History of Geophysics

Volume 5

C. Stewart Gillmor
Series Editor

The Earth, the Heavens and the Carnegie Institution of Washington

Edited by
Gregory A. Good

With a Foreword by Maxine Singer,
President of the Carnegie Institution of Washington

American Geophysical Union
1994

Library of Congress Cataloging-in-Publication Data

The Earth, the heavens and the Carnegie Institution of Washington / Gregory Good, editor.
p. cm. — (History of geophysics : v. 5)
Includes bibliographical references.
ISBN 0-87590-279-0
1. Carnegie Institution of Washington. 2. Mount Wilson Observatory. 3. Geophysics—Research—United States. 4. Space sciences—Research—United States. 5. Astrophysics—Research—United States. I. Good, Gregory. II. Series.
QC801.E36 1994
550'.7209753—dc20 93-46865
CIP

ISSN: 8755-1217
ISBN: 0-87590-279-0

Printed in the United States of America

CONTENTS

Sounding the Ionosphere

International Spin-offs of Carnegie Research

Post-War Geophysics at the CIW

Resources for Historical Research

Foreword

History and Science at the Carnegie Institution of Washington

In this very grand building that houses the Carnegie Institution of Washington, portraits and photographs of Vannevar Bush and the other leaders who served in these rooms remind us of our history. Only a few miles away is the campus of our Earth and planetary scientists on Broad Branch Road, where a superb model of the Good Ship *Carnegie* is displayed. Likewise, the halls of our Observatories in Pasadena are lined with photographs of the great Carnegie astronomers from early in the century--a daily message to our young astronomers that the Institution expects them to continue that greatness. But our sensitivity to our history goes beyond such symbols.

We are very satisfied and pleased that historians interested in our early days have shown that Carnegie Institution contributed importantly in preparing the way for the explosion of science in America after 1940. The Institution early in the century filled a role as the leading patron of the nation's research. We recognize and honor this aspect of our past, though it offers very few parallels to our present situation. (We are no longer a "big fish in a small pond," but rather the opposite.)

But if our place in the overall world of science is smaller, we sense that the Institution still has a special role to play. We are favored by a healthy endowment, although it is not large enough to make it possible to do everything we would like. As neither a university nor a government or industrial entity, we are largely without obligation to others except to society at large. Thus we have an enviable independence to choose our own directions, one that few others enjoy. In return, of course, we have an obligation to use it wisely.

How then should we use this freedom? Should we compete in the arenas that are widely perceived to be at the frontier's edge, or should we choose relatively unpopulated scientific pathways? Are we risking our uniqueness if we allow our reliance on federal monies to increase? How precious is our historic leadership in astronomy when measured against the costs for a new, very-large-diameter telescope in Chile?

In facing such questions, absolutely fundamental to us is the bedrock set of ideas that we inherit from our past. These derive from Mr. Carnegie's founding guidance as reinforced and shaped by our subsequent leaders and scientists. Together, these ideas from our history establish our identity.

Thus we hold fundamental that basic science ultimately benefits humankind. We defend the investigator's freedom of enterprise in research, and we believe in the potential of the largely independent single investigator, one who often interacts closely with others in small and loosely organized groups. We deem it crucial that our scientists be able to take new and previously unplanned scientific directions at little notice.

Our Trustees--Mr. Carnegie's representatives today--take seriously our heritage and our obligation to reinterpret that heritage in the light of present circumstances. When, for example, earlier this year the Trustees began an extended examination of our posture and future directions in plant biology, their perspectives were enhanced by a serious historical review of the ancestry of our Department of Plant Biology, presented both orally and in booklet form. The case exemplified our conviction that our history is fundamental.

Carnegie's Historians

Some may be surprised to learn that Carnegie Institution, which is now wholly engaged in the physical and biological sciences, was once a major practitioner of the historical profession. For many years, the Institution operated a Department of Historical Research, based in Washington. Its leader from 1905 to 1928 was the renowned historian J. Franklin Jameson--a giant in the evolution of the American Historical Association and its professional journal, and the principal figure in establishing the National Archives. Much of the Carnegie Department's work under Jameson was not in the preparation of interpretive histories but rather in the exploring and opening up of material useful to future researchers--preparing guides to documents in foreign archives bearing on U.S. history, for example.

Historian Morey Rothberg has shown that Jameson's reluctant resignation from Carnegie was preceded by strong differences in outlook between Jameson and the then Carnegie president, John C. Merriam. Jameson firmly believed that prior to offering interpretations, the historian should develop incontrovertible bedrocks of evidence; Jameson's was, thus, a seeming "scientific" view of history-writing; but it was Merriam, the scientist, who urged Jameson and his colleagues to develop conclusions early and publish them. It was a position Merriam often took in dealing with the Institution's scientists, out of concern that

Carnegie's investment in research could be lost if results went unpublished. (Historians and scientists seem to be a lot alike; both must decide when to punctuate data gathering by offering interpretations and syntheses.)

After Jameson's resignation, the Carnegie work in U.S. history was gradually replaced by a growing program in middle-American archaeology.

Meanwhile another giant--the pathfinding historian of science, George Sarton--for more than three decades received a full salary as a research associate of the Institution. Sarton came to America as a refugee from war-torn Belgium in 1915 and, after a series of overtures to Carnegie president Woodward, joined the Institution's payroll in 1918. His place of work was Harvard University's Widener Library, which offered a superb collection of materials for his work. Sarton paid for his library privileges by giving lecture courses at Harvard; but he received no monetary stipend from Harvard for many years.

Sarton spent most of his time gathering the knowledge that became his encyclopedic survey *Introduction to the History of Science*, which was published in five covers between the years 1927 and 1948 as Carnegie Institution publication 374. He wrote that "my appointed task, the task which justifies my existence, is to prove inductively the essential unity of knowledge and the essential unity of mankind." Thus, the massive text moved back and forth among the civilizations of Greece, Rome (and western Europe), Islam, and the Far East; the grand endeavor ended with the 14th century, having tapped Sarton's last reserves of stamina. His treatment was strongly chronological because, Sarton once explained, although one cannot be sure that a later event is influenced by a earlier one, the opposite cannot have occurred. Few things can be known so assuredly.

Sarton was a heroic scholar, driven by a vision to establish the history of science as a discipline. He created, edited, and on occasion kept alive with his own money the journal *Isis*, which he regarded as the necessary complement to the volumes of the *Introduction*. Sarton knew Greek, Latin, and the modern European languages, and had some reading ability in the oriental languages. At first he relied on assistants to read Arabic texts, but while working on the first volume of the *Introduction*, he came to realize that the dominance of Arabic as the language of science in the 9th-11th centuries dictated that he himself obtain command of that language. Thus in mid-life, amid an incredibly busy workload, largely self-imposed, Sarton systematically learned, and later regularly corresponded in, Arabic.

I might add that on recent occasions I have been involved with panels or groups addressing multiculturalism in the curricula of American universities. In trying to find ways to help scientists deal with that topic, I have in general addressed the history of science and in particular talked about the many roots worldwide of modern scientific thinking including an emphasis on Arab contributions; for our knowledge of that history, we are of course indebted to George Sarton.)

Sarton's intensity brings to mind Andrew Carnegie's early charge to the Institution--to seek out the "exceptional" individual and to give that person opportunity for a life's work in research and discovery. Sarton, who wrote often to the successive presidents of Carnegie Institution thanking them for support *and* asking for extra funds, in a 1923 appeal to president Merriam seemed to be reminding Merriam of Mr. Carnegie's words (something that present-day Carnegie scientists also do to their president):

> Sometimes men appear, extremely specialized ones, who could do very well a certain kind of thing, but there is no demand for it. These men are not ready to work so many hours a day; they must needs give every thought of their mind, every beat of their heart, every dream of their sleep to that special work; in fact, they are slaves to it.... Very little would be needed to help them, but no help comes.

Merriam actually never doubted that Sarton was indeed an exceptional scholar and that his work merited the Institution's support. Even during the 1920s and 1930s, when the purchasing power of the Institution's largely fixed income was diluted by the doubling in price levels after World War I and by the general financial conditions, monies were found to pay a full salary to Sarton and provide some clerical assistance. In those days, such clerical assistance was provided by the hiring of Radcliffe students to type copies of documents and notes. Starting in 1929, the Institution also paid for several professional research assistants, and the effort was officially designated "Section of the History of Science." A Sarton assistant for many years was Alexander Pogo, whose tasks included the gathering of notes for a never-published volume of the *Introduction* on the 16th century. Pogo much later became a familiar figure caring for the library and rare books at Carnegie's Mount Wilson Observatory. When he died a few years ago, he left a very generous bequest to the Institution, to everyone's surprise and delight. We will probably honor him by naming some portion of the to-be-refurbished offices of the Observatories in Pasadena in his honor.

Another, younger Sarton associate was I. Bernard Cohen, who later succeeded Sarton as the editor of *Isis*.

THE HISTORICAL RECORDS OF THE INSTITUTION

We at Carnegie are trying to enhance the condition of our historical records and their accessibility to historians. Our archival files have been greatly improved in recent years under the care of Susan Vasquez. Susan has been helped each summer by visiting worker John Strom. Shaun Hardy, the new and very talented and energetic librarian at Broad Branch Road, is plunging into the large task of properly organizing the archives of the Department of Terrestrial Magnetism and Geophysical Laboratory. Meanwhile in Pasadena, the files of the Observatories have been transferred to the Huntington Library under the attention of the trusted archivist and historian Ron Brashear. Much remains to be done, but we are pledged to continue moving in the needed direction within the limited resources that we can commit.

I am especially grateful, as a relative newcomer to the Institution, to Susan Vasquez and Ray Bowers, who not only tend the archives but are themselves living resources on the Institution's history. In particular, I want to note that Ray has been instrumental in the preparation of this foreword, which is adapted from an introductory talk I gave at the start of the symposium "The Earth, the Heavens, and the Carnegie Institution of Washington: Historical Perspectives after Ninety Years," held in Washington, D.C., June 15-17, 1992.

Maxine Singer
President
Carnegie Institution of Washington

The "Founding" Meeting of the Board of Trustees of the Carnegie Institution of Washington, January 19, 1902, at the offices of the U.S. Geological Survey. Seated (left to right, • = Executive Committee member, ○ = Finance Committee member): D. B. Henderson, ex-officio (Speaker of the U.S. House of Representatives); William P. Frye, ex-officio (President of the U.S. Senate); ○Lyman J. Gage; •Elihu Root; John Hay; Andrew Carnegie. Standing: •Caroll D. Wright; •S. Weir Mitchell; S. P. Langley, ex-officio (Secretary of Smithsonian Institution); ○D. O. Mills; • John Shaw Billings, Vice-Chairman of the Board; •Abram S. Hewitt, Chairman of the Board; William W. Morrow; Wayne MacVeagh; Charles L. Hutchison; William N. Frew; •Daniel Coit Gilman, President of CIW and Chairman of the Executive Committee; Henry L. Higginson; Henry Hitchcock; William Lindsay; •C. D. Walcott, Secretary.

Preface

This volume examines the history of the Carnegie Institution of Washington and its role in the evolution of the geosciences and astrophysics: important science in an important setting. The Carnegie Institution would certainly be on any short list of the two or three institutions most critical to the founding of these sciences. Moreover, the sciences of the Earth and of near-space have grown increasingly important in the 20th century, yet their history is not nearly as well known as that of, for example, theoretical physics or evolutionary biology. This volume intends to refocus attention on these essential aspects of science in our century.

The genesis of this volume took place in May 1988 at the Spring Meeting of the American Geophysical Union, during a brainstorming session about ways to spotlight the history of geophysics. The idea of a thematic meeting, separate from the usual press of an AGU meeting or a History of Science Society meeting, was suggested. Later, while working in the archives of the Carnegie Institution, I was introduced to Maxine Singer, then the relatively new president of the Institution. She was interested in whether it would be possible to focus more attention by historians of science on the various activities undertaken by Carnegie scientists over the years: not just geophysics and astrophysics, but also genetics, biophysics, and much more. In July 1990, with President Singer's support, about 20 scientists and historians of science gathered for a planning meeting in Washington. We developed several lists of themes for a symposium. One was arranged topically by areas that have been investigated by Carnegie scientists: seismology and the structure of the Earth, tectonophysics, high pressure physics, cosmic rays, petrology, the age of the Earth, solar-terrestrial relationships, the Sun's magnetic field, and so on. A few themes also suggested other perspectives: the development of the infrastructure of science, styles of research, administration of science at frontiers, the roles of Carnegie scientists in identifying and developing initiatives in the geo/space sciences, the building of institutions and disciplines, the development of private foundation support of science, and the functioning of the Institution as a surrogate for a national science policy body early in this century. These themes and topics, while not immediately evident in the table of contents, recur throughout the volume.

From that point, it took two years to invite speakers to the conference, form a coherent program, and find the funding to make it happen, in June 1992. The articles presented here are the result of further work, review and revision, and are a tribute to the significant sacrifices made by the authors and other scholars.

I wish to express my thanks first to those organizations that supported me while I acted as clearing house for this community effort. Much of the organizational work was sandwiched between research forays during my sabbatical year from West Virginia University, 1990-1991, which I spent among congenial colleagues at the Institut für Geschichte der Naturwissenschaften, University of Hamburg, Germany. That year as a fellow of the Alexander-von-Humboldt Stiftung was critical to the success of the conference. The most important financial support for the conference itself came from the Carnegie Institution, which defrayed the expenses of most of the speakers. Several speakers were helped, too, by parts of West Virginia University. W. D. Parkinson made the second half of his trip from Tasmania courtesy of the WVU Institute for the History of Technology and Industrial Archeology, and Wang Shen was supported in his trip from the People's Republic of China by the WVU Council of International Programs. It should also be noted that many speakers paid their expenses out of pocket, an often unappreciated contribution to scholarship.

Lastly, though, I express my appreciation to the many fine people at the CIW's administrative offices, at the Geophysical Laboratory, and at the Department of Terrestrial Magnetism who nudged me and the process along at critical moments. They, along with the secretaries and graduate students at WVU who did so much, deserve special credit, and should take much satisfaction in seeing this volume in print. My heartfelt thanks to you all.

Gregory A. Good
West Virginia University
Morgantown
Editor

Introduction

The Breadth, Height, and Depth of the Geosciences and Space Sciences at the Carnegie Institution of Washington

GREGORY A. GOOD
History Department, West Virginia University, Morgantown, West Virginia 26506

The Carnegie Institution of Washington (CIW), founded in 1902, was among the first privately funded scientific-research institutions in the United States. It was an experiment in the organization of scientific research and provided a model for many later organizations around the globe. Among the areas strongly supported by the CIW were investigations of the Earth and its relations to the Sun and planetary system.

This volume is the result of a 3-day symposium that met in June 1992. The symposium's goal was to explore the history of the geosciences and astrophysics during the first 60 years of the Institution. We wanted to stimulate discussion of several historical questions. First, how did the complex of the geo/space sciences develop in this century? Second, what kinds of environments and strategies supported its development? Third, what have the relations been between science and its sponsors, and how have new disciplines and research fields developed in interdisciplinary settings? The main goal of the symposium was to contribute to understanding the historical development of the geo/space sciences at the Carnegie Institution of Washington: important science in an important setting. We had no illusion, and this volume has no pretension, to answering these questions in a general way. But we hope to have moved in this direction.

The symposium centered on the development of research programs in the geo/space sciences at this one institution, and on the evolution of the institutional forms and policies within which these developments occurred. The sciences of the Earth and of near-space have roots in earlier centuries. But during the first half of the 20th century many of these sciences witnessed significant instrumentational and theoretical developments. During this period, geophysics and geochemistry became widely recognized words and, perhaps, disciplines. In this period, major discoveries were made regarding the interior architecture of the Earth, the cause of the Earth's magnetic field, atmospheric dynamics, the ionosphere, and solar-terrestrial relations. This was a period of great activity, of exchanges or borrowing of methods between disciplines, and of changes of approach that occurred across many sub-disciplines almost simultaneously. These events set the stage for the dramatic era of the International Geophysical Year, the plate-tectonic revolution, and space-based studies of the Earth. The first half of the century was critical in the history of the geo/space sciences, and yet it has been the subject of little historical scrutiny.

The development of the institutions that support the geo/space sciences has been just as important as the development of the instruments, methods, and theories associated with them. The geo/space sciences are prototypical big sciences. They demand world-wide data networks, big and expensive instruments, sizeable staffs, and commitment of relatively large budgets over long periods of time. Their products may be either quite abstract and useless or of direct commercial utility. Institutional support cannot be ignored in the history of these sciences. Between 1900 and 1950, the types of sponsors of geo/space science and their relative importance changed significantly: private research institutions, foundations, national and state governments, oil and mineral companies, and the military. In 1900, private research institutions were critical. Among these, the CIW was one of the best endowed. For a variety of reasons that have never been adequately explored by historians, a large fraction of the CIW's support of science in its early decades was directed to the geo/space sciences.

From the Institution's founding, grants were awarded to outside researchers studying the Earth. Soon thereafter, three departments at the CIW became primary loci for geo/space sciences: the Department of Terrestrial Magnetism (DTM) and the Geophysical Laboratory (GL), and the Mount Wilson Observatory. The DTM's efforts were focused very closely for the first few decades on a global geomagnetic survey, to make possible a mathematical

The Earth, the Heavens and the Carnegie Institution of Washington
History of Geophysics Volume 5

analysis adequate for physical theory. Beginning in the 1920s the DTM broadened its researches to include, gradually, cosmic rays, nuclear research, paleomagnetism, the ionosphere, and other phenomena.

The Geophysical Lab concentrated on laboratory techniques for investigating how rocks might behave at great depths in the Earth. That is, its scientists perfected techniques of high-pressure and high-temperature studies. Especially influential were the researches of N. L. Bowen on fractional crystallization. The Geophysical Lab, like the DTM, broadened its researches in the 1920s and cooperated in the development of seismology at the Caltech Seismological Laboratory.

One other Carnegie department was a main location of geo/space science: Mount Wilson Observatory in California. George Ellery Hale and visiting researchers explored such questions as the Sun's magnetic field, the evolution of stellar and planetary systems, sunspots, and planetary temperatures. This department, like the GL and the DTM, still exists and still conducts research in geo/space science.

The authors in this volume discuss the history of geo/space science at these three departments and at the Institution generally. A number of themes are evident in the volume. Some authors look more closely at the dynamics of research programs, at the interplay between theory and evidence. Others are more concerned with the relative roles of personality and policy, or with strategies of institutional support, or with questions of social history. Others focus on the history of disciplinary development or of the relations between private institutions and the government. The volume invites a comparison of the relative merits of these approaches.

The articles in the first section examine how the original research programs of the Carnegie Institution were chosen. Ellis L. Yochelson asks how the interaction of C. D. Walcott and Andrew Carnegie affected these events. H. S. Yoder and Gregory A. Good discuss the elaboration of early program ideas at two of the permanent departments: the Geophysical Laboratory and the DTM, respectively. Questions of the international and national relations of the CIW arise in these three discussions.

Early CIW grants to outside scientists and cooperation with other agencies are discussed by R. S. Brashear (on cooperation of Smithsonian Astrophysical Observatory with Mount Wilson regarding the solar constant), David DeVorkin (on Henry Norris Russell's appointment as a CIW Research Associate at Mount Wilson), N. S. Hetherington (on T. C. Chamberlin's CIW-supported cosmo-geological work), Ralph Jewell (on Vilhelm Bjerknes' work on atmospheric dynamics under CIW sponsorship), and Peggy Champlin (on Raphael Pumpelly's geoarcheological expeditions in Turkestan). John Lankford asks what roles women have played in the "great factory observatories" and why they are invisible in the history of the geo/space sciences. Owen Gingerich suggests ways in which these case studies may be placed in the context of major changes taking place in astronomy worldwide at the beginning of the century.

Several shifts in research programs attract close attention. These include C. Stewart Gillmor's discussion of ionospheric investigations by M. A. Tuve and Gregory Breit at the DTM in the 1920s and Homer Le Grand and Thomas D. Cornell on post-World War II geophysics there.

Several instances of "cooperative research" as a conscious policy are also featured. Naomi Oreskes discusses Felix Vening Meinesz's important S-21 gravimetric expedition, Sam Silverman and Marion Edmonds Smith use Roald Amundsen's famous Northwest Passage traverse and his interaction with Harry M. W. Edmonds (of the DTM) to dissect the nature of expeditionary science. Bruce Hevly analyzes a Washington "research network" in atmospheric physics involving Robert Goddard's rockets and scientists at the DTM, the Bureau of Standards, and the Naval Research Laboratory. These three papers illustrate both the centripetal and centrifugal aspects of very different institutions and individuals cooperating across disciplinary lines.

The penultimate theme explored is the role of Carnegie researchers and programs in defining research areas. This can alternatively be viewed as the establishment of "research initiatives" which then were carried on by others beyond the institutional walls. Hence many of the papers concern how an interdisciplinary project becomes established, how it is incubated, and how this project then fares beyond the institution which engendered it. This theme is especially prominent in papers by Elburt F. Osborn, R. W. Home and W. D. Parkinson, J. E. Kennedy, W. O. Kupsch, Wang Shen, and L. T. Aldrich, which discuss basic research at the CIW in igneous petrology, geomagnetism, ionospheric physics and geochronology that led to continuing research programs far beyond the Institution.

The last thematic section of the volume concerns the resources on which history of geo/space science depends. It is about, especially, scientists' papers, institutional records, pictorial records, and ephemeral printed sources. Just as scientists cannot conduct research without data, historians need these resources. These records are essential to understanding science history. D. Day's article calls for preservation of the archival record. The articles by J. Strom, R. S. Brashear, and R. E. Doel demonstrate that the CIW and some other institutions have recently made some significant progress toward this desideratum. The historical articles in the volume show clearly what these records can reveal of our history. One hopes that other scientists and other institutions will follow the example of the Carnegie Institution of Washington. Take care of your records.

A great number of researches in the geo/space sciences at the CIW are not addressed in this volume. They include: the CIW Advisory Committee on Seismology, X-ray crystallography, oceanography in the 1920s and 1930s, L. H. Adams' work on elastic constants, military work in both World Wars, volcanology, L. V. Berkner and the ionosphere, cosmic ray research, recent work by Ho-Kwang Mao and others on planetary interiors, the pressure effect on the Curie temperature, phase diagrams of silicate systems, stability relations of hydrous and iron-bearing minerals, putting petrology and mineralogy on a sound physico-chemical basis, and much more. Clearly, Carnegie scientists have been instrumental in this century in defining and extending the frontiers of the Earth and space sciences.

An equally long list of institutional and social historical questions is possible. Who were the people who did the research? Lankford is correct that we know little about the roles of women researchers in these sciences, but indeed, beyond a few prominent individuals, we know little about the social backgrounds of *any* Carnegie scientists. This question is of interest not just to American historians, but also in current policy discussions regarding diversity and the encouragement of a new generation of researchers. Likewise of historical and contemporary importance are questions of how the research was administered. What decisions led to its funding or re-direction? Moreover, while the Carnegie Institution contributed significantly to the theoretical and observational foundations of the geo/space sciences, one must stress the equal stature of its contributions to the institutional basis for these sciences. Many scientists and administrators in the United States and around the world have interacted with the CIW, and many have looked to it for examples of how to organize their endeavors.

To understand science is ultimately to understand it as a whole. A history that ignored the humanity of science, both its glories and its ambiguities, would be incomplete. While the researchers and their investigations are at the center of science history, they are embedded in contexts ripe with humanity. Understanding those contexts is part of understanding science. One hopes that other scholars will follow the examples of the authors in this volume. The historical breadth, height, and depth of the geosciences await much more investigation.

Andrew Carnegie and Charles Doolittle Walcott: The Origin and Early Years of the Carnegie Institution of Washington

Ellis L. Yochelson
Research Associate, Department of Paleobiology, National Museum of Natural History, Washington, D.C. 20560 and United States Geological Survey (retired)

Introduction

The founding and early history of the Carnegie Institution of Washington (CIW) already have been carefully studied and documented several times [Madsen, 1969; Miller, 1970, p. 166-181; Reingold, 1979; Servos, 1983]. As with any other accurate historical investigation, these works provide information for further reflection on the history of CIW, and it descendent part, the Geophysical Laboratory. Manuscript sources suggest that in the formation of both the main structure and that one of its working components, Charles Doolittle Walcott played far more of a central role than has been credited to him by earlier studies of these organizations. As a secondary theme, these sources also indicate that although Andrew Carnegie did not involve himself directly in the affairs of the Institution once CIW was set up, Walcott made certain that the philanthropist was fully aware of all major developments during the early days of the organization.

Andrew Carnegie in 1901

It is well known that after a bruising battle with Childs Frick, in 1901 Carnegie (1855-1919) retired from the steel industry and plunged full-time into philanthropy [Wall, 1970]. It is also well-known that Andrew D. White, former President of Cornell University, and then Ambassador to Germany, was his advisor [LC, Carnegie Papers, White to Carnegie April 30, 1901; May 18, 1901]. White visited Carnegie late that summer at Skibo Castle, his home in Scotland to persuade the philanthropist to support the academic community, possibly in the form of a national university [Miller, 1970, p. 170]. White had hoped that Daniel Coit Gilman [Franklin, 1910], just retiring as President of Johns Hopkins, would also come to Scotland, but Gilman had made other plans.

At this stage, Carnegie actually preferred to deal directly with White alone. Apparently one of the notions being considered was that upon President Gilman's retirement, Johns Hopkins University might be moved to Washington, DC, and increased in size and stature to become the major regional university in the middle Atlantic and the south. Nothing of obvious substance resulted from White's meeting in Scotland. Ambassador White returned to America in September on the same ship as Carnegie and tried again quite unsuccessfully to convince him to support a university. Thus, an academic institution in the traditional sense was ruled out before CIW was conceived.

Carnegie was a tough businessman and one who battled to become primary in the field, regardless of the enterprise. Speculations as to Carnegie's motives in setting up the Washington institution have been discussed [Madsen, 1969, p. 155]. One may add to those reasons already on record that support of American universities might have held little appeal for Carnegie because of the prior involvement of John D. Rockefeller with the founding and support of the University of Chicago. Carnegie Institute of Technology in Pittsburgh, founded in 1900, indicated that Andrew Carnegie was capable of carving out his own particular niche in the academic field. On the other hand, the assets of Big Steel could not match those of Big Oil, and perhaps a national university was too big a bite. Carnegie was not the sort of man who would move into a field knowing beforehand that he would never achieve a preeminent position.

Carnegie had been the first to employ a trained chemist in American steel manufacture [SIA, Walcott Papers, Walcott to Carnegie, February 1, 1915; Carnegie to Walcott, February 3, 1915], and he knew that research could produce information that yielded practical results. Rockefeller had recently started support of medical research, but there was a vast field for scientific inquiry that was still untapped. No one will ever know exactly what Carnegie had in mind, but these are considerations toward an interest in research that were present in the fall of 1901.

Again, it is well-known that on November 16, 1901, John Shaw Billings (1838-1913) and Daniel Coit Gilman (1831-1909) met with Andrew Carnegie at his mansion in New York City and that Gilman was designated to be president of the new enterprise, whatever it was to be

The Earth, the Heavens and the Carnegie Institution of Washington
History of Geophysics Volume 5

[Madsen, 1969, p. 156; Reingold, 1979, p. 117]. Animosity between these two members of the "brain trust" is a central theme to histories of CIW, yet Dr. Gilman dined at the Billings household prior to their second meeting with Carnegie [NYPL, Billings Papers, Gilman to Billings, November 20, 1901] a few days later. They submitted a joint report to Carnegie, admittedly one slanted more toward research and post-graduate training than towards education.

Lieutenant-Colonel Billings did not use his title in his civilian, second career, but he was not an academic in any sense of the word [Garrison, 1915; Henry, 1964, p. 78-82; Mitchell, 1919]. He and Gilman met occasionally in Washington or Baltimore and had been intermittent correspondents for at least three decades. Billings had assisted Gilman in the organization of The Johns Hopkins Medical School. There is no question of the forcefulness of his character, yet Gilman knew him well, and if the president emeritus of Johns Hopkins University felt seriously threatened by Billings, he easily could have withdrawn at this early stage when the future course of the proposed institution was still nebulous.

CHARLES DOOLITTLE WALCOTT IN 1901

At this time, Walcott (1850-1927) was both Director of the Geological Survey and effectively in charge of the forest reserves of the United States [Yochelson, 1967]. Correspondence at Johns Hopkins University shows that he had known Gilman since at least the early 1890s—and probably much earlier—for his cousin the geologist George Huntington Williams was one of the stars of the new university in Baltimore. When Williams died unexpectedly in 1894, Walcott arranged for members of the Geological Survey to teach at the university. A reasonably close connection with Gilman persisted and in 1902, at the 25th anniversary of the university and the formal investiture of Ira Remsen as Gilman's successor, Walcott received an honorary doctorate from Johns Hopkins.

Walcott was equally well-known to Billings. Casual contact between the two in Washington prior to Billings' retirement from the Army cannot be documented, but the Army Medical Museum and the United States National Museum were adjacent buildings on the south side of the Mall. Further, both Walcott and Billings frequented the Cosmos Club, and both had served as its president; a few years later, Walcott was to write Dr. Billings "I recall full well the strong leadership you exhibited as both Treasurer and President of the club . . ." [NYPL, Billings Papers, Walcott to Billings, November 17, 1903]. In 1898, two years after Walcott was elected to the National Academy of Sciences, Billings resigned as Treasurer of the Academy and recommended Walcott for that office, which he then held until 1902.

Despite his governmental administrative load, Walcott was involved in many other activities. He was first president of the Geological Society of Washington in 1893, and more importantly for this discussion, President of the Washington Academy of Sciences, an office he held for more than a decade, almost from its inception in 1898. Walcott had recognized that the National Academy of Sciences was, if not moribund, at least exceedingly slow moving. One aim of the local academy was to give science a voice in Washington. Although this distressed some members of the National Academy, Ira Remsen, the new president of Johns Hopkins wrote Billings that: "I entirely agree with you. We have nothing to fear from the Washington Academy of Sciences—or if we have, there must be something wrong with us" [NYPL, Billings Papers, Remsen to Billings, March 22, 1900].

More or less coincident in time, the George Washington Memorial Association had been founded; Mrs. Walcott was active in that group. The aim of this organization was to help start a National University in honor of the first president. Walcott soon arranged for committees of both groups to meet and join forces. As a result, one of the first actions of the Washington Academy was to lobby successfully for Congressional legislation opening government laboratories to graduate and post-graduate students.

After that bill became a law in the winter of 1901, a third group was conceived, formed partly by a fusion of committees from each organization [Anonymous, 1901]; (correspondence in Cattell Papers of the Library of Congress from Walcott show that he was the author of this statement). The immediate aim was to construct a building that would serve as administrative headquarters for a national university, or at least for a university symbolically in the nation's capital. Walcott also saw this building as a potential permanent headquarters for both the National and Washington academies. In mid-May, Walcott wrote Gilman "At a meeting of the joint committee of the G.W.M.A. & W.A.S. held today it was decided to incorporate the Washington Memorial Institution at once" [JHU, Gilman Papers, Walcott to Billings, May 15, 1901]. The Institution was incorporated forthwith, the Board of Trustees met, and as finally organized in early June, Walcott was president of the Board of Trustees—not a surprise—and Gilman was Director of the Institution [JHU, Walcott to Gilman, June 3, 1901]; Nicholas Murray Butler of Columbia University was Secretary of this nascent organization. Gilman wrote his friend Andrew White:

> Curiously enough a Board of Trustees has been initiated in Washington to promote the opening of the museums,

Daniel Coit Gilman (upper left), Andrew Carnegie (upper right), John Shaw Billings (lower left), and Charles Doolittle Walcott (lower right). Photographs courtesy of, respectively, Johns Hopkins University, CIW, National Library of Medicine, and the author.

libraries, etc. to students, and yesterday I was made chief officer of the Board. There are no funds, buildings, regulations, or laws, —only ideas to be worked out. The movement which has the approval of capital men in Washington and throughout the country, can, I think be brought into great service, —but just how, I do not yet see [Franklin, 1910, p. 338].

Throughout his scientific career, Walcott had a keen business sense. He knew how to make organizations work. In the early 1890s, he rewrote the bylaws for the Biological Society of Washington. He helped found the Geological Society of Washington and played a role in writing its constitution and bylaws. He helped assemble the Washington Academy out of a chaotic Joint Commission representing a number of local societies [Gilbert, 1899]. Quite apart from this, Walcott was actively engaged in the construction and managing of three apartment houses in the city. As a result of these various activities, he had first-hand knowledge of the legal consequences of incorporation and the concepts of governance. Even more importantly, he knew his way around the Washington local government.

Walcott had higher education on his mind and was concerned as to what the Federal bureaus might contribute to a mix of government and university, as indicated by his speech upon receiving an honorary degree from the University of Chicago in the spring of 1901 [Walcott, 1901]. By summer, Walcott was on a hunt for money. Within a short time of his assuming its presidency, the Washington Academy of Sciences had sufficient funds, mainly through his efforts, to print "Proceedings." The Washington Memorial Institution, however, needed a far larger amount if it were to succeed. That fall, he communicated to Gilman:

I have just written to Dr. Hadley [President of Yale University] asking him if he knew Mr. Carnegie sufficiently well to ask for an appointment, at which you, Dr. Hadley and I could present the question of his becoming the financial founder of the institution.

I shall also talk with the Washington gentleman who was instrumental in securing a large sum for the City Library here, and who has offered to call Mr. Carnegie's attention to the Washington Memorial Institution if it is thought desirable" [CIW, Historical File 1, Walcott to Gilman, November 15, 1901].

Among Walcott's papers is a 7-page memorandum "Origin and Development of the Carnegie Institution of Washington, D.C.." An earlier 6-page "Organization of the Carnegie Institution of Washington" is also present, along with an even earlier 4-page "Memorandum on the conditions that led to the Carnegie Institution of Washington," both heavily edited in pencil by Walcott. There are also six small sheets of "Notes" on the subject. None of this material is dated, and the best guess one can make is that the final draft may have been prepared during the closing days of the Gilman era at CIW for the benefit of his successor. Where these documents can be checked against other sources, they are accurate, and there is every reason to believe that the story occurred exactly as Walcott recounted it.

From these papers, the end of the academic concept can be traced. After his return to America in the fall of 1901:

Dr. White went to Washington and met at an afternoon tea at Gardiner Hubbard country house, "Twin Oaks," Dr. D.C. Gilman and Mrs. Charles D. Walcott. He spoke of his talk with Mr. Carnegie about endowing a National University and his failure to convince him of the necessity for it. Dr. Gilman and Mrs. Walcott then told him of the scheme for a research institution in Washington, and Dr. White said that he would speak to Mr. Carnegie about it as it seemed to be a fine proposition. He evidently did so as Mr. Carnegie sent for Dr. Gilman [SIA, Walcott Papers, Origin, p. 5].

This meeting was in October, 1901, and Mrs. Archibald Hopkins, President of the George Washington Memorial Association, was also present [SIA, Walcott Papers, Memorandum, p. 3].

Even though Walcott did not meet Carnegie until December 2nd, a slightly earlier contact was a letter, so far untraced, which was sent to both Rockefeller and Carnegie. In preparing his draft memorandum, Walcott noted Carnegie was to donate money " . . . as a foundation for a scientific Institution for Research along lines outlined to him in my letter of Nov. 1901" [SIA, Walcott Papers, Notes, p. 4].

In Walcott's dealings with Congress, he had shown a remarkable ability to shift focus when necessary and seize an opportunity. The Washington Memorial Institution was constructed in a broad enough manner that its emphasis could be moved almost entirely from education to research, now that it was evident that Carnegie was not interested in a national university. It would be interesting to know just what Walcott had proposed in this Carnegie-Rockefeller letter.

The import of the joint report of Gilman and Billings should be reconsidered in this light. If the memorandum of Walcott's is correct, Gilman was not meeting Carnegie as an advocate for an academic institution. Rather, he was there to suggest some Washington organization that was aimed much more toward research. Accordingly, the basic differences, previously assumed, between the research ideas of Billings and the academic slant of Gilman may have been more apparent than real. Walcott noted that Gilman's main contribution was to serve as an advocate of research when he met with Carnegie in November, 1901, but as this comment was indirectly a harsh criticism of Gilman, Walcott did not include the remark in his final version of the history [SIA, Walcott Papers, Memorandum, p. 4].

Gilman wrote to Andrew White about the events in November when he had met with Carnegie and Billings. In reply, after expressing some harsh words about Billings, White added:

> In the beginning, Dr. B. was certainly opposed fundamentally to the plan we had at heart, but he may have seen the light since, especially if Mr. C. shows him that he has himself come to a decision. If so, the doctor ought to give you valuable aid. What I am most concerned about is the right selection of the main site in due time . . . [CIW, Historical File 1, White to Gilman, November 30, 1901].

One apocryphal story in Washington is of Carnegie having breakfast with Walcott and being asked to tell him about the Englishman Smith. After correcting Carnegie that the name was Smithson, Walcott extolled the virtues of the Smithsonian Institution to such an extent that Carnegie decided he wanted a comparable organization bearing his name. However the original notion for a research organization might have been planted, this was not the route.

THE START OF CIW—DECEMBER 1901

After his meetings with Billings and Gilman, Carnegie moved quickly. On Thanksgiving Day, he wrote to President Theodore Roosevelt [LC, Carnegie Papers, Carnegie to Roosevelt, November 28, 1901] about a gift to the nation. He made no secret of the fact that the funds were in the form of United States Steel 50-year bonds. Carnegie also indicated his willingness to construct any necessary buildings in Washington, even though later he was opposed to permanent facilities for CIW.

Despite the short time between the meetings of Gilman and Billing with Carnegie, and this letter, some preliminary discussion must have taken place with the White House, for two days in advance of Carnegie's formal letter proposing the bequest, Andrew Carnegie's secretary wrote "Dear Dr. Billings, Messers Gilman, Carnegie & you are to lunch with the President 1-30 Monday, Dec 2nd" [NYPL, Billings Papers, Bertram to Billings, November 26, 1901].

The luncheon was held, Carnegie announced his gift, and Walcott entered the picture.

> Dr. Walcott returned to Washington in November and on the afternoon of December 2, 1901, Dr. Gilman called at the office of the Geological Survey and said that Mr. Andrew Carnegie would like to see Dr. Walcott. He went down to the hotel [The Willard on Pennsylvania Avenue] with him and met with Mr. Carnegie for the first time. Mr. Carnegie said that he had had luncheon with President Roosevelt and he thought it right to tell him first that he intended to give ten million dollars ($10,000,000) as foundation for a scientific research institution in Washington along the lines outlined to him in a letter which had been sent to him and further explained by Dr. Gilman. After a short conversation Mr. Carnegie led Dr. Walcott to a bay window and said that he wished his dear old friend, Dr. Gilman, to be the first president of the new institution, as he had done a great work at Johns Hopkins from which he was about to retire. Mr. Carnegie spoke of Dr. Gilman being seventy years of age and that not much initiative could be expected from him. He then asked Dr. Walcott to act as Secretary and to take an active part in the organization and conduct of the institution in its first years at least. This Dr. Walcott agreed to do as far as other duties permitted, as he was then Director of the U. S. Geological Survey" [SIA, Walcott Papers, Origin, pp. 5-6].

Walcott kept a pocket diary in which he jotted a few lines each day. It confirms the various dates and meetings cited by Walcott in his unpublished memorandum. At the same time, there must have been some inkling or at least speculation associated with Carnegie's Washington excursion, for prior to Walcott's 4:00 PM visit to the New Willard Hotel, in his office at the Geological Survey, there was a "Meeting of Dr. J. S. Pritchett, C. D. Wright, C. Hart Merriam & C. D. W. 9-10:30 on Washn. Memorial Institute." [SIA, Walcott Papers, Diary, December 2, 1901]. From this same source, one also finds out that Gilman, Pritchett, Billings, and Walcott, dined with Mr. Carnegie that night. After noting the size of Carnegie's gift, Walcott added " . . . for research work & postgraduate work in Washington."

It seems clear from Walcott's memorandum that Carnegie and Walcott were immediately attracted to each other. At 51, Walcott was fourteen years younger than the steel baron, and 12 years younger than Billings; perhaps, upon reflection, Carnegie's advisers for this new venture appeared just a bit too old for Mr. Carnegie's taste. Walcott was known for having a commanding presence, and, incidentally, he was a foot taller than the 5'2" philanthropist. Whatever the reasons, they seemed to trust each others' motivations implicitly.

That same afternoon, Carnegie dropped a note to Roosevelt: "Messr. Gilman, Billings. & Walcott and Pritchard [sic] Heads of Geological Survey & Boston Polytechnic—respectively, have approved enclosed— . . . [LC, Carnegie Papers, Carnegie to Roosevelt, December 2, 1901]; apparently a statement on the bequest, no longer available. In the final details for an official statement to be released a few days later appears the comment: "The other directors shall be Andrew D. White, Daniel Coit Gilman, John S. Billings & Charles D. Walcott, who have been my advisors, . . ." [LC, Carnegie Papers, Gilman to Roosevelt, December 5, 1901]. In his letter of November 28 to Roosevelt, Carnegie had mentioned only the first three names. This confirms that Walcott had instant impact on Carnegie.

At the White House luncheon, Nicholas Murray Butler was also present. He remained behind trying to convince Roosevelt that a quasi-governmental research institute run by the President and Congress simply would not work. Eventually that evening they telephoned Secretary of War Elihu Root who unwillingly came to the White House. After getting an outline of the scheme, he retorted "What damn fool suggested this idea?" [Jessup, 1938, p. 489-490].

Root was not the only one upset. Carnegie's largess was a great news story. As the reporters dug into it, they pounced on the point that the donation was not to be cash, but the 50-year bonds in the United States Steel Corporation. Newspaper clippings in the library of the Carnegie Institution of Washington indicate widespread concern as to the propriety of the government even tacitly endorsing a gift from United States Steel. Peculiar stock transfers and corporate mergers in the steel industry were current events, and memories of the Homestead Strike lingered in many minds. Walcott continued the story:

> During the week of December 9, 1901, Dr. Walcott talked with President Roosevelt in relation to the proposed Institution, and he requested him to consult with some of the leaders in the Senate and the House of Representatives in relation to the possibility of an act of incorporation being obtained from the Congress. Dr. Walcott consulted with Mr. Cannon and several members of the House, Senator Allison and other Senators, and told the President that as far as he could learn there was no probability of obtaining a charter at that session of Congress, owing to the hostility then existing toward the United States Steel Corporation. This was strongly manifested as Mr. Carnegie had stipulated that he would present ten million dollars ($10,0000,000) in United States Steel Corporation bonds as the foundation of the new institution.
>
> This was reported to President Roosevelt, who had also consulted with Senators and Members, and he decided to send for Mr. Carnegie and inform him that it was not practicable to obtain the desired act of incorporation. Mr. Carnegie had set his mind on having a "National Charter," and it was feared that he might decline to have anything further to do with it. On December 18th President Roosevelt sent for him and told him that he could not secure a National Charter. Mr. Elihu Root was present and both urged Mr. Carnegie to get a charter elsewhere and carry out his intention. Mr. Carnegie gave no indication of what he would do about it. In the afternoon Dr. Gilman and Dr. Walcott talked with Mr. Carnegie as to what could be done. Mr. Carnegie said that as Congress "declined to grant a charter, it might be just as well to let it go". Dr. Walcott told him that he thought the work was worth doing and that he would draw up the articles of incorporation under the general incorporation law of the District of Columbia which was a national law and submit it to him. Mr. Carnegie said: "All right; go ahead, and send it over to me" [SIA, Walcott Papers, Origin, pp. 5-6].

Although today Washington D.C. has a city government, this is a relatively new development. Local government was in force for a few years after the Civil War, but thereafter the Congress had direct control of the District of Columbia. For some legal purposes, the city was treated as a Federal territory, and Walcott was quite correct that incorporation in Washington was effectively national and was second only to a Congressional charter. That particular legal point may well have been the factor which persuaded Carnegie to continue with his scheme and to base the organization in Washington. At the same time, because of the Federal presence in local affairs, working one's way though the city bureaucracy was a complex affair. It was here that Walcott's real estate activities and chartering of non-profit organizations became the key to success.

FOUNDING THE INSTITUTION—1902

As a result of his meeting with Walcott, Carnegie recovered totally from the Congressional rebuff. He wrote the Principal of St. Andrews University:

> Very busy just now planning the new body which is to control the Ten Million Dollars which I am giving here to be used for securing the exceptional man for the work for which he is intended, and supplying the necessary apparatus for experiments and research. We aim to find the geniuses of the Republic and set them to work on the higher problems. Glad to say that the President takes the deepest interest in it, and I have

> been twice already lunching with him at the White House [LC, Carnegie Papers, Carnegie to Donaldson, December 20, 1901].

The theme of finding and supporting the exceptional man caught the fancy of reporters, but the problem of how to identify such an individual caused difficulties [Reingold, 1979, p. 318-320].

Having been given a "go ahead" on the afternoon of December 18, Walcott, who was always a fast mover when the timing was right, acted swiftly. By December 20, he had dispatched a letter to New York communicating the draft of the articles of incorporation which he had written the evening before and an extract of the pertinent regulations of the District of Columbia. By his second paragraph, Walcott was resolving the point that the majority of incorporators had to be citizens of the District of Columbia.

> In view of the greater scope and character of the institution, I suggest that you come here when the articles are decided upon, and unite with Secretary [of State] John Hay, Justice [of the Supreme Court] Edwin Douglas White, Dr. Daniel Coit Gilman, and, if you wish, myself in signing the same. These incorporators can at once meet after the articles of incorporation are filed, elect the trustees, and call a meeting for organization of the board, adoption of by-laws, etc [LC, Carnegie Papers, Walcott to Carnegie, December 20, 1901].

Andrew Carnegie, man of action, had met a kindred soul. Carnegie liked what he read, and Walcott became one of the original incorporators of CIW.

In the same letter, Walcott emphasized the need for organization of the Board of Trustees in such a way that each trustee would serve for a limited time and would be replaced on a rotational basis. He brought up the issue of Carnegie as a member of the Board. "I regard this as the greatest of your many acts to benefit mankind, and as a member of that Board you would be in more thorough touch with all that is going on, and be able to give counsel and suggestion that would be of great value and service to the institution" [LC, Carnegie Papers, Walcott to Carnegie, December 20, 1901]. At the conclusion of this long letter, he wrote, "I have been studying and working for the establishment of a [sic] institution for research for several years, which I trust will be sufficient reason to you for my making the many suggestions contained in this letter and the accompanying papers." As an even more personal note, Walcott scribbled "Out of the hands of the dentist and all right again—Sorry to be so used up when you were here." There is irony on a cosmic scale to be offered millions while suffering the anguish of a tooth ache!

Carnegie did not come to Washington for the incorporation, his place at that event being taken by Carroll D. Wright, Superintendent of Labor since 1885.

> As Carroll D. Wright, of the Department of Labor, is suggested as a member of the executive committee, will you not send for him and inspire him with your conception of the purpose of the new Institution. He is an unusually clear-headed, suggestive man, and worth taking into your confidence. All members of the suggested executive committee will then have talked with you and they will do better work as a result of it" [LC, Carnegie Papers, Walcott to Carnegie, December 28, 1901].

In 1901 the Bureau of Labor was under the Department of Interior so that Wright and Walcott had official contact, as well as more personal ones; Wright was also deeply involved in the Washington Memorial Institution plan. From this, it is evident that Walcott played a role in determining the composition of the executive committee; it was through Wright that CIW ventured into the "softer" sciences.

Three days after his first long letter, Walcott wrote again giving the name of a possible candidate for the Board from the west coast, and conveying and supporting Gilman's suggestion that the bylaws be changed to hold the annual meeting on November 25, Carnegie's birthday; Carnegie declined that honor. Walcott concluded, "I expect to be in New York on Sunday, the 29th inst., and again on my return from Rochester, January 3rd. I should be very glad to call and pay my respects if agreeable to you" [LC, Carnegie Papers, Walcott to Carnegie, December 23, 1901]. Mindful of the proprieties and, like Carnegie, a deeply religious man, Walcott added in pen "on Sunday."

Walcott may not have planted the seed for a research institution in Washington, but after the sprout broke ground, he forced its growth. The botanical metaphor may be extended to the old saying that money does not grow on trees, a point better known to the government scientist even than to those in academia. Walcott's speed in preparing a draft, even while suffering a toothache, is remarkable. To that must be added the point that in Washington the time of year before Congress went into session was extremely busy. Within a few months, the Newlands Act would be passed and Walcott would also be in charge of the Reclamation Service. Reclamation was a pet project of Roosevelt's and Walcott was actively involved on Capitol Hill shepherding the legislation. Add to that, Walcott's trip to Rochester was to give an address as outgoing president of the Geological Society of America; for the average person, this duty alone would have been enough to occupy one fully.

The 29th was not convenient for Mr. Carnegie, so Walcott and Dr. Gilman went to New York earlier than Walcott had planned.

> On December 27th, [I] met Mr. Carnegie at his home in New York and discussed the proposed articles of incorporation with him, Dr. D. C. Gilman, Abram S. Hewitt, and John S. Billings. Mr. Carnegie approved of my draft of the proposed articles of incorporation, and decided the name should be: Carnegie Institution. I suggested that it might be well to give it a National name and state that the Institution was founded by Andrew Carnegie. After thinking it over, he said "No, Carnegie Institution is good enough" [SIA, Walcott Papers, Organization, p. 4].

Walcott was among friends at this meeting. At the 25th anniversary dinner of the U. S. Geological Survey in 1904, he credited Abram Hewitt with the founding of the organization when Hewitt was in the House of Representatives and pointed to him as a long-time friend of the Bureau. Still, the speed of events was a bit more than Mr. Hewitt liked. He suggested that the bonds be turned over to a Board of Trustees who at their leisure could decide when and where to incorporate [LC, Carnegie Papers, Hewitt to Carnegie, December 30, 1901]. There was no interest on Mr. Carnegie's part in any slowdown. Walcott's diary entry provides the additional information that he had coffered with Nicholas Murray Butler the night before, that Walcott had a private hour with Andrew Carnegie prior to the 10:30 gathering, and that this meeting went so smoothly that there was plenty of time afterwards for a luncheon with Mr. and Mrs. Carnegie [SIA, Walcott Papers, Diary, December 26 and 27 1901]. In assembling the history, Walcott was quite clear that: "My draft of proposed articles of incorporation of research institution was approved by Mr. Carnegie & gentlemen present" [SIA, Walcott Papers, Notes, p. 2].

As soon as the morning's meeting was concluded, Carnegie wrote to prospective trustees. Walcott formally accepted three days later [LC, Carnegie Papers, Walcott to Carnegie, December 30, 1901], and with the exception of Grover Cleveland, all those contacted agreed to serve. The day following the New York meeting, when Walcott was back in Washington, packing for his trip to Rochester, he still took the time: "As per request of Mr. Carnegie talked with President Roosevelt, John Hay, Justice White and Carroll D. Wright. Hay, White & Wright agreed to serve" [SIA, Walcott Papers, Notes, p. 2].

Carnegie was not thinking small and even with Hay and Root, two cabinet officers, already involved, he was extremely keen to have President Roosevelt on the Board. Roosevelt was equally willing, writing: "I will serve with the greatest of pleasure" [LC, Carnegie Papers, Roosevelt to Carnegie, December 31, 1901]. On January 3, 1902, Walcott "Talked with President Roosevelt 6. P. M. & asked him to become a Trustee of the proposed Carnegie Institution for Research. He declined for official reasons" [SIA, Walcott Papers, Notes, p. 3]. Cooler heads had prevailed.

The problem of Carnegie's wish for Roosevelt on the Board of Trustees, or an earlier draft attempt to emulate the Smithsonian Institution, or both, led to the addition of five *ex officio* members of the Board. In one way, this action appeared useful in adding to the prestige of the emerging organization, but it was an action that would cause difficulty less than a year later. Indeed, indirectly this created a problem that would influence the future course of CIW. Several different lists of additional prospective members of the Board of Trustees are in the Carnegie collection [LC, Carnegie Papers, volume 86], but they are undated and it is impossible to determine their sequence. On all but one list, Walcott's name appears. One list has the Secretaries of Agriculture and Interior as *ex officio* members of the Board and that could only have come from Walcott's knowledge of government laboratories in Washington.

Ex officio positions were prominent in Carnegie's mind. Walcott noted this additional category of Trustees on December 30, though it is not entirely clear whether it was he or Carnegie who had instituted the idea. The draft bylaws were modified accordingly, and old Washington hand Walcott suggested incorporation be on Saturday because Friday was "cabinet day." All of the incorporators became members of the Board of Trustees.

Walcott was still concerned with a basic problem in building this new organization, on which he had touched earlier:

> As I recall, something was said about having the trustees serve for life. I sincerely trust that this will not be done, as it means the accumulation of men in the board who can not attend the meetings, or if present will take very little active interest in what is going on. A number of our most important institutions in the country are now suffering from dry rot as a result of permanent membership in the boards of trustees.
>
> In the draft of the By-laws which I sent you, it was provided that there shall be a change of one-fifth the membership each year, after the first four years. I think this is little enough infusion of new blood [LC,

> Carnegie Papers, Walcott to Carnegie, December 30, 1901].

Walcott did not specify the institutions, but he was familiar with the Smithsonian and the National Academy, neither of which in 1901 could be termed scintillating organizations. His idea of limited terms for trustees is not in the original bylaws.

Meanwhile, however, the future seemed rosy and CIW was incorporated, January 4, 1902, the day after Roosevelt finally declined. Marcus Baker, a faithful Geological Survey employee, rushed the signed Articles of Incorporation to the Recorder of Deeds and rushed back with the information that the organization officially existed. A few moments later, the Trustees were elected and CIW began. Three days later, Carnegie conferred with Gilman, Hewitt, Billings, Wright, and Walcott [SIA, Walcott Papers, Notes, p. 2]. With the exception of Secretary of War Root, and the Philadelphia physician S. Weir Mitchell, a long-time friend of Billings, this was to be the Executive Committee. All was now arranged for the new organization. Just to keep chronology accurate, this gathering was 12 days before the Board formally met to elect the Executive Committee.

By mid-January, Walcott was writing to Carnegie on CIW letterhead stationary. He was concerned about how vacancies on the board would be filled, but enthusiasm was running high. Even the refusal of the Chief Justice of the Supreme Court to serve as a Trustee—it would have been a conflict of interest with his position as Chancellor of the Smithsonian Institution—did not dampen ardor. The organization was set and there was no need to patronize Carnegie. Thus, there was no insincerity in Walcott concluding a letter: "Your great gift to the Scottish Universities will be of lasting service, but I think the new institution here will be of great benefit to mankind as time goes on" [LC, Carnegie Papers, Walcott to Carnegie, January 18, 1902].

1902—The First Steps

After one more conference between Walcott and Carnegie and three meetings between Dr. Gilman and Walcott, the gathering of the Board of Trustees was set. At 2:30 PM on January 29, 1902 in the office of the Secretary of State, CIW was launched. The Trustees met again briefly the following day and, thereafter until late November, all action was taken by the Executive Committee. Henry Adams, historian-cynic, was a close friend and neighbor of Trustee John Hay, but that did not prevent him from making acerbic comments on CIW.

> Our distraction of the week has been the meeting of the Carnegie Trustees, twenty-seven contemporaries of mine doddering like me. I could not believe Carnegie so wretched an organizer. Any elderly literary frump would have done better, but I suppose Carnegie gained his object by harnessing respectability to his coach. He harnessed one too many when he got Wayne [McVeagh], who gave all the trouble he could. Hewitt drifted in through the snow to see me, and really carries his eighty years with considerable affection of success, though he had little to say that sounded youthful. Henry Higginson stayed with me till Friday, very pleasant, and quite youthful in mind, so that I really enjoyed his company, though he could tell me nothing novel [Leverson, et al., 1988, Adams to Cameron, February 2, 1902].

Adams' view of the Board may have been a little harsh, though admittedly for a group whose aim was to foster research, science was inadequately represented among the Trustees.

Walcott was elected Secretary of the Board. He reported to the patron:

> I have thoroughly enjoyed being associated with the men forming the Board of Trustees, and greatly appreciate the honor of being elected Secretary. I think, however, that if it were all balanced against the pleasure of meeting and knowing you, I should choose the latter. I have rarely met a man whose personality and whose ideas of life were more pleasing to me [SIA, Walcott Papers, Walcott to Carnegie, January 31, 1902].

Mr. Carnegie was equally delighted with the Board of Trustees, with their actions at the organizational meeting in January, with the auspicious start of his enterprise, and especially with Walcott. He travelled down to Florida a happy man and a month later wrote Walcott, calling him "My dear First Authority" [SIA, Walcott Papers, Carnegie to Walcott, February 28, 1902]. There was ice in Florida, and Carnegie became interested in the issue of climatic change. Walcott dug out a report made by the Weather Bureau a few years earlier and sent it along.

Fortunately for the future of CIW, Walcott also served as Secretary of the Executive Committee. One of the first actions of that Committee was to grant him a salary of $100 per month, an amount he earned! They also awarded, President Gilman $10,000, a princely sum [CIW, Minutes of Executive Committee, 1902, p. 2]. A home had to be found for the new organization and Walcott located it first in a private house at 1439 K Street and later in the Bond Building, on the southwest corner of 15th and New York Avenue, both relatively close to his Geological Survey office. Marcus Baker became Assistant Secretary.

If the Board of Trustees was weak on research interests, the Executive Committee, presided over by Dr. Gilman who generally did not take part in discussions, was better, but was peculiarly constituted. Billings and Mitchell were both trained as physicians and were active in that field. Abram Hewitt was a business man who knew the steel industry. Caroll D. Wright was mainly a statistician. John Hay may never have even been aware of what research implied. Root, one of the busiest of Roosevelt's Cabinet officer, was not a scientist, but he was keenly aware of what scientists did. Unfortunately, he was seldom at the early Executive Committee meetings, especially those in New York. In trying to reconstruct past events and personalities, one should never overlook the importance of what is called the old boys' network. Root and Walcott both grew up in central New York. Young Walcott, out collecting fossils at age 18, met Root's father, "Cube" Root, the mathematician and mineralogist at Hamilton College. When Walcott, who never went to college, received his first honorary doctorate from Hamilton College in 1898, Elihu Root was on the Board of the college.

Despite his limited participation at first, Root was active and greatly interested in the organization. Overall, he was an important member of the Executive Committee for decades; the Elihu Root Auditorium in the headquarters of CIW is an appropriate memorial to his service. Wright, Root, and Walcott would have agreed on virtually every policy matter, and probably Hay would have followed Root's lead. Billings and his friend Mitchell had little in common with Gilman or Hewitt. Walcott labored hard, but in 1902 the Executive Committee never developed a consensus on the issue of support for individuals or for larger projects. In the end, they agreed to search for the "exceptional man" and support a few modest-sized investigations which Billings particularly desired. Inside of a few months, cordiality within the meetings declined.

As part of his duties, Walcott prepared a draft of "Aims of the Carnegie Institution," and appended to it are a few thoughts that he either used in speeches or in later documents. They show Walcott's concept of the organization and are worth repeating for the benefit of future administrators: "To lead and help—not to dominate and suppress." "Aid to the individual of promise, whether he comes from the farm, the workshop, or the university." "Aid to objects in the front of the line of progress of science rather than to those left behind in the advance." "The successful manufacturer replaces his plant every half decade if conditions demand. The research institution must be prepared to do the same." "Simple, effective, business-like administration, with as much concentration of energy as the success of the enterprise will permit" [SIA, Walcott Papers, appended to Aims].

Almost immediately, CIW was inundated by letters. Any number of persons had good ideas as to how to spend Mr. Carnegie's money. For example, Walcott sent his fifth summary of letters received to Carnegie slightly more than month after his election as Secretary [LC, Carnegie Papers, Walcott to Carnegie March 6, 1902]. Walcott seems to have been determined that Carnegie be fully informed of the progress of the organization bearing his name.

The Ex Com, as Walcott abbreviated the Executive Committee, had the responsibility of announcing the CIW to the scholarly public; this in no way reduced the number of inquiries from those who wanted Mr. Carnegie's money, but it did add some wheat to the chaff. President Gilman reported informally that "Some admirable suggestions have been made, and I should not be surprised to find that the aggregate of requests for aid worthy to be granted would exceed the income of funds" [LC, Carnegie Papers, Gilman to Carnegie, March 5, 1902]. Walcott saw matters a bit differently and wrote an historian: "I have looked over most of the communications received and find that yours has more real practical suggestions in it than nine-tenths of all that has come in." [LC, Jameson Papers, Walcott to Jameson, March 7, 1902].

Notwithstanding that its membership was scattered among Washington, Baltimore, Philadelphia, and New York, the Ex Com did work very well at the beginning. A key point is that Abram Hewitt participated, despite his age. At its very first meeting, after the Board had adjourned, the Ex Com instituted the concept of small advisory committees to prepare reports for the Board as to the needs of their respective fields. Even more significantly, most of the members of these committees were designated before that meeting was completed. Although there was discussion in regard to a few of the later appointees, there was no dissension. This notion of advisory committees and their immediate staffing suggests that during January a great deal of thought had been given to the matter of how to apportion Mr. Carnegie's largess.

Some concern has been expressed regarding early use by CIW of the terms "minor" grants and "research assistants," as being inappropriate, or at least an unfortunate choice of wording [Reingold, 1979, p. 319]. As to the latter, even before the Board of Trustees met, Ira Remsen wrote to Billings, "The idea of research assistants commends itself to me more & more [NYPL, Billings Papers, Remsen to Billings, January 16, 1902]. It was never intended as a demeaning term. Major and minor are not so clear, but the Board agreed to appropriate $75,000 for administration and for use of the Ex Com between their first and second meeting [CIW, 1903, p. xxiv]. They would make no large investment of funds, that is major amounts, until after advisory committees had been appointed and had reported. On the other hand, the Ex Com did have the authority to spend limited sums for worthy projects. In that light, major and minor do not appear inappropriate.

A few small sums were appropriated by the Ex Com; for example, Dr. Mitchell was interested in venoms and although only three grants were made in 1902, among them was one for a study of these poisons. By far, the main effort was directed toward the advisory committees. The problem of finding appropriate advisors, contacting them, setting up meetings, and getting reports in time for the fall meeting was formidable.

In a talk that fall, Walcott gave his version of history. In this he outlined a progression through the Philosophical Society of Washington, the Joint Committee, and the Washington Memorial Institution:

> . . . until plans for research were formulated clearly enough to attract Mr. Carnegie's attention.

He said sixteen advisory committees, including forty-six members, had been appointed, whose confidentially printed reports, filling over 200 pages, had been presented to the trustees at the last meeting. Some general principals, both of exclusion and inclusion have been adopted, special - emphasis being laid on the man responsible for any research [Hayford and Wead, 1902, p. 2]. That spring Walcott had advised Carnegie:

> In a general way Dr. Billings is looking after all marine biological questions; Dr. Mitchell, physiology; Col. Wright, economics; and Dr. Gilman, the general field of research, especially historic archaeology, history, philology, art, and education. The Secretary (Walcott) will give special attention to geology and physics, and endeavor to keep all of the committees at work so that there will be a large amount of data upon which to base recommendations to the Trustees in November for a general plan of operations [LC, Carnegie Papers, Walcott to Carnegie, April 3, 1902].

By early May, fourteen advisory committees were at work. The Ex Com held five meetings before Dr. Gilman left for a European trip in April. One authority has at least implied that by absenting himself, Gilman lost what chance he might have had to control the institution [Masden, 1969, p.166-167]. Yet there really was little more to be done at this stage. Even though Gilman returned in late August, after seeing Mr. Carnegie in Scotland, the Ex Com did not meet again until October.

Actually, President Gilman had lost control of events at CIW long before his European tour.

> He entered most enthusiastically into the subject of the organization of the Institution up to early in March, 1902, when he presented certain views in relation to the conduct of the administration. These did not fully meet with the approval of the Executive Committee, and Doctor Gilman very rarely after that brought any matter to the attention of the committee, as he felt that he was not in harmony with it. This feeling developed still further during the fall of 1902, Doctor Gilman confining himself almost entirely to act as the Chairman of the Executive Committee without suggesting matters for the consideration of the Committee" [SIA, Walcott Papers, Origin, p. 6].

The lack of harmony—animosity may be too strong a word for the gentle Dr. Gilman—was not directed at Walcott, for Gilman wrote him three cordial letters from Europe, reporting on his trip [CIW, Historical File 2]; all have the friendly form of salutation to "My dear Walcott."

A foretaste of problems came in May. Dr. T. A. Jaggar, Jr., a promising young man, sent Walcott a telegram "Planning to go Martinique at once. Is there any chance of support from Carnegie Institution [CIW, Geophysical Lab, Jagger, Thomas. A. Dr., May 10, 1902 and attached later correspondence]. In his capacity as Secretary, Walcott wired back that same day: "Your telegram received. Dr. Gilman is in Europe, the Executive Committee scattered, and there is no authority for incurring such expenditures at present. I hope you will find some other way of going." In his capacity as Director of Geological Survey, Walcott immediately sent Jaggar letters of introduction, indicating that in part he was representing the Survey.

Jaggar was able to get on the Navy cruiser "Dixie" and was appropriately grateful for the letters. He indicated that he proposed to write a memoir on the eruption which could be published by CIW. "The investigation will necessarily be modified in its scientific efficiency according to the source of funds. At present I am undertaking it at my own expense."

In his capacity as Secretary, Walcott circulated the letter of ". . . . Jaggar, Jr., of Harvard University, who has gone to Martinique to investigate the volcanic eruption now going on there. Doctor Jaggar is eminently well qualified to make such a study and to prepare a memoir on the subject." Walcott suggested an allotment of $500 for expenses, allowing the issue of publication to wait and ended: "Personally I am willing to vote for the allotment."

Root and Wright were also in favor of Jaggar's request. The elderly Hewitt pointed out that the American Museum had sent an investigator and concluded that: "If the other members of the Committee think well of the appropriation I will cheerfully assent, but I do not feel qualified to pass any judgement as to its expediency."

Jaggar did not get the $500, for there was no majority of the Ex Com. Gilman and Mitchell were unavailable, but Billings left no question as to his opposition.

> It appears to me that all the investigations into the present condition of affairs in connection with the volcanic eruption in Martinique should be made by geologists and others already on the ground or who will be sent there by the French government,-and without further information of the kind of investigation which it is proposed that Dr. Jaggar should make, I should not be willing to vote for the funds to carry on the work. So far as I can see there is no need for the Carnegie Institution to collect records of what is past and as for the future work involving the taking of seismological records, I should wish to know what the French Government will do before voting that the Carnegie Institution should undertake the work.

This was an exceptionally strong reply to a modest request for support.

1902—THE LATTER PART

While Gilman was away, Walcott prepared a statement of guidance for the advisory committees. Billings responded to a draft:

> I have gone over these notes and return the first part, being those which it is proposed to send to the various scientific advisors. My advice is to omit from this everything except that which relates to original research. The question of higher education is one upon which we do not want any advice at present, and we are likely to get a great deal more than we require [CIW, Historical File 2, Billings to Walcott, May 20, 1902].

It would be interesting to have clearer insight into what Dr. Billings considered research. As noted above, a few days later he viewed an ongoing eruption with disfavor as something of the past. At the same time, Billings was arranging for CIW to support publication of *Index Medicus*, a fundamental bibliography, but one that hardly fits any definition of original research.

While the CIW itself was being organized, Walcott was also pursuing a laboratory for geophysical research within the organization. That may be set aside for the moment, but a laboratory does require a site and on May 1, Walcott mentioned land to Carnegie [SIA, Walcott Papers, Walcott to Carnegie, May 1, 1902]; even though Carnegie was now at his castle in Scotland, this did not inhibit Walcott's correspondence. By the middle of the next month, Walcott had options on 110 acres of prime real estate. Before taking out the options, Walcott had the properties examined by trustees Root, Hay, Wright, and Henderson, an *ex officio* trustee by virtue as his position as Speaker of the House of Representatives; it is difficult to see what further steps he might have taken to insure this was a prudent action. Walcott reported his real estate activity to the founder and interweaved some philosophy of administration.

> During the past three months I have been at work, as a subcommittee of one, looking up possible sites for the Institution that would provide for present and future needs. Before starting out I carefully read your Deed of Gift and remarks, and also took into consideration the suggestions that have been made from time to time by various persons interested in the organization. The more I thought the matter over the more thoroughly convinced I became that it would be necessary to have a vigorous central administration. An experience of twenty years in connection with the administration of scientific organizations, and observations of others having similar objects, lead me to think that the energies of any institution will be more or less dissipated, and that but a limited degree of success will attend it unless it have an organization where the standard of thought is set and maintained. I do not believe that the Carnegie Institution can become a great research organization unless it have a strong effective direction from a central office. Men engaged in research work are very like other human beings. The most exceptional and gifted men need to be looked after if waste of time, energy, and money is to be avoided. I find that some men will go on year after year with very little supervision or suggestion, but where there is one such, there will be ten more or less impractical men who require consultation and suggestion. The same will also be true of committees of experts who will be placed in charge of the special committees. This is clearly demonstrated by the committees which have been placed in charge of investigations under other scientific organizations. Some committees are effective, others are not.
>
> It should be the duty of the executive officers of the Carnegie Institution to see that the work is carried forward effectively, no matter by whom it is being done. If all is going well there will be no need for suggestion or change. If it is not going well, it should be looked after promptly.
>
> With these views in mind I come to the conclusion that it was desirable to secure as a site, not only for the present but for centuries to come, a considerable tract of ground in a readily accessible suburb of Washington, where the future development of the city would not

interfere with the purposes of the Institution. This renders it essential that the immediate environments of the central administration building and laboratory should be under the control of the Institution" [LC, Carnegie Papers, Walcott to Carnegie, June 19, 1902].

Shortly after writing this, Walcott and his family migrated to Woods Hole for the summer. A stenographer came with him and he carried on an extensive correspondence for CIW in addition to his official duties. Even when Dr. Gilman returned from Europe, Walcott handled almost all activities.

> It was the custom of Doctor Gilman as President of the Institution to visit Washington for a few hours once each week and talk with me in relation to matters connected with the institution. This he did during his three years as President. He also attended the various meetings of the Executive Committee" [SIA, Walcott Papers, Origin, p. 6].

Walcott remarked: "I find that the best time for me to work at the office of the Institution is from 9 a.m. to 10.30 a. m., and from 3 p. m. to 4.30 p.m. This enables me to be at the Survey office during the hours of the day when public men are most apt to drop in; also to attend to the work of this office" [CIW, Historical File 2, Walcott to Gilman, December 2, 1902]. Prior to an Ex Com meeting in late October where the plan of work that was to be submitted to the Board was to approved, Walcott almost pleaded with Gilman to come to Washington earlier so that they might confer prior to the meeting [CIW, Historical File 2, Walcott to Gilman, October 24, 1902].

By November 7, the various reports had been gathered and sent off to the printer. The next step was the second meeting of the Board of Trustees.

NOVEMBER-DECEMBER, 1902—MAJOR SETBACK

As late as October 6, Walcott was still writing to Billings concerning the examination of title to the land under option [NYPL, Billings Papers, Walcott to Billings, October 7, 1902]. As late as early November, Abram Hewitt was still taking a part in affairs of CIW. He suggested in a letter to President Gilman that Walcott receive an honorarium of $5,000 for his services; Gilman was delighted to pass this suggestion along to the members of the Ex Com [NYPL, Billings Papers, Gilman to "Dear Sir," November 7, 1902]. These two separate events make a letter from the astronomer Norman Boss all the more interesting.

> Gossip concerning the affairs of the Institution both in Baltimore [at the fall meeting of NAS] and in Washington seems to be quite active. I found a receptive attitude all that was necessary in order to learn something about the supposed tendencies at issue. Sentiment seemed to be almost unanimous against what the Baltimore men term the "Government Bureau" idea of organization for the Institution, as opposed to the "University plan." The former means, of course, concentration upon a large scale of buildings, laboratories, and officials in Washington.
>
> From my communications both with Dr. Walcott and Dr. Gilman it did not seem to me that there was much concert of action between them; yet I purposely avoided discussion of personal issues and probably missed information thereby. [NYPL, Boss to Billings, November 17, 1902].

Boss went on for several more pages mentioning an idea of Dr.Gilman's for an advisory board. He then went back to Alexander Agassiz as one who would be strongly in favor the "University plan" and mentioned Simon Newcombe as a "government bureau" type. It is impossible to say if this letter influenced the later actions of Dr. Billings, but if nothing else, it reinforced his opinions.

There was more than just gossip behind this letter. Six months earlier, Boss had taken it upon himself to write something of a self-serving letter to Carnegie [LC, Carnegie Papers, Boss to Carnegie, April 7, 1902], pointing out how much he had accomplished for astronomy at the Dudley Observatory in Albany, laboring for decades with virtually no support in either funding or staff. A point emphasized by Boss was that: "Nothing should be done which might have a tendency to cramp individual effort, to discourage the exceptional man, or to reduce scientific investigation to a machine-like organization, or to establish scientific dictatorship." Walcott had built the Geological Survey into a powerful instrument for science, and in no small measure some of the non-government scientists in America were both fearful and jealous of his accomplishments.

As is well known, Abram Hewitt was too ill to preside when the Board of Trustees assembled again in November. Equally unfortunate for Walcott, that fall Carnegie became ill in London and did not return to America until December. He might have been at the annual meeting and he might have been further influenced if Walcott could have written to him. On an undated discussion of the site, Walcott scribbled that "All of executive committee approved except Billings" [LC, Carnegie Papers, Volume 93].

With Hewitt confined to his home, Vice-Chairman Billings was in the Chair. After the various formalities and minor reports were disposed of in a first session, the report of the Ex Com was taken up. Billings pointed out that the Committee was unanimous on all recommendations save one, the purchase of land. Billings then deliberately went on to put this item at the head of his agenda. He mentioned that support for the action was nearly unanimous and that: "If we were to have a laboratory business, if it were a matter of ten or fifteen acres, I should not have objected." [CIW, Transcript, First Meeting Board of Trustees, p.17]

Walcott answered a question as to when the option would expire. He had invested a great deal of time in finding the acreage, in having it examined by several parties—including several members of the board—and in keeping the Ex Com and Andrew Carnegie fully informed. It was for naught. What happened next is best described by a term known to poker players; Walcott was sandbagged. Alexander Agassiz, President of the National Academy, and therefore an *ex officio* member of the Board, spoke up.

> I feel very much the way Dr. Billings does, that it is a very dangerous thing for the Institution to pledge itself, before knowing what we do now, to the equipment of a large number of laboratories situated in Washington. That would mean that the Institution will naturally change itself into a University if it is to run a series of laboratories. . . . It means the establishment of a large number of offices and a large number of officers, and after establishing two or three laboratories the Carnegie Institution will find itself exactly in the position which many universities are, without a cent to do what they have been established to do. They will have funded their whole capital, they will have mortgaged their income and have nothing to do for the primary object for which the institution has been established—assisting individuals in their research or special lines of research—and for that reason I hope we will put off, at any rate, the purchase of any thing which might bind the Carnegie Institution to the policy of having laboratories on an extensive scale in Washington. [Minutes, First Meeting of Board of Trustees, p. 18-20].

Alexander Agassiz was the Nemesis of government scientists, though it was never clear what this group had done to deserve his wrath. He had assaulted John Wesley Powell in the 1880s, and now he gave his successor more or less the same treatment. In Agassiz' view, research should not be done by the government. It was to be a private affair of the academic community and the sole function of the government was simply to place funds and equipment at his disposal. Alexander Agassiz had made a fortune, yet even he not afford a ship such as the Fish Commission steamer "Albatross," which the government essentially provided at his beck and call; his continual attacks on government science are indeed curious.

At this point, there ensued an argument concerning the need to establish a reserve fund, and the debate on land wandered away from the subject. Dr. Gilman sat silent until virtually the end of the meeting when, after it was far too late, he produced a note from Elihu Root, who had been present at the morning session, strongly supporting the purchase. Walcott wrote little privately about his major defeat: "The Institution starts in a conservative manner" [SIA, Walcott Papers, Diary, November 25, 1902].

There is the old doggerel that: "For want of a nail, a shoe was lost, for want of a shoe . . ." Had CIW purchased 110 acres at $2,500 per acre, the entire course of the Institution would have been different, for it would have developed centralized laboratories. Had Hewitt been well enough to preside, the problem might never have arisen, for although he was originally opposed to the purchase, Walcott had won him over. Had Root been present at the afternoon session, he might have swung the tide in favor of the purchase, for he was a Trustee that the others would listen to. Had Carnegie not wanted Roosevelt on the Board, *ex officio* member Agassiz would not have been at the meeting, and Walcott might have been able to carry the day.

FIGHTING AGAINST DRIFTING—1903

Notwithstanding this major disruption of his plans, Walcott carried on. An Ex Com meeting was scheduled for mid-January at Mr. Hewitt's house [JHU, Gilman Papers, January 3, 1903]; the Chairman of the Board may have been infirm, but his mind was still clear. Early in January, a cryptic note appears in Walcott's diary. After attending to correspondence in the CIW office, he wrote: "Had a talk with Dr Gilman about the C.I. and his relations to it" [SIA, Walcott Papers, Diary, January 5, 1903]. Perhaps Walcott was advising him to be more forceful. Two days later, Carnegie was in Washington for the opening of the public library he had endowed and Walcott conferred with him [SIA, Walcott Papers, Walcott to Carnegie, January 8, 1903]. Shortly thereafter, Walcott and Charles R. Van Hise visited Carnegie in New York; a few days later, Mr. Hewitt died. One result of these two events was another long letter from Walcott to Carnegie.

> Of the many things that I have to thank you for, the privilege of knowing Mr. Hewitt, and being thrown into close contact with him, is probably the greatest. . . .
>
> My great object in this matter is to make the Carnegie Institution the great research institution of the world. I do not think that this can be done by making small grants to A., B., and C., in connection with university laboratories. That is a proper and wise thing to do, but

it will not carry forward the ideas indicated by you in your Deed of Gift. If it were in my power today, I would secure Dumbarton, [one of the estates which had been under option] and as time goes on, establish there a group of Carnegie laboratories and ample groups for carrying on all research work that could be undertaken in this region. On the hill overlooking to Rock Creek National Park, and the city, there would be erected an administration building, where rooms would be provided for temporary research work such as that of Professor Newcomb, and where space for computers, books, etc. is needed but not laboratory equipment. In a short time a considerable amount of such work will have to be provided for [LC, Carnegie Papers, Walcott to Carnegie, January 19, 1903].

It was a grand dream but it was not to be. Lest Walcott be accused of too much dreaming, space for computers meant a place for those who compute to sit, in the same way a typewriter was one who types with a machine. Language does change over the years.

Walcott saw Carnegie again at Hewitt's funeral and they exchanged views. It is generally accepted that after CIW was founded, Carnegie—with two minor exceptions [Reingold, 1979, p. 314]—made a point of not meddling in its affairs. To a large extent this is true. However, at this critical junction for CIW, with the Chairman of the Board dead, and the President ineffectual, Carnegie had definite ideas and moved to implement them. Shortly after the funeral, Walcott wrote him:

I wish to make it clear to you in view of our conversation yesterday, that I have no immediate desire to give up my position as Director of the Geological Survey. If, in the progress of events, it seems evident that the Carnegie Institution will develop into the research institution I think it should be, I would if it were your choice, and that of the Trustees, do all in my power to direct it, even if it involved severing my connection with the Survey" [SIA, Walcott Papers, Walcott to Carnegie, January 22, 1903].

This is an interesting letter to ponder. Walcott had only to crook a finger and Carnegie would have thanked Gilman for his services, whispered to the Board, and Walcott would have been President. Equally, Walcott could easily have persuaded Gilman to retire and the Board would have elected Walcott, if only because he was the only one who had a grasp on what was being done. As Director of the growing Geological Survey, Walcott was receiving $6,000 a year. Six months into the Reclamation Service for which he was also Director, he was in a new policy area fraught with potential political difficulties—all of which he managed to avoid—and some complex engineering problems. Apart from personal loyalty to the Geological Survey and respect for Gilman, there are no reasons why he should not have taken a position which paid $4,000 more, required only modest effort on his part, and would have allowed him considerable time for his research. As another play in the great historical game of "what if . . . ," had Walcott replaced Gilman in 1903, the move of CIW toward larger projects would have accelerated.

Walcott's personal loyalty is shown in another way. He suggested to Carnegie that a national collection be made in technology and named after Hewitt. Carnegie was dubious, [SIA, Walcott Papers, Carnegie to Walcott, February 10, 1903], but Walcott persisted. Although this was a failed attempt to honor a memory, it was one of the strands which eventually led to the founding in 1910 of the Bureau of Mines.

If Walcott had most of the burden for CIW in 1902, he had even more in 1903. By virtue of the way the organization was set up, most checks had to be sent to New York for Billings' signature and then returned to Washington; this was in addition to the load of correspondence. The Ex Com produced paper, but no clear direction for CIW. Walcott was away for two months in the spring on Reclamation Service business, and a month after his return, he reported to Henry Fairfield Osborn of the American Museum of Natural History:

Carnegie Institution matters are very quiet. We have rented a few rooms in an office building, where Mr. Baker and three clerks attend to different matters. I hope as soon as time permits to take up the consideration of work to recommend to the Trustees in 1904. The Executive Committee have not outlined any plans, and I presume Dr. Gilman will spend the summer as usual on the Maine Coast [AMNH, Walcott to Osborn, June 13, 1903].

During the summer, Walcott moved to Canandaigua Lake in central New York to escape the Washington heat, but again a stenographer was along. All the decisions on quality of paper for the yearbook, the binding, use of seal, and other details were made by him and approved by Billings. The correspondence involved in such housekeeping was staggering and added to this were applications for research assistantships and general correspondence. Just to make life even more complicated, Roosevelt made Walcott chairman of a committee to review the placement of all scientific work in the Federal establishment.

Walcott later had another three months in the west in 1903, again mostly on Reclamation Service affairs. Though he does not quite write what was on his mind, the direction in which CIW seemed to be going was distressing to him. "Nearly every man to whom a grant was made last year has renewed his application for another grant, usually asking for much more than the previous year. If these increases are made, it will of course prevent any new grants for next year" [AMNH, Walcott to Osborn, October 16, 1903].

On top of this load, Walcott had to deal with John Shaw Billings face-to-face at the Ex Com meetings. "By many, perhaps most, of his friends, Billings was held to be a cold-blooded personage" [Garrison, 1915, p. 354]. That is a quaint way to describe a friend! It is not clear what sort of interests Dr. Mitchell was pursuing, but Billings and Walcott were the dominant members of the Ex Com, and there were some interesting discussions between them. On the other hand, after some months the overall situation was not quite so grim as it must have been in the earlier months of 1903. Walcott wrote Billings:

> I greatly enjoyed the last meeting of the Executive Committee, as things were done in a businesslike way and matters of importance agreed upon that I think will have a strong influence on the future development of the Institution.
>
> While I do not agree with you in all things, I greatly admire your clear and logical method of working out and presenting matters, and the vigorous way in which you champion that in which you believe and fight that which you do not believe [NYPL, Billings Papers, Walcott to Billings, November 25, 1903].

The Board meeting in December, 1903, went well. There is an interesting comment on the meeting to Mr. Carnegie from Senator Henry Cabot Lodge, who was not elected to fill a vacancy on the Board:

> I was not at all surprised at the result among the Trustees, for I felt very sure that unless active steps were taken beforehand our friend Walcott would have everybody on the Board committed to his list, and that he would take pains to have me shut out if he possibly could. My selection, however, is not a matter of very great moment, my only value being that I take a great interest in the fund and sympathize very strongly with your feelings in regard to it and that I live here in Washington; the last fact, I fancy, being a strong disqualification in the eyes of Mr. Walcott. But I do think it would be a very serious matter if Walcott should succeed in preventing the election of Pritchett, or should secure the place for himself [LC, Carnegie Papers, Lodge to Carnegie, December 11, 1903].

Carnegie responded: "Glad to receive your letter, which you may be sure I have taken to heart. The gentleman you suspect had nothing to do with the matter" [LC, Carnegie Papers, Carnegie to Lodge, December 14, 1903]. Lodge shrugged it off, but it was apparent that being a Trustee was a plum, and that he was annoyed at not being appointed.

A day or two after the annual meeting of the Board, it was public knowledge that Dr. Gilman had submitted his resignation. This was hardly a bombshell and it was to take effect in a year; Dr. Henry S. Pritchett was being considered as a candidate for President even before Dr. Gilman was officially on his way out. That December, a more serious problem for CIW than the loss of their first president was the death of Assistant Secretary Marcus Baker who had done yeoman service in all senses.

GETTING ORGANIZED FOR BIGGER SCIENCE—1904-1905

Early in January of 1904, Walcott wrote a long letter to Carnegie. About all he received in reply that was tangible was: "Another point—The organization at present, as you know, is not well settled. The President's resignation requires reconstruction" [LC, Carnegie Papers, Carnegie to Walcott, January 7, 1904]. Carnegie was a bit behind the times for reconstruction had already begun. By late 1903, the early concerns of Congress were quieted and the Ex Com was moving for a Congressional charter long before Carnegie suggested any modifications. Walcott prepared the groundwork with the Speaker of the House and reported:

> I send you herewith a copy of the bill introduced by Mr. Cannon, to incorporate the Carnegie Institution of Washington. The bill was prepared by Mr. Root, and I understand met with your approval.
>
> Matters at the Institution are well in hand" [LC, Carnegie Papers, Walcott to Carnegie, March 19, 1904].

Actually the draft of the new incorporation was prepared by J. L. Cadwalander [NYPL, Billings Papers, Root to Cadwalander, January 22, 1904], a noted lawyer on the Board, who had a long preamble and still included some provision for educational activity. The House cut the preamble, and a month later Walcott wrote Mr. Carnegie:

> The charter for the Carnegie Institution passed the House and the Senate Committee, but is held up just at present by Senator Lodge, who stated that he wished to propose an amendment to it. I have just wired Mr. Root to communicate with Senator Lodge and get him to move quickly, as Congress will probably adjourn by the 28th.

I wrote Dr. Pritchett some time ago that the next time he was in Washington I wished to have a talk with him. He was here a few days since, and I learned that something that had given me an unfavorable impression was not the result of any action of his. I told him that if he was selected as President of the Carnegie Institution, I would assist him in every way I could, either actively or as a looker on, as seemed best [LC, Carnegie Papers, Walcott to Carnegie, April 23, 1904].

Lodge was not fully nudged into movement by Root, and John Hay also had to be called up to get the Senator to act. For the record, it was Lodge who removed the last vestige of advanced instruction and support of higher education from CIW. This was an action advocated by Walcott to prevent any future educator from changing the character of CIW to that of an "ordinary university" [NYPL, Billings Papers, Walcott to Billings, March 20, 1904]; the concept of a national university had vanished.

As he had done in 1902, Walcott was one of the incorporators of the rechartered CIW. He saw to the transfer of its assets, reelection of the Board of Trustees, and myriad other details. That summer was again one of frantic activity for Walcott, the Reclamation Service taking increasing amounts of time.

Simultaneously, the CIW took his time and a string of letters flew to Billings. Among other matters, Walcott noted that President Gilman declined even the minor task of drafting a new application form until there could be an Ex Com meeting in the fall to consider it. He added:

This letter does not need a reply. You are the only one with whom I have the opportunity of talking over Carnegie Institution matters, which explains, in a measure, why many apparently trivial things are sent to you" [NYPL, Billings Papers, Walcott to Billings, July 27, 1904].

In one account of CIW history, there is at least a hint that Robert S. Woodward wrested control, first from Walcott and then from Billings [Reingold, 1979, p. 117]. This seems quite unlikely for almost immediately after the election of the new president, Walcott transferred activities to him. Further, Woodward was certainly the choice of Billings [CIW, Billings Trustee File, Billings to Woodward, June 13, 1904], though admittedly he might not have liked all of his subsequent actions. Carnegie was kept fully informed of the search for a new president. Even though he seemingly favored Pritchett, Carnegie made no move to prevent Woodward's election.

Walcott was delighted with the choice in December, 1904. . . . I have never been at a meeting of a board of trustees where the was a better feeling and a more general satisfaction with the result than at the meeting on the thirteenth. Personally, I am delighted with the election of Dr. Woodward. We have been personal friends for many years, and I as mentioned to you two years ago, when talking about geophysical work, he is one of the clearest headed, most straight forward men engaged in scientific work.

I have no personal dislike toward the candidate [Pritchett] of whom you spoke several times, but I was not satisfied that he was the man for the place.

Dr. Woodward has taken vigorous hold, and I have turned over to him all matters that I have in charge, and have assured him that it will give me the greatest pleasure to assist him until he has full knowledge of all that has been done or is pending as a result of the work of the past three years. Now that the administration of the Institution is in such capable hands, I hope it will advance rapidly to the position that we looked forward to its occupying when the work was begun" [LC, Carnegie Papers, Walcott to Carnegie, December 23, 1904].

Carnegie replied: "Delighted to receive yours of the 23rd. It is most gratifying to me that President Woodward is hailed by all as the right man in the right place. Especially am I glad that he was your candidate" [SIA, Walcott Papers, Carnegie to Walcott, December 28, 1904].

If anyone was being outgeneraled, it was Billings. The Ex Com was reorganized by adding a member and changing the chairmanship, with Carroll Wright being the new Chairman. Gilman, Hay, Mitchell and Root were full committee members; Woodward, Billings, and Walcott were *ex officio* by virtue of their offices on the Board. It was a committee that Billings could not dominate and the way was clear for the founding of a geophysical laboratory. Walcott had laid a great deal of groundwork to reach this point.

CONCLUSIONS

This investigation indicates that the general narrative history of CIW needs hardly any significant change. Nevertheless, the documentation presented should demonstrate that Walcott played a much more significant role in the founding and early organization of CIW than has been accorded to him. His title may have been that of Secretary but for the first three years at least, it was he who kept the organization moving forward. Under no definition can John Shaw Billings be classed as a villain, yet it was he who bulldozed President Gilman, and it was he who pushed support for the individual investigator rather than for the large well-equipped laboratory. Walcott struggled to keep Drs. Billings and Gilman working together. When given the opportunity to succeed President Gilman at great personal gain in both salary and time for research, Walcott declined the opportunity.

Another point that emerges is that Andrew Carnegie did not simply donate his money and move on to other things. On the record, he never interfered with the Board, but it was the Executive Committee and not the Board that was critical, and it seems extremely unlikely that he and Billings never discussed matters informally in New York. Walcott kept Carnegie fully informed on all major points that arose, and tried to influence him on a course of action, away from simple grant giving to individual projects, no matter how worthwhile, toward significant support of a few fields. It does seem reasonable that no one puts down ten million dollars and then ignores it and Walcott's persistent writing may have caused Carnegie to drop a word to Billings.

During the summer of 1903, Charles Doolittle Walcott began to steer CIW toward larger projects and away from small grants; Woodward completed this process. Walcott served as Secretary until the end of 1905, when CIW was finally under strong leadership. He continued to serve on the Executive Committee until 1922. Without him, the course of CIW would have been different. If Walcott had not met Carnegie early in December, 1902, there may not have been a Carnegie Institution in Washington, or even a Carnegie Institution.

The extensive correspondence between Carnegie and Walcott, sampled herein, is interesting in its own right. From the inception it went beyond cordial to friendly. Four months after they first met when Carnegie was leaving for Scotland, Walcott appended to a letter: "Bon Voyage, a happy summer, and a safe return" [LC, Carnegie Papers, Walcott to Carnegie, April 3, 1902]. Years later, Walcott had no hesitancy in adding a handwritten postscript to another letter: "How did you come to get the option on so much of the Lake Superior ores?" [SIA, Walcott Papers, February 10, 1915]. One consequence of the continued correspondence which kept Carnegie aware of CIW was the later donation by him of another $10,000,000 to CIW for support of astronomy. That is another story and one which would never have come about if Walcott had not shepherded CIW through its difficult formative years.

ACKNOWLEDGEMENTS

The New York Public Library generously granted permission to quote from the correspondence of John Shaw Billings; the Joseph Hamberger Archives, Johns Hopkins University, and the American Museum of Natural History, were equally gracious with correspondence of Daniel Coit Gilman and Henry Fairfield Osborne, respectively. Those at work at the Carnegie Institution of Washington (CIW), the Manuscript Division, Library of Congress (LC), and the Smithsonian Institution Archives (SI) were uniformly helpful in suggesting sources and in filling requests. For many years, Dr. Hatten S. Yoder, Director Emeritus, CIW Geophysical Laboratory, has encouraged my interest in Walcott, has been tolerant of a paleontologist on his "turf", and has materially improved my draft.

ARCHIVAL SOURCES

The papers of Andrew Carnegie are mounted in a series of numbered volumes in the Manuscript Division of the Library of Congress. The majority of these papers are of letters received, though some copies of letters sent are included. Most of the material of interest is in chronological order and is scattered through volumes 83-103. A card catalogue of correspondents is an invaluable adjunct to the volumes. The Jameson letter mentioned is in Box 134 of the Jameson Papers; Cattell letters mentioned are in box 44 of the Cattell Papers.

The papers of Charles Doolittle Walcott are in Record Group 7004 of the Smithsonian Institution Archives. A few duplicate items of Carnegie-Walcott correspondence occur also in the Library of Congress, but there is actually little overlap between the two collections. Most of the items mentioned are in boxes 32-34. These archives also have a duplicate set of minutes for early meetings of both the Board of Trustees and the Executive Committee of CIW. The CIW archives maintain folders, named as indicated in the text, as well as unpublished minutes of the Executive Committee and Board meetings; a duplicate set are in the Walcott papers.

REFERENCES

Anonymous, Washington Memorial Institution for post-graduate study and research in Washington, *Science*, new series *13*, 921-924, 1901.

Carnegie Institution of Washington, *Year Book*, No. 1, 305 pp., Carnegie Institution of Washington, Washington, DC, 1903

Franklin, F., *The Life of Daniel Coit Gilman*, 446 pp., Dodd, Mead and Company, New York, 1910.

Garrison, F. H., *John Shaw Billings, a Memoir*, 432 pp, G. P. Putnam's Sons, New York, 1915.

Gilbert, G. K., First annual report of the Secretary, *Washington Academy of Sciences, Proceedings*, *1*, 1-14, 1899.

Hayford, J. H., and Wead, C. D., Report of meeting of December 6, 1902. *Philosophical Society of Washington, Bulletin 14*, p. 2.

Henry, R. S., *The Armed Forces Institute of Pathology: the First Century 1862-1962*, 422 pp., U. S. Government Printing Office, Washington, D.C., 1964.

Jessup,. P. C., *Elihu Root, Volume 2, 1905-1937*, 586 pp., Dodd, Mead, and Company, New York, 1938.

Leverson, J. C., Samuals, E., Vanderser, C. and Winner, V.H. (Eds.) *Letters of Henry Adams, volume 5, 1899-1905*, 726 pp., Belnap Press, Harvard University, Cambridge, Massachusetts, 1988.

Madsen, D. A., Daniel Coit Gilman at the Carnegie Institution of Washington, *History of Education Quarterly*, *9*: 154-186, 1969.

Miller, H. S., *Dollars for Research: Science and its Patrons in Nineteenth-Century America*, 258 pp., University of Washington Press, Seattle and

London, 1970.

Mitchell, S. W., Biographical Memoir of John Shaw Billings 1838-1913, *National Academy of Sciences, Biographical Memoirs*, *8*, 375-383, 1919.

Reingold, N., National Science Policy in a Private Foundation: The Carnegie Institution of Washington, in *The Organization of Knowledge in Modern America, 1860-1920*, edited by Alexandra Oleson and John Voss, pp. 313-341, The Johns Hopkins University Press, Baltimore and London, 1979.

Servos, J. W., To explore the borderland: The foundation of the Geophysical Laboratory of the Carnegie Institution of Washington, *Historical Studies in the Physical Sciences*, *14*, 147-185, 1983.

Walcott, C. D., Relations of the National Government to Higher Education and Research, *Science*, new series, *13*, 1001-1015, 1901.

Wall, J.F, *Andrew Carnegie*, 1137 pp., Oxford University Press, Oxford and New York, 1970.

Yochelson, E. L., Charles Doolittle Walcott 1850-1927, *National Academy of Sciences, Biographical Memoir*, *39*, 471-540, 1967.

Development and Promotion of the Initial Scientific Program for the Geophysical Laboratory

H. S. YODER, JR.
Geophysical Laboratory, Carnegie Institution of Washington
Washington, D.C.

INTRODUCTION

The perception that geological processes were amenable to quantitative analysis and experimental demonstration by applying the laws of physics, physical chemistry, and chemistry developed after the 1860s. There were no illusions about the difficulty of deducing a natural process having many variables when only the end products, the rocks, were in hand. Nevertheless, it was reasoned that some understanding of the processes of rock formation might be achieved if the products could be duplicated in laboratory experiments. The principal problems involved carrying out the experiments at the actual conditions presumed to exist in the Earth. The compositions of the rocks had been determined for the major elements with sufficient accuracy, but reproducing the high temperatures and high pressures in the laboratory was yet to be accomplished in a measurable way.

The concept of a physical laboratory in which to carry out these experiments quantitatively was established by Clarence King in 1880 after the founding of the U.S. Geological Survey in 1879 [Rabbitt, 1980]. His logical mode of analysis of geological problems is recorded in an 1878 paper entitled "Systematic Geology" that deals with fusion, genesis, and the classification of volcanic rocks. He assigned George F. Becker (Figure 1), a field geologist with a strong background in physics, chemistry, and mathematics, to direct the new laboratory. While Becker was in the west on other assignments, the laboratory studies were carried out by Carl Barus, first at the American Museum in New York, then in a house in New Haven (1882), and later in the U.S. National Museum (1884) under the direction of F. W. Clarke. The experiments included measurement of the electrical properties of rocks [1891], the electromotive force of various thermocouples [1892], the melting temperature of rocks [1892], and the volume change on melting of diabase [1893]. The work was abandoned in 1892 as a result of the reorganization of the USGS precipitated by a serious cut in funding [Rabbitt, 1980].

The Earth, the Heavens and the Carnegie Institution of Washington
History of Geophysics Volume 5

A second physical laboratory was established, also under Becker, in the USGS by its third director, Charles D. Walcott, in July, 1900. In October 1900 Arthur L. Day (Figure 2), an American staff worker of the Physikalische Technische Reichsanstalt in Berlin, was appointed to the Geological Survey to organize the new laboratory because of his experience in high-temperature research [Cahan, 1989]. The plan laid before the director by Becker stressed the study of silicates at high temperatures to ascertain if they exhibited isomorphic or eutectic properties. The belief that some mineral groups were continuous functions of composition, that is isomorphous, was first established by Tschermak [1864]. But the idea that magmas were eutectic-like yielding definite mixtures of minerals arose when Guthrie [1884] first defined the term, an idea strongly supported by Teall [1888], and later by Becker [1901], as a basis for rock classification.

Because the plagioclase feldspars appeared to be the most important isomorphous group among the common rock-forming minerals, Becker set Day and E. T. Allen, a chemist, the task of investigating the melting characteristics of that isomorphous mineral group. Day obtained the necessary laboratory equipment, some of which was made at the Reichsanstalt. The working conditions on the fourth floor of the new building at 1310 F Street were far from ideal, and the special equipment required a large measure of patience and persistence on the part of the investigators. Walcott apparently perceived the growing needs of the laboratory and particularly the limitation of public funding for such an enterprise "remote from industry" and leaped at the opportunity that arose when he learned of the largess to be bestowed by Andrew Carnegie in support of basic research [Yochelson, this volume]. Early in December, 1901, Walcott asked Becker to prepare quickly a statement for an independently endowed geophysical laboratory.

THE MAGNIFICENT SCHEME

On December 16, 1901, Becker delivered his outline for a geophysical laboratory to Walcott under the title "Concerning a Geophysical Laboratory." A handwritten draft with corrections and a typed version are preserved

Fig. 1. George Ferdinand Becker (1847-1919).

Fig. 2. Arthur Louis Day (1869-1960).

Fig. 3. Charles Richard Van Hise (1857-1918).

[LC, Becker Papers, Box 26]. The first paragraph of the typescript is as follows:

> It is difficult to conceive of a more magnificent scheme than the founding of a generously endowed laboratory, devoted to researches into the physical and chemical conditions affecting the history of the globe. Very little work has been done in this direction for several reasons. As a rule physicists and chemists know too little of geology to appreciate the applicability of these sciences to the elucidation of the history of the earth, while few geologists have the training in exact science which would fit them to undertake such researches independently. Again, the investigations required are so laborious and expensive that no institution now in existence is in a position to undertake them, systematically, on an organized plan. The nearest approach to such work is that now going on in the division of Chemical and Physical Research of the Geological Survey, but there is little ground for hope of adequate appropriations from Congress for this purpose.

The two-page statement was presumably too brief in scientific substance for Walcott to use in persuading the CIW Board of Trustees, incorporated on January 4, 1902, for which he had been elected secretary, and a more detailed report was requested.

Becker submitted the desired scientific program on March 21, 1902, to Walcott. [LC, Becker Papers, "Project for a Geophysical Laboratory," Box 26]. The program emphasized the need for experiments on the physical properties of rocks as they applied to "terrestrial density, upheaval and subsidence, and vulcanism." The principal projects were listed under 1) Mechanics, 2) High-temperature work, 3) Solutions and their relations, 4) Thermodynamics, and 5) Constitution of matter. The scientific program was published in 1903 with detailed plans for staff, building design, budget, and organization in an Appendix to the Report of Advisory Committee on Geophysics [Becker, 1903].

THE ADVISORY COMMITTEE

The Advisory Committee on Geophysics for CIW [Walcott, 1903] consisted of three geologists, T. C. Chamberlin (Univ. of Chicago), C. R. Van Hise (Univ. of Wisconsin), C. D. Walcott (USGS), and three physicists, R. S. Woodward (Columbia Univ.), C. Barus (Brown Univ.), and A. A. Michelson (Univ. of Chicago). The Advisory Committee chaired by Woodward submitted a list on September 23, 1902, also published in the first *Year Book*, of 16 specific problems involving the broader scope of geophysics, as viewed today (atmosphere, oceans, and lithosphere), and explicitly including geochemistry. Suggestions for specific research projects and support had been obtained by letter from Lord Kelvin, E. Suess, F. Becke, O. Kohlrausch, J. H. van't Hoff, G. H. Darwin, and W. Nernst. Their list of specific problems, abbreviated here, included:

- Heat transfer in the atmosphere.
- Determination of gases in magmas, rocks, and meteorites.
- Function of the ocean as a reservoir of atmospheric constituents.
- Physical chemistry of natural solutions as related to ore deposits.
- Alteration and recrystallization of minerals under varying conditions.
- Heat of formation of natural compounds.
- Deformation of rocks.
- Effect of pressure on the melting of minerals, including volatiles.
- Thermal conductivity of rocks.
- Elastic constants of rocks under varying conditions.
- Sources of internal heat on earth.
- Relationship of heat distribution to deformation and volcanism.
- Tidal deformation.
- Moon-earth tidal relationship.
- Density and mass distribution in the earth.
- Gravity in oceans and continents.

Because "the trustees were not prepared to act," [CIW *Year Book*, 1904, p. xxxv] further study of the subject of geophysical research, especially in Europe, was assigned to Van Hise (Figure 3). On the basis of those discussions he laid out a four-part program considered by geologists as "most pressing" that included the relation of liquid and solid rocks, minerals and rocks from aqueous solutions, deformation of rocks, and physical constants of rocks [Van Hise, 1904a, 178-181]. In addition, the opportunity to systematize the seismological investigations of the world was recognized. These areas were backed up with a detailed and definitive outline of experiments for the investigation of igneous and metamorphic rocks prepared by eight petrologists in the interest of promoting the "Science of Petrology."

THE COMMITTEE OF EIGHT

The report submitted on October 10, 1903, by a Committee of Eight provided even greater detail for the initial program of research for the proposed geophysical laboratory. The Committee consisted of Whitman *C*ross (USGS), Joseph P. *I*ddings (Univ. of Chicago), Louis V. *P*irsson (Yale), and Henry S. *W*ashington (Private Laboratory, Figure 4)—the group famous for the ***CIPW*** system of rock classification (1902)—and Frank D. Adams

Whitman Cross, 1854-1949

Joseph Iddings, 1857-1920

Louis V. Pirsson, 1860-1919

Henry S. Washington, 1867-1934

Fig. 4. Four members of the Committee of Eight who also participated in the formation of the CIPW system of rock classification

Alfred Church Lane, 1863-1948

James Furman Kemp, 1859-1926

Frank Dawson Adams, 1859-1942

John Eliot Wolff, 1857-1940

Fig. 5. Remaining four members of the Committee of Eight.

(McGill University), James F. Kemp (Columbia), Alfred C. Lane (Michigan State Geological Survey), and John E. Wolff (Harvard, Figure 5). It was indeed a distinguished group: Six of these petrologists became members or foreign associate of the National Academy of Sciences. Their suggestions outlined in some detail are tabulated in brief in Table 1.

TABLE 1. Geophysical Investigations Suggested by CIW Committee of Eight, 1903

Igneous Rocks	*Metamorphic Rocks*
1. Physical properities, f(P,T)	1. Physical properties, f(P,T)
2. Mutual solution of minerals	2. Thermal properties
3. Diffusion in liquids and solids	3. Rock-water interaction
	4. Hydrothermal mineral solubility
4. Crystallization from liquids	5. Chemical reactions and crystallization
A. Liquidus of simple systems	
Rates of crystallization	6. Hydration and dehydration
B. Gas solubility in magmas	7. Crystal growth
Hydrous mineral stability	8. Solution of stressed crystals
Crystal size, habit, texture	
	9. Rock deformation
5. Chemical analysis	10. Development of foliation
6. Thermal properties of minerals	11. Effects of stress on composition
	12. Origin of graphite in metamorphic rocks

There appears to be no written record on who organized the committee or how the group was assembled. (It may be presumed that their suggestions resulted from discussions held in Washington when at least six of the Committee of Eight met at the Washington Meeting of the Geological Society of America between December 30 and January 2, 1902-03. Neither Pirsson nor Wolff was registered, but Wolff is listed as having given a paper on 2 January.) The emphasis on physics espoused by Becker and the CIW Advisory Committee on Geophysics thereby evolved toward physical chemistry as promoted by Van Hise and the Committee of Eight.

GRANTS IN GEOPHYSICS

As the plans developed for a geophysical laboratory, CIW provided grants to initiate some aspects of the program. Apparatus was to be built and the feasibility determined of carrying out experiments at temperatures beyond the calibrated scale and at "extreme pressures." The first CIW grant in geophysics in 1903 was to F. D. Adams at McGill for study of the deformation of rocks. Another grant went to G. K. Gilbert at the USGS for exploring the feasibility of a deep drill hole in which to measure the geothermal gradient. The world's record at that time was a hole 2,800 feet deep, so the proposal to diamond drill granite and retrieve core to 10,000 feet was indeed a pioneering venture. It was an initial step toward a Mohole-type project, that is, deep drilling for scientific purposes. Because of the estimated high costs that project was abandoned. Becker received a grant in 1904 from CIW for experiments on the elasticity and plasticity of solids, and Day received funding for an investigation of mineral fusion and solution under pressure. The "under pressure" part unfortunately depended on new apparatus that was not developed because of the demands of other basic problems. The accepting of grants for research by a government agency from a private institution was indeed innovative and deemed an "enlightened" policy by Becker.

MULTIPLE NUCLEATION

These individual and uncoordinated efforts in geophysics by Adams, Becker, and Day were matched by similar investigations around the world. For example, rocks and minerals were being studied experimentally in Austria by C. Doelter, in France by F. Fouqué and A. Michel-Levy, in Russia by F. Loewinson-Lessing, P.Tschirwinsky, P. Lebedev, A. E. Logario, in Poland by J. Morozewicz, in Norway by J. H. L. Vogt, and a large number of groups in Germany. The "Science of Petrology" was indeed progressing rapidly in many laboratories as pioneering methods were being tested.

After the thin-sectioning of rocks became readily available and the petrographic microscope was developed, descriptive texts of rocks and their classification were published, and the chemical analyses of rocks and minerals were accumulated in tables. By 1900 Iddings had already defined the crystallization course of a magma that yielded igneous rocks; Becker outlined the concept of fractional crystallization; and Barus had measured the volume change on melting of gabbro and calculated the change of melting point with pressure. In addition, Bakhuis Roozeboom was applying Gibbs's thermodynamic theory to multicomponent systems; Teall laid out the iron-enrichment concept, recognizing the sequence of magmas from basic to siliceous; and the first phase diagram for the Stassfurt salt deposits had been published by Löwenherz, a student of van't Hoff.

Furthermore, a temperature scale up to 1160°C had been established at the Reichsanstalt, and the manganin gauge for pressure measurement had been invented by Lisell at Upsala. The electric wire-wound furnace was available and the use of platinum crucibles established. Hydrothermal pressure vessels were available in France, Germany, England, and in Washington at the USGS. By the incorporation of the CIW, January, 4, 1902, the basic ideas

ANDREW CARNEGIE
2 EAST 91ST STREET
NEW YORK

December 19th, 1905.

My dear Mr. Walcott,

Yours of December 16th received. I had heard of the money appropriated for administration building, but had not heard of the large physical laboratory. You know my own opinion is that no big institution should be erected anywhere, but the exceptional men should be encouraged to do their exceptional work in their own environment..

There is nothing so deadening as gathering together a staff in an institution. Dry rot begins and routine kills original work. At least that is the opinion of

Yours very truly,

Andrew Carnegie

Dr. Chas. D. Walcott,
Department of Interior,
Washington, D.C.

Fig. 6. Letter from Andrew Carnegie to Charles D. Walcott dated December 19, 1905.

Geophysical Laboratory
CARNEGIE INSTITUTION OF WASHINGTON

ADDRESS
1330 F STREET
WASHINGTON, D.C.

GEORGE F. BECKER
ARTHUR L. DAY

October 25, 1905.

Dr. Arthur L. Day,
Director Geophysical Laboratory.

Sir:

As noted in your letter of October 16th, the result of a verbal consultation has been that the research on elasticity under my charge has gradually incurred an indebtedness to the machine-shop under your charge, amounting approximately to $500.00. I enclose herewith my check to you for that amount in satisfaction of this debt.

I entirely concur in your suggestion that hereafter an account be kept of work done for me, and that I be charged with the actual time spent upon it and with the actual cost of materials used.

Very respectfully,

Geo. F. Becker
Geophysicist.

Fig. 7. Letter from George F. Becker to Arthur L. Day dated October 25, 1905.

and apparatus were available for a major coordinated effort in experimental geology.

THE UNIQUE PHILOSOPHY

One must ask, then, what was the need for a diversified staff in a geophysical laboratory? The problems were well defined, but the critical ingredients that were missing were accuracy, reproducibility, and coordination. The philosophy of approach was clearly laid out by Becker: "We must patiently begin with the simplest problems that can be devised and, aided by the most perfect appliances known, study them exhaustively before proceeding to more difficult and complex cases" [Day et al., 1905, p. 6]. In short, the concept of examining the effects of a single variable while holding all others constant was the cornerstone of his approach. It influenced the decision to use only the purest chemicals available, calibrated temperature and pressure scales, and the component-by-component method of studying phase equilibria. That the geophysical laboratory began its work in the USGS in 1904 on the isomorphism of the plagioclase feldspars and some simple eutectic systems, two of the most important concepts of the day, was in fact Becker's dream already presented to Clarence King in 1882. Day [1910, p. 142, footnote 2] acknowledged Becker as the leading force in "the introduction of the quantitative methods of physics and chemistry in the service of geology. . . ."

THE PROMOTION

Walcott was a very effective politician in generating support for a geophysical laboratory with this new philosophy. He tried to persuade Andrew Carnegie not only by personal meetings, and with letters from many of the outstanding scientists abroad, but also by engineering a resolution through S. F. Emmons at the International Congress of Geologists in Vienna in 1903. The resolution passed by the Council of the Congress Géologique International is quoted by Van Hise [1904a, 177-178] in his "Report on Geophysics." In addition, Walcott arranged for extended discourses by both Van Hise [1904b] and Becker [1904] at the famous International Congress of Arts and Science of 1904 held in St. Louis. Becker gave a detailed list of critical problems, whereas Van Hise gave a more general discussion of geological issues.

In spite of all of Walcott's efforts, Carnegie was never persuaded, but the result could have been foretold when Woodward, chairman of the Advisory Committee on Geophysics, was elected president of CIW late in 1904. At the December 12, 1905 meeting of the Trustees, Woodward and Walcott persuaded them to establish *the* Geophysical Laboratory. To say the least, Carnegie was not pleased and quickly wrote to express his opinion (Figure 6). He clearly was opposed to extracting "exceptional men" from their own environment and especially to the erecting of buildings. Nevertheless, the decision was in the hands of the Board of Trustees, not the donor of the endowment.

Obviously, Woodward and Walcott were so sure of the outcome that major decisions had been made well in advance of the December 1905 vote. What skillful politicians they were! In a letter dated October 25, 1905 (Figure 7) with the letterhead already prepared—well in advance of the vote—in the same script used today, the apparent appointment of A. L. Day, then assistant to Becker at the Survey, as the first director had been made; and yet the old address of the USGS was retained. This simple letter has quite a story behind it [Servos, 1983, pp. 174-175], especially in regard to the apparent bypass of Becker as director, presumably because of his rigid stance on his proposed budget considered excessive by Walcott. Becker has been referred to erroneously as the director by Williams [1932, vol. 1, p. 309] and in recent editions of the *Encyclopaedia Britannica* (e.g., 1977, vol. 1, p. 918) as the first director of the Geophysical Laboratory. Becker, age 58, was a geologist with a strong physics background, whereas Day, age 36, was a physicist with essentially no geological background. The official appointment of Day as director was made several months into 1906; however, Day had been receiving from CIW a small "honorarium" from April 1, 1904 and a "salary" from April 1905, so the appointment was not unexpected. The work of Day, Allen, and Iddings [1905] on the plagioclase feldspars was published by CIW in 1905 with the consent of Walcott, CIW secretary, in recognition of the support from CIW. It was considered to be the first publication of the Geophysical Laboratory. Day [1907] reiterated Becker's unique philosophy of approach, but reduced the number of amenable projects to calorimetry, extension of the calibration of the high-temperature scale, the action of water on minerals, and high-temperature mineral equilibria.

A TIMELESS BLUEPRINT

The outline of basic research problems prepared by Becker, Van Hise, the Advisory Committee, and the Committee of Eight is still valid today. Their perceptiveness in listing the major issues in experimental petrology is indeed incredible. The descriptive terms have been changed into the fashionable language of the day, but the relevancy and priorities of those critical questions remain in spite of the progress made on the complex multivariate problems of the Earth. Almost all of the critical questions outlined in their proposals have been investigated at the Geophysical Laboratory at some time in its history [Yoder, 1989].

Because of the population explosion, attempts to raise the

quality of life worldwide, and the coming demand for natural resources, it is essential that consideration be given now for another laboratory of similar scale with the same philosophy of approach devoted not to equilibria in closed systems, but to the nonequilibria, irreversible, open systems characteristic of most ore deposits. The type of experimental work briefly mentioned by Van Hise [1904a, p. 180] and by Becker [1904, p. 550] for an understanding of ore deposits will soon be a matter of urgency. In spite of the stigma of commercialism, the need for basic research in that area, still awaiting a coordinated, systematic effort, is indeed great.

REFERENCES

Barus, C., On the Thermo-Electric Measurements of High Temperature, *Bull. U.S.G.S.*, No. 54, p.313, 1889.

Barus, C., The Viscosity of Solids, *Bull. U.S.G.S.*, No. 73, p. 139, 1891.

Barus, C., Thermoelectrics of Platinum-Iridium and of Platinum-Rhodium, *Phil. Mag.*, *15*, 376-381, 1892.

Barus, C., Fusion Constants of Igneous Rocks, *Phil. Mag.*, *16*, 1893

Becker, G. F., Fractional Crystallization of Rocks, *Am. J. Sci.*, *4*, 257-261, 1897.

Becker, G. F., Report on the Geology of the Philippine Islands, *U.S. Geological Survey Annual Report*, *21*, pt. 3, 493-625, 1901.

Becker, G. F., Project for a geophysical laboratory. *Carnegie Institution of Washington Year Book for 1902*, No. 1, 44-58, 1903.

Becker, G. F., Present Problems of Geophysics, *Science*, *20*, No. 513, 545-556, 1904.

Cahan, D., *An institute for an empire: The Physikalisch-Technische Reichsanstalt, 1871-1918*, 315pp., Cambridge Univ. Press, Cambridge, 1989.

Cross, C. W., J. P. Iddings, L. V. Pirsson, and H. S. Washington, A Quantitative Chemico-Mineralogical Classification and Nomenclature of Igneous Rocks, *J.Geol.*, *10*, 555-690, 1902.

Day, A. L., Geophysical Research, *CIW Year Book for 1906*, No. 5, 177-179, 1907.

Day, A. L., Some Mineral Relations from the Laboratory Viewpoint, *Bull. Geol. Soc. Amer.*, *21*, 141-178, 1910.

Day, A. L., E. T. Allen, and J. P. Iddings, The isomorphism and thermal properties of the feldspars, CIW Publ. No. 31, 95pp., 1905.

Guthrie, F., On Eutexia, *Phil. Mag. 17*, 462-482, 1884.

King, C., Systematic Geology, *U.S. Geological Exploration of the Fortieth Parallel*, 696-725, U.S. G.P.O., Washington, 1878.

Rabbitt, M. C. Minerals, lands, and geology for the common defence and general welfare, 1879-1904, 2, 407pp. U.S. Geological Survey, U.S. G.P.O., Washington, 1980.

Servos, J. W., To Explore the Borderland: The Foundation of the Geophysical Laboratory of the Carnegie Institution of Washington, *Hist. Stud. Phys. Sci.*, *14*, 147-185, 1983.

Teall, J. J. H., *British Petrography with Special Reference to the Igneous Rocks*, Dulan & Co., London, 469pp, 1888.

Tschermak, G., Chemish-Mineralogische Studien-I: Die Feldspargruppe, *Sitzber. Akad. Wiss. Wien*, *50*, 566-613, 1864.

Van Hise, C. R. , Report on Geophysics, *CIW Year Book for 1903*, No. 2, 173-184, 1904a.

Van Hise, C. R., The Problems of Geology, *J. of Geol. 7*, No. 7, 589-616, 1904b.

Walcott, C. D., Proceedings of executive committee, *CIW Year Book for 1902*, xxx-xl, 1903.

Williams, A. F., *The Genesis of the Diamond*, Ernest Benn, Ltd., London, 2 vol., 636p, 1932.

Yoder, H. S., Jr., Scientific highlights of the Geophysical Laboratory 1905-1989, *Annual Report of the Director Geophysical Laboratory, Carnegie Institution of Washington, 1988-1989*, 143-197, 1989.

Vision of a Global Physics: The Carnegie Institution and the First World Magnetic Survey

GREGORY A. GOOD
History Department, West Virginia University, Morgantown WV 26506-6303

INTRODUCTION

In 1900, a letter appeared in the journal *Science*, claiming that "The present time can well be likened to the years when Gauss inaugurated a Magnetic Association . . ." and that a "veritable boom" was taking place in the study of the Earth's magnetism [Bauer, 1900, pp. 651-652]. The author of this letter was Louis Agricola Bauer, then the chief of the Division of Terrestrial Magnetism at the U.S. Coast and Geodetic Survey (USCGS). Bauer listed in this letter government-sponsored magnetic work in the U.S., Britain, Egypt, Germany, India, New Zealand, Australia, Italy, and Mexico. He also pointed to recent theoretical advances by Adolf Schmidt, instrumentational innovation by Max Eschenhagen, and the efforts of Wilhelm van Bemmelen as indications of a renewed quest to understand the global phenomena of magnetism. Within five years, Bauer was to establish an agency dedicated to pursuing a magnetic survey of the entire globe, an agency known as the Department of International Research in Terrestrial Magnetism—the DTM.

Before looking at the DTM and the world magnetic survey Bauer conducted through it, a brief summary of Bauer's life is in order. Bauer was born in Cincinnati in 1865, of German-American parents. He was trained as a civil engineer and worked as a computer for the Coast and Geodetic Survey from 1887 to 1892. His supervisor there was Charles Anthony Schott, who had been with the Survey since the 1850s and who, as chief of the Computing Division, directed the Survey's geomagnetic work. Bauer spent 1892-1895 at the University in Berlin, where he attended classes in physics and geophysics by, among others, Max Planck and Wilhelm von Bezold. His doctoral dissertation, finished in 1895, was a mathematical analysis of the secular variation of the Earth's magnetism. He also worked at the Geomagnetic Observatory at Potsdam, where he came to know Germany's most prominent magneticians. On his return to the U.S., Bauer developed a dual career, teaching physics and geophysics at several universities while also directing magnetic surveys. He became the first chief of the Coast and Geodetic Survey's Division of Terrestrial Magnetism when it was founded in 1899. He increased the density of the geomagnetic survey grid in the U.S. significantly and established five magnetic observatories. He also founded the international journal *Terrestrial Magnetism* in 1896 and thus began to build an institutional basis for promoting his main research interest [Fleming, 1932; Reingold, 1970].

THE PIONEERING STAGE: RAISING THE STATUS OF TERRESTRIAL MAGNETISM

The ideas of promoting the physical study of the Earth and its cosmical relations were circulating at the Carnegie Institution of Washington (CIW), founded in 1902, independently of Bauer. The Advisory Committee on Geophysics, in that year, had claimed a broad territory, from the body of the Earth and its "great envelopes" to its motions and external relations [Woodward et al., 1903, pp. 28-29]. This included geochemistry, rock elasticity, seismology, and gravimetry, but also terrestrial magnetism. The report insisted that the latter topic belonged more to geophysics than it did to geodesy and topography, which was where it was placed in the Coast and Geodetic Survey [Ibid., p. 32]. It concluded that

> in general, the deeper terrestrial questions lead out in the end into the realm of cosmology, where the studies of the geologist, astronomer, physicist and chemist blend. . . . the phenomena of the Earth are most important factors in cosmology [Ibid., p. 33].

Bauer could not have presented his ideas to a more sympathetic body.

Indeed, his proposal fit with other events as the Institution's interest in geophysics was defined. A second report on geophysics, presented in 1903 by C. R. Van Hise, narrowed the mission of the CIW Geophysical Laboratory to problems related to geology and capable of solution by physical and chemical means then available [Van Hise, 1904, p. 178]. He specifically recommended exclusion of problems dealing with the atmosphere, the interior of the Earth, and cosmology, as they were not useful to the geologist [Van Hise, 1904, p. 182]. These were exactly the sorts of questions that Bauer wanted his Department to concentrate on. Historian John Servos shows clearly that the Geophysical Laboratory quickly narrowed

The Earth, the Heavens and the Carnegie Institution of Washington
History of Geophysics Volume 5

its mission to geochemistry [Servos, 1983]. Whether this was meant to make room at the Institution for Bauer's plan, or whether his plan came along fortuitously and this narrowing was an independent process, is an open question.

Bauer's original mission statement for the DTM indicates the breadth of his vision. According to his published report, the first project of the Department would be a magnetic survey of areas not being studied by others. When this proposal was first presented in 1902, this meant the oceans, South America, Africa, and parts of Asia, since several polar expeditions then underway included magnetic research, as did government-sponsored research in Europe and the U.S. Secondly, the DTM was to be an international coordinating bureau, especially for promoting uniform study of geomagnetic variations. It would guarantee observation and reduction of data by inter-comparable methods, a need made painfully obvious by the languishing of the vast data collected during the International Polar Year, still unpublished after two decades. This second project initially included the coordination of both observatory and field research. Lastly, Bauer envisioned a project that would extend to the depths of the ocean and upward through the atmosphere, since the distribution of the Earth's magnetism and electricity could never be known only from surface observations. His general goal was to trace "the intimate relationship between terrestrial magnetism and other sciences, such as meteorology, geology, and astronomy . . ." [Bauer, 1904a, pp. 203-204].

By the time the proposal was accepted by the Board of Trustees in December 1903, the specifics of Bauer's plan were altered somewhat. The first year of operation was to concentrate on office work, especially the analysis of currently available secular variation and other data. The budget provided for a director (Bauer), several computers (people), two magneticians, and clerical assistance [CIW, file DTM.Name, Scope, Plan, Walcott to Bauer, September 14, 1904]. No extensive field work was authorized. Still, Bauer ordered instruments for future surveys and sought approval to observe "magnetic and electric phenomena of momentary occurrence . . ." [Ibid., Bauer to Walcott, February 24, 1904]. His purpose was the study of magnetic storms and solar eclipses. He thought that geomagnetic and geo-electric observations could answer basic questions about solar radiation and about the upper reaches of the Earth's magnetic field. He wrote:

> It is becoming more and more evident that we are receiving radiations from the Sun similar in effect to cathode rays—possibly radium radiations—which when they come within reach of the Earth's lines of magnetic force extending into outer space about one fourth of the distance of the Moon from the Earth, are deflected by these lines of magnetic force and thus in a large measure are prevented from penetrating to the lowest regions of our atmosphere and escape detection in any amount near the Earth's surface. They remain however, in the region above us where they may be of prime influence in affecting the magnetic, electric and meteorological conditions of the Earth and hence the very conditions of life [Ibid.]

This less known aspect of the DTM's research program led to a series of worldwide cooperative observations of solar eclipses, atmospheric electricity, and cosmic rays over the next several decades. Altogether, Bauer wrote more than fifteen articles on eclipses and solar-terrestrial relations, and published many more reports of results [Harradon, 1932].

Bauer was crusading, and the Carnegie Institution was but one of the stages for his crusade. As he told an audience of scientists at the St. Louis Exposition in 1904, he wanted to raise geomagnetism to the "plane of independent investigation", level with astronomy and meteorology [Bauer, 1907c, p. 1]. He insisted that the subject suffered by being studied only as a sideline to geodesy and other subjects. Avocation was not good enough. Complete concentration was required.

> The magnetician must struggle to have accorded him equal privileges and recognition with the astronomer, the astrophysicist, the geologist, or the meteorologist [Ibid., p. 2].

Only when magneticians specialized could one expect advances in theory. Until then, Bauer lamented, some scientists would continue to present theories previously rejected.

The first task before magneticians, then, was to raise the status of geomagnetism. This process, Bauer announced, had been started by the establishment of the DTM. This Department could focus on its subject without distraction and embrace the whole globe. With this single purpose, magneticians could turn to the next critical task: to convince theoreticians that both the data and the mathematical tools were inadequate for decisive tests of theories. All aspects of terrestrial magnetism were, he said, at the pioneering stage, from the study of magnetic storms and the distribution of the Earth's field to those of secular and diurnal variation. In all of these areas, theory was not well developed. The study of diurnal variation alone had adequate data. Bauer claimed that little more was known of the distribution of the Earth's field than had been in the time of Edward Sabine, around 1850. In even the most prominent case of geomagnetic theory, Gauss's harmonic analysis of the distribution, the coefficients were not any better known than then. With geomagnetic surveys restricted mainly to Europe, the U.S., and a few widely spaced oceanic voyages, basic questions could not be answered. Bauer concluded that his generation's special problem was to "bring together the great facts concerning the Earth's magnetism" so that either his generation or a later one could make "decisive deductions of theory" [Ibid., pp. 2-3].

Bauer's hopes for his subject were certainly grand. His journal *Terrestrial Magnetism* was to promote the research area's international stature, to provide a sense of identity for its researchers, and to act as a catalyst by making them aware of data-gathering activities, new instruments, developments in theory, and new mathematical tools. It would also encourage uniformity. Likewise, his new Department at the CIW was to encourage activities already underway. The best indications of this in the early years of the DTM were its grants to Adolf Schmidt for analysis of magnetic storm data and its efforts to coordinate magnetic observations during a solar eclipse in 1905 [Bauer, 1905a, pp. 73-74; Bauer, 1906, pp. 266-267].

Bauer's new Department was also to go beyond what other researchers were doing. This coincided with the intent of the Institution at large. Its first president, Daniel C. Gilman, stated it would avoid "ground already occupied," by which he primarily meant subject areas [Gilman, 1903]. The DTM first determined the state of knowledge of the global distribution and secular variation of geomagnetism, to determine where more data was needed, and thus where the DTM should send expeditions [Bauer, 1905a, pp. 69-70]. Existing data was, for example, inadequate for determining whether the total magnetization of the Earth was decreasing or not. The DTM's world magnetic survey grew directly from this project.

THE DTM'S WORLD MAGNETIC SURVEY

One way to obtain a worldwide survey was suggested by Alexander von Humboldt in his *Kosmos* in the early 19th century. He called for a survey through the cooperation of the scientific academies and governments of the maritime nations.

> Four times in every century an expedition of three ships should be sent out to examine as nearly as possible at the same time the state of the magnetism of the Earth, so far as it can be investigated in those parts which are covered by the ocean. . . . Land expeditions should be combined with these voyages. . . . [Humboldt, 1868, 2, pp. 719-720; Bauer, 1909, p. 198].

Humboldt set 1850 as the epoch for the first international effort. But the governments weren't listening and no coordinated effort occurred. As Bauer noted, the problem was not that governments were completely unsympathetic to magnetic research, since in fact they sponsored a moderate amount of magnetic work in the late-19th century. The problem was that these were "more or less isolated and incomplete surveys, independently undertaken . . . and distributed over a great number of years." [Bauer, 1909a, p. 198]. The resulting lack of uniformity limited the usefulness of the data obtained.

Bauer maintained that Humboldt's approach was doomed, at least in the case of a world magnetic survey. A magnetic survey must be conducted with dispatch to avoid the deleterious effects of secular variation. He wrote:

> The time-element in the Earth's magnetism, even for a period of a few years, is of such moment as to completely mask the fine, hair-splitting points which would necessarily and rightly have to be raised on some international mode of action, to say nothing of the painful and cumbersome method which would have to be employed to conform with the rules of official correspondence between nations [Bauer, 1909a, p. 199].

It was exactly to avoid these problems that Bauer approached the CIW to sponsor his vision. With the enthusiasm of America in that expansionist age, Bauer saw a private institution as the appropriate agency to coordinate international research. With no official connection to any government, Bauer argued that he could approach the appropriate scientific officials in different countries directly, without being obliged to go through their bureaucratic superiors. He could tell them what the DTM planned to do, ask if they had similar plans afoot, ask if they had enough funding for such a project, and ask them to suggest "some friendly basis of cooperation." In this way, he said, international research could be accomplished without offense, jealousies, duplication, or overlap. He imagined that scientists could approach each other at ease privately, no matter what their positions in their respective governments and no matter what the diplomatic relations between those governments. This was a rather rosy view of the situation, but Bauer's approach at least minimized these difficulties enough that in most cases the DTM did or stimulated the research it wanted. Thus it did no survey work in Europe, India, or the U.S., and even limited its activities in Argentina where preliminary efforts were being made. Bauer also encouraged systematic magnetic surveys in other countries, such as Canada [Good, 1991].

The encouragement of magnetic surveys in other countries and colonies presented Bauer with endless problems. The pace and quality of work by scientists outside the DTM structure did not always meet Bauer's standards, yet he was often unable to force the issue. There were also difficulties obtaining official permission for travel to some colonies and occasionally revolutionaries or bandits interfered with the Department's travelling scientists.

Several examples will help form clearer images of the DTM's land magnetic surveys. Early efforts in Africa are especially interesting, as they reveal not only how logistics and cooperation were arranged, but also something about the political and cultural backdrops of this work. Bauer unflinchingly referred to Africa as the Dark Continent. His discussion of the itinerary of an expedition in 1909 strongly reminds one that this was Africa's colonial era. This expedition started in the Cape Colony and proceeded through German Southwest Africa, Rhodesia, British Central Africa, German East Africa, and on to Egypt, where a magnetic survey was being conducted by the British government [Bauer, 1909b]. Bauer arranged for the cooperation of the British and German governments, which granted passports, special credentials, and even free passes on railroads. One remote section of the expedition, which required the hiring of 100 porters, was financed by two English colonial gentlemen. The research was done not by regular DTM employees, but by two "research associates" given special grants: Dr. J. Carruthers Beattie (Physics Department, South African College) and Dr. J. T. Morrison (Physics Department, Stellenbosch College). The DTM participated in later joint ventures in Africa, too, including one with the South African government in the late 1920s. As a later researcher reported, "The routine usually followed was that laid down by the Carnegie Institution of Washington for their magnetic observers" [Grindley, 1947, p. 254]. In 1934-35, the DTM sent its own observer R. H. Mansfield to make secular variation measurements around Africa [DTM, Mansfield file].

The first survey through China and over the Himalayas into northwestern India was conducted by D. C. Sowers, a Fellow of the Royal Geographical Society in London, as an employee of the DTM [Bauer, 1913, pp. 524-525]. Other observers in China included another British national, a Swedish missionary, and the American president of the Canton Christian College [DTM, General Files to 1934, Edmunds file and China file]. This alliance between the DTM and missionary work was partially accidental, and lasted from the teens into the 1930s. By then, Chinese individuals conducted the survey work and other research. Some travelled to the U.S. for research and training visits. One Chinese physicist who came in the 1930s for training in ionospheric research was Paul C.T. Kwei [Wang, in this volume]. As Bauer stated early on, one subsidiary aim of the world survey was to encourage "native" observers.

The rapidity of the land magnetic survey is surprising, considering the difficulties overcome and the relatively small staff. During its first decade, Bauer estimated that two thirds of it was completed. Land expeditions had established about 2500 magnetic stations in 94 countries or colonies worldwide. They had travelled over 800,000 miles. Ultimately, the DTM sent out or cooperated with over 200 expeditions. He lauded the observers for their "devotion, zeal, enthusiasm, and ability" in pursuing their duties

> in strange countries, amid strange people with strange customs and speaking a strange language; often over infrequently traversed roads and even at times in regions either rarely or never before . . . reached by white man [Bauer, 1914, pp. 497-498].

There was a decided sense of crusade and white man's burden in this ambitious scientific program.

AT SEA: THE *GALILEE* AND THE *CARNEGIE*

The one opportunity for a significant international effort by the DTM with a minimum of diplomatic activity was an oceanic magnetic survey. This was sorely needed, especially since the oceans comprise so much of the globe's surface. Although magnetic observations had been taken at sea since at least Columbus' time, no systematic global survey had been undertaken beyond coastal waters. The famous expeditions of Edmond Halley, J.C. Ross, *La Perouse*, and the *Challenger* had obtained much important data, but they did not approach the needs of geomagnetism in the 20th century.

To redress this deficiency, Bauer shifted the DTM's activities in the second year (1905) away from analysis of available data and toward the accumulation of new. Bauer decided to tackle the North Pacific Ocean first [Bauer, 1905b]. He did this with the advice of Ettrick W. Creak of the British Hydrographic Office, who had been on the *Challenger*, and of the physicist Arthur Schuster. But Bauer's intention from the start was clearly to extend this survey to all the world's oceans.

This oceanic work was planned meticulously. After a careful search, a wooden sailing vessel, the brig *Galilee*, was found which had relatively little magnetic material on board. The steel rigging was replaced with hemp, the iron blocks and tackle were removed, and an elevated observing deck was built to get the magnetic instruments as far as possible from the iron bolts in the main structure. Ultimately the vessel had the lowest magnetic constants of any ship then conducting magnetic research [Bauer, 1907a, pp. 607 and 611].

Equal care went into selecting the track the *Galilee* was to follow. Bauer consulted with George W. Littlehales, of the U.S. Navy Hydrographic Office, who advised about currents and prevailing winds in different seasons and parts of the Pacific [Bauer and Littlehales, 1905]. They laid out a route consisting of three spiraling loops, which would carry the ship through each of the five-degree quadrangles into which they had divided the northern Pacific. The plan was to sail this course of over 70,000 miles during three eight-month seasons spread over three years. The cost was estimated at $12,000 per year [Bauer, 1905b, p. 389]. This initial oceanic survey would provide a density of magnetic stations about one-fortieth that of the land survey conducted in the U.S. by the Coast and Geodetic Survey (one station every 30 miles versus one every 200 miles) [Bauer, 1907b, pp. 174-175; Bauer et al., 1917, p. 96].

The development of improved instruments for magnetic measurements at sea was an important part of Bauer's program. Motion of the ship was not as much a difficulty as was devising an easy way of determining azimuth. It was also important that each instrument measure two of the three magnetic elements. This allowed a maximum use of the minimal observation space on ship. It also allowed cross-checking of every measurement by two instruments. Bauer and his colleagues began with the instruments available: the Ritchie liquid compass, a Kelvin compass, and the Lloyd-Creak dip circle. New instruments gradually evolved out of these during the three cruises of the *Galilee*. By the end of the last cruise in 1908, declination could be read within five tenths of a degree, compared with one tenth of a degree obtained with the same instrument on land [Bauer et al., 1917, p. 19]. In some cases readings at sea were as good as those on land [Ibid., p. 93]. Horizontal intensity at sea was measured within about one three-hundredth of its value [Bauer, 1907b, p. 186]. By the end of these cruises, the limiting factor was not instrumental error but the accuracy of the correctional factors required because of the remaining iron in the ship.

The survey of the North Pacific had some important results: for science, for commerce, and for Bauer's ambitions. The 442 new magnetic stations throughout the region now gave a significantly improved distribution of data on the magnetic field, and ultimately were important for studies of secular variation [Bauer et al., 1917, p. 93]. These results led almost immediately to new magnetic charts which showed that the best declination charts then in use had systematic errors of between one and two degrees. Likewise, the best charts of inclination were one to three degrees off, and the best of horizontal intensity were about 4% too high [Bauer, 1907b, p. 186].

But from Bauer's perspective, perhaps the most important result was the following. The *Galilee* demonstrated that even when all precautions were taken with the best available vessel, the deviations caused by shipboard iron were troublesome in every way. They introduced corrections, which required onerous labors by computers. More important, however, the corrections had to be redetermined every three or four days. The deviation was due to soft iron, the magnetization of which changed readily depending on the heading of the ship and how long it stayed on that heading. The procedure by which the corrections were determined—called "swinging the ship"—wasted precious time useful for magnetic research. Moreover, because of the unpredictable variation of this deviation between determinations, the error was not controllable. Bauer saw a clear solution: build a totally non-magnetic ship.

In 1909 the magnetic survey yacht *Carnegie* was launched in Brooklyn, N.Y., before a crowd of 3500 people. It was built by the master builder responsible for Kaiser Wilhelm's yacht the *Meteor* and for the winner of the 1906 Atlantic cup race [Bauer, 1910a, p. 197]. Its overall length was 155 feet, with a mean draught of 12 feet, 7 inches, and a displacement of 568 tons when fully loaded. It had a brigantine rig with almost 13,000 square feet of sail. It was a sailing vessel as had not been seen before. It was made of white oak, yellow pine, Oregon pine, and teak, held together by locust tree nails, copper and bronze bolts, and composition spikes. Its four anchors were each 5500 pounds of bronze, and in place of anchor chains, it had 11-inch diameter hemp cable. Bauer revelled in the irony that this ship was named after a person who made his fortune in steel, and yet, "every effort was made to avoid steel" in its construction [Bauer, 1910b, pp. 236-237]. Bauer stationed one of his most experienced magnetic investigators at the shipyard during the seven months of construction to test all metals used in the *Carnegie* [Bauer, 1910a, p. 198].

As he noted in an annual report to the Institution, the approval of the construction of the *Carnegie* provided a "source of encouragement and stimulus to every investigator in magnetic science" [Bauer, 1909c, p. 167]. At a cost of $115,000, this was certainly the most expensive scientific instrument created to that time for geomagnetism, or in fact for most any science. Or perhaps this ship should be compared not to scientific instruments, which it carried, but to observatories or laboratories. It was one of the CIW's largest special appropriations to date, comparable to those for various telescopes at Mt. Wilson Observatory and to those for the construction of the Geophysical Laboratory and the later building for the DTM.

The most important part of the ship for its research was the deck house, which consisted of an observation room with a circular observatory at each end. The observatories each had a revolving dome made of a bronze framework and glass plates, several of which could be removed for astronomical or azimuth observations. The main virtue of the domes was that they kept observers dry while making magnetic measurements during foul weather. They also kept sailors and errant iron at least eight feet from the instruments.

The magnetic instruments were mostly derived from those used on the *Galilee*, except for the marine Earth inductor, which replaced the Lloyd-Creak dip circle in 1912 [Bauer et al., 1917, pp. 190-203]. The sea deflector, which originally consisted of an attachment designed by Bauer to be added to a standard compass, allowed the measurement of horizontal intensity. After a decade of development it was an instrument made entirely in the DTM instrument shop. As Bauer said, a more appropriate name would have been sea magnetometer, since it measured declination and horizontal intensity with equal facility. In the latter function it used the sine-deflection method. A deflecting magnet was mounted above the center of the compass card. The angle through which the card rotated out of the magnetic meridian readily yielded the value of horizontal intensity [Bauer et al., 1917, pp. 190-195]. The marine Earth inductor differed from those used in land surveys mainly because of mechanical problems presented by the gimbal mounting. Unlike on land, the marine Earth inductor did not use a null method, but instead determined inclination by interpolation of galvanometer deflections taken either side of this position [Bauer et al., 1917, pp. 196-202].

The first assignment of the *Carnegie* was a cruise of the North Atlantic, from Newfoundland to England. Setting sail in October 1909, the ship encountered rough seas with westerly gales nearly every day. Nevertheless, magnetic measurements were taken on 11 days of the 12-day trip. The results of "swinging the ship" before and after this cruise confirmed that no deviation corrections were made necessary by the remaining magnetic material aboard [Bauer, 1910a, pp. 198-199]. As we would say today, she was magnetically clean. The chief advantage of this was that no time was ever again wasted swinging the *Carnegie*. Another advantage, however, was that its magnetic results were available immediately at the end of every cruise, and even at its ports-of-call in mid-cruise. The results of this first cruise revealed systematic errors in declination of one degree, in inclination of half a degree, and in horizontal intensity of ten percent [Bauer, 1910a, p. 199].

During the next twenty years, the *Carnegie* conducted seven cruises, a total distance equivalent to twelve circumnavigations. Its crew of fourteen sailors and seven or eight scientists was repeatedly in the news and popular magazines around the world, largely because of the romance and mystery of their mission. One of its most famous voyages was the first single-season circumnavigation of Antarctica, completed in what was for a sailing vessel the short time of 118 days. It continued to draw attention because of the errors it found in charts of declination—which reached a maximum of fifteen degrees in the Indian Ocean [Bauer, 1920, pp. 300-301]. Although Bauer referred more than once to navigational utility as a "sordid" justification of his ship's "splendid" research, he was not above trooping out this argument time and again.

During the First World War the work of the world magnetic survey was necessarily curtailed, as might be expected for a project of international scope. The *Carnegie*, on a cruise in the South Atlantic, docked in Buenos Aires for the duration rather than risk being sunk. Many of the scientists with whom the DTM cooperated were Germans and were now out of reach, as were the German colonies. Moreover, the DTM converted almost to an arm of the U.S. Navy. It provided compasses and other navigational instruments, and it conducted research on the detection of submarines and the development of magnetically triggered mines. Its pure scientific research took second place to military functions [DTM, WWI Navy files].

AFTER THE GREAT WAR: THE COMPLETION OF THE FIRST WORLD MAGNETIC SURVEY

After the War, the mission of the DTM came into question, as a new president, John C. Merriam, took over the reigns of the CIW. Merriam convened a one-day "Conference on Fundamental Problems of the Earth's Magnetism and Electricity and Most Effective Methods of Research" in 1922 [CIW, DTM, Fundamentals of Earth's Magnetism 1922]. One of the chief conclusions was that the magnetic survey was nearing completion and that "future work need not be so intensive." Secular variation would be the main concern of future survey work. Emphasis would shift to observatory work, and laboratory investigations. Merriam conducted many private conversations with Bauer and other DTM staff members about the "problem" of the DTM. W. F. G. Swann recommended more attention to analysis and less to observation, especially regarding the *Carnegie* [CIW, Merriam Memos, DTM, Memorandum of Conference with Dr. Swann, December 19, 1922]. Swann urged the hiring of a mathematical physicist. In a conference with Bauer and his main administrator, J.A. Fleming, in 1923, Merriam advised the DTM to concentrate on the Earth's crust and atmosphere [Ibid., Memorandum Regarding Conference with Dr. Bauer and Mr. Fleming, May 11, 1923].

Clearly, he wanted a different kind of leadership from Bauer than was forthcoming. He wanted the DTM to cooperate more with other CIW Departments working in seismology, solar-terrestrial relations, and other geophysical problems. He also wanted a broader range of basic research on the constitution of matter at the DTM, conducted independently of Bauer. In short, Merriam thought that it was time to move beyond the world magnetic survey, and that Bauer's control of the DTM had to be lessened [Ibid., Copy of Handwritten Notes by J. C. Merriam at Executive Committee Meeting of May 18, 1923]. The very strength of Bauer's vision, to raise terrestrial magnetism to the status of an independent branch of science, also limited the DTM's program. With the world magnetic survey drawing to a close, the DTM required a new vision.

CONCLUSION: NEW LEADERSHIP FOR THE DTM

The story beyond this date becomes much more complex, as the DTM searched for a new mission and new leadership. Under new directors, the program was broadened to include the experimental research of Merle Tuve in high-energy physics, the theoretical research of Gregory Breit, the cosmic ray work of R. A. Millikan and Scott Forbush, and others. But the DTM neglected neither terrestrial magnetism nor geophysics more generally in the coming decades. It continued geomagnetic survey work through the 1930s and 1940s, although more effort was made to encourage such work by governments, including the U.S. In the 1930s the DTM brought in Sydney Chapman from England and Julius Bartels from Germany, among other important theoreticians, as Research Associates for extended stays. They wrote their classic text *Geomagnetism* under DTM sponsorship [Chapman and Bartels, 1940].

Among the other scientists to become prominent in geophysics who had a tie with the DTM during these years were Teddy Bullard, Walter Elsasser, Harry Vestine, and James Van Allen. Bullard conducted a magnetic survey for the DTM in Africa in the 30s [Wallis and Green, 1947, pp. 20-21]. One of Elsasser's first discussions of his dynamo theory of the Earth's main field was at a DTM symposium, "The Interior of the Earth," in 1940 [Teller and Tuve, 1940]. The critical work of Elsasser and Bullard on the production of the Earth's field presumed the observational base provided by Bauer's vision. Vestine carried that vision forward. As chief of the DTM's Land Magnetic Survey section in the 1940s, he directed the analysis of the largest body of magnetic distribution data available in the world. This analysis, largely supported by the U.S. Naval Ordnance Laboratory during World War II, resulted in an authoritative publication [Vestine et al., 1947].

Perhaps ironically, though, the culmination of Bauer's vision coincided more closely with Humboldt's original call for a survey sponsored not by a private institution, but by the cooperation of many nations. Vestine was one of the essential organizers of the World Magnetic Survey conducted in the 1960s through the auspices of the International Association of Geomagnetism and Aeronomy. Bauer might have been pleasantly surprised to see such an international effort succeed.

In conclusion, I wish to ask directly, did Bauer see terrestrial magnetism as an emerging and truly independent discipline, with all of the intellectual and institutional implications that come with that view? I think not, or at least not at first. Although his research concentrated on geomagnetism, his training included theoretical and experimental physics and astrophysics. I think, rather, that he saw specialization in geomagnetic research as equivalent to specialization in problem areas of more standard physics, such as electromagnetic radiation. Such areas required concentration on particular theoretical issues and experimental techniques, certainly. But this concentration did not imply that electromagnetism, for example, was a new discipline. Perhaps the one tactical difference Bauer faced was that the study of geomagnetism, unlike electromagnetism, was firmly tied in many people's views to activities such as geodetic surveying. Bauer sought to sever this tie, so that more useful associations might be made, especially to physics and astrophysics. His creations of the DTM and of the journal *Terrestrial Magnetism* were steps toward this realignment.

ARCHIVAL SOURCES

All archival material for this article is in either the main CIW archives at P Street or in the DTM archives. Unless stated otherwise in a citation, material in the main CIW archives is in the files there for the DTM. The file names referred to are those that appear on the file folders.

REFERENCES

Bauer, L. A., Some Recent Contributions to Terrestrial Magnetism, *Science*, N.S. *11*, 651-655, 1900.

Bauer, L. A., Proposed International Magnetic Bureau, *CIW Year Book [for] 1903*, *2*, 203-212, 1904a.

Bauer, L. A., Report of the Department of International Research in Terrestrial Magnetism, *CIW Year Book [for] 1904*, *3*, pp. 68-74, 1905a.

Bauer, L. A., A Contemplated Magnetic Survey of the North Pacific Ocean by the Carnegie Institution, *Nature*, *70*, pp. 389-390, February 23, 1905b.

Bauer, L. A., Report of Department of Research in Terrestrial Magnetism, *CIW Year Book [for] 1905*, *4*, pp. 264-274, 1906.

Bauer, L. A., The Work in the Pacific Ocean of the Magnetic Survey Yacht *Galilee*, *National Geographic*, *18*, pp. 601-611, September 1907a.

Bauer, L. A., Recent Results of Terrestrial Magnetic Observations, *Technical Quarterly and Proceedings of the Society of Arts [Boston]*, *20*, pp. 170-186, June 1907b.

Bauer, L. A., The Present Problems of Terrestrial Magnetism, *International Congress of Arts and Science, Universal Exposition, St. Louis*, 8 vols., edited by Howard J. Rogers, *4*, pp. 1-7, Houghton, Mifflin and Co., Boston and New York 1907c.

Bauer, L. A., The Instruments and Methods of Research, *The Popular Science Monthly*, *74*, pp. 184-201, 1909a.

Bauer, L. A., The Magnetic Survey of Africa, *The National Geographic*, *20*, pp. 291-292, 1909b.

Bauer, L. A., Department of Terrestrial Magnetism, *CIW Year Book [for] 1908*, *7*, pp. 163-174, 1909c.

Bauer, L. A., Department of Terrestrial Magnetism, *CIW Year Book [for] 1909*, *8*, pp. 194-202, 1910a.

Bauer, L. A., The Most Curious Craft Afloat: The Compass in Navigation and the Work of the Non-Magnetic Yacht Carnegie, *National Geographic*, *21*, pp. 223-245, 1910b.

Bauer, L. A., The Magnetic Survey of the Oceans, *The Geographical Journal*, *42*, pp. 517-530, 1913.

Bauer, L. A., The General Magnetic Survey of the Earth, *Bulletin of the American Geographical Society*, *46*, pp. 481-499, 1914.

Bauer, L. A., The Cruises of the 'Carnegie', *World's Work*, *39*, pp. 280-301, 1920.

Bauer, L. A., and G. W. Littlehales, Proposed Magnetic Survey of the North Pacific Ocean, *CIW Year Book [for] 1904*, *3*, pp. 269-273, 1905.

Bauer, L. A., W. J. Peters, J. A. Fleming, J. P. Ault, and W. F. G. Swann, *Ocean Magnetic Observations, 1905-1916, and Reports on Special Researches*, in series Researches of the Department of Terrestrial Magnetism, Carnegie Institution of Washington Publication No. 175, *3*, 447 pp., Washington, D.C., 1917.

Chapman, S. and J. Bartels, *Geomagnetism*, 2 vols., 1049 pp., Clarendon Press, Oxford, England, 1940.

Fleming, J. A., Louis Agricola Bauer (1865-1932), *Science*, *75*, pp. 452-454, 1932.

Gilman, D. C., *CIW Year Book [for] 1902*, *1*, pp. xli-xlii, 1903.

Good, G. A., Scientific Sovereignty: Canada, the Carnegie Institution, and the Earth's Magnetism in the North, *Scientia Canadensis*, *38*, pp. 3-37, 1991.

Grindley, E. N., The Earth's Magnetic Field in Southern Africa at the Epoch, 1 July 1930, *Philosophical Transactions of the Royal Society of London*, Series A, *240*, pp. 251-294, 1947.

Harradon, H. D., Principal Published Papers of Louis A. Bauer, *Terrestrial Magnetism and Atmospheric Electricity*, *37*, 220-224, 1932.

Humboldt, A. von, *Cosmos: A Sketch of a Physical Description of the Universe*, 5 vols., transl. by E.C. Otté, Harper and Brothers, New York, 1868.

Reingold, N., Louis Agricola Bauer, *Dictionary of Scientific Biography*, 18 vols., edited by C. C. Gillispie, *1*, pp. 521-522, Charles Scribner's Sons, New York, 1970.

Servos, J., To explore the borderland: The foundation of the Geophysical Laboratory of the Carnegie Institution of Washington, *Historical Studies in the Physical Sciences*, *14*, 147-185, 1983.

Teller, E. and M.A. Tuve, The Sixth Washington Conference on Theoretical Physics, March 21-23, 1940, *Science*, *91*, pp. 621-623, 1940.

Van Hise, Report on Geophysics, *CIW Year Book [for] 1903*, *2*, 173-184, 1904.

Vestine, E. H., L. Laporte, C. Cooper, I. Lange, and W. C. Hendrix, *Description of the Earth's Main Magnetic Field and its Secular Change*, 532 pp., Carnegie Institution of Washington Publication No.578, Washington, D.C., 1947.

Wallis, W. F. and J. F. Green, *Land and Ocean Magnetic Observations, 1927-1944*, in series Researches of the Department of Terrestrial Magnetism, Carnegie Institution of Washington Publication No. 175, *8*, 243 pp., Washington, D.C., 1947.

Woodward, R.S., C. Barus, T. C. Chamberlin, A. A. Michelson, C. R. Van Hise, and C. D. Walcott, Report of Advisory Committee on Geophysics, *CIW Year Book [for] 1902*, *1*, 26-43, 1903.

Vilhelm Bjerknes's Duty to Produce Something Clear and Real in Meteorological Science

RALPH JEWELL
Department of Philosophy, University of Bergen
Sydnesplass. 9, N-5007 Bergen, Norway

My subject today is the Earth's atmosphere and how in the first three decades of this century the sense of duty and honor of the Norwegian physicist Vilhelm Bjerknes (1862-1951) crucially affected the way it came to be viewed as more or less distinct air masses separated by the now-familiar atmospheric fronts. Bjerknes inspired into existence the Bergen School of Meteorology and its characteristic "Polar Front Meteorology," which set new patterns of work which Meteorology has adopted and adapted ever since.

When I consider the theme of Bjerknes's striking sense of duty and honor in science, I have in mind a dramatic image that is rooted in Norwegian history. It is given in a ballad told by the Norwegian poet Per Sivle, about a battle in Norwegian history as famous in Norway as the battle of Hastings is among Britons, and which happened at roughly the same time—the battle of Stiklestad. The ballad tells of a great leader in this battle who became mortally wounded. He summoned up all his strength and resolve, and thrust the staff of his banner into the ground, crying out the words: "And this is what is great and truly wonderful: though the man may fall, the banner shall stand!" [Sivle, 1925 (1909-10)]. These words are words of defiance, of stubbornness, of duty, of honor. In 1929, directly alluding to this ballad, Bjerknes used these very same defiant words. He was then 67 and feeling regretful that he had not yet reached the most cherished promised goal of his scientific work. At the same time he hoped soon to make a final breakthrough towards the great prize, the theory, that had so far eluded him. He felt committed to a binding scientific duty to gain it. For Bjerknes the "banner" that was to remain standing and outlive him was to be this *theory* that seemed imminent or impending—a theory that would be capable of standing by itself and that could make its declarations about reality without the necessity of his continual personal presence and intervention. To understand the force Bjerknes felt in this image of a banner standing free and by itself without his continued support, would take us to the heart of science as Bjerknes's personal calling, towards seeing the special attraction that theory has for scientists more generally, and towards seeing the crucial role that the Carnegie Institution of Washington (CIW) played in the way Bjerknes and the Bergen School made such an impact on dynamic meteorology and the meteorological forecasting services of the world.

In 1929 his wave-theory of cyclones seemed on the verge of being formulated to his satisfaction. This theory was intended to give a proper grounding for the empirical discoveries concerning the atmosphere's dynamic structure that his Bergen School had been making during the previous eight years. Only with sound grounding in proper theory would these empirical discoveries stand fast and stand firm. This was a matter of honor for him, for only in that case did he believe that he could satisfy his obligation to the CIW for the considerable financial support it had given him, and for its trust in his meteorological initiatives. He felt honor-bound to provide in return something in science that was clear, real, and whole—"a complete victory" in science, and not just a half. How did Bjerknes come under the force of such an obligation in meteorological science, to produce something that would last? And how did the CIW's support for his work form some of the basic conditions for it? These are the questions that I consider here.

At the time the CIW of Washington was founded and its first plans for supporting science were being formulated, meteorology was something of an embarrassment to science. It was in such an ailing state that leading scientists variously criticized it for being "in the Slough of Despond," "a miserable science—not even believed in by its own folk," and "the sore spot of the conscience of natural science." The range of actual criticisms and complaints that meteorology drew forth makes for interesting reflections on the origin of the thought that a science is not quite a mature science, and conversely also of what affords a particular science a good defense against such strictures. The CIW was formed partly to do in science what, without it, might not be done. Clearly, in meteorology something was crying

The Earth, the Heavens and the Carnegie Institution of Washington
History of Geophysics Volume 5

out to be done; but what?

When the CIW looked for the "exceptional man" to benefit meteorology, they sought advice from experts at home and abroad. Bjerknes became the chosen man. How the decision to support his research was made, and how this support was maintained for over thirty years form important parts of the story here. The details involve surprises of the kind that we expect from real dramatic stories.

Although Bjerknes's name is now linked mainly with the twentieth-century reform of meteorology through the work of his Bergen School, that connection should not mislead us. Bjerknes always regarded himself as first and foremost a theoretical physicist. It was theoretical physics that drew from him his deepest loyalty and his strongest commitment personally to do something really worthy in life. In both the highly theoretical physics that he cultivated and in his commitment to duty in his work, he followed his father, the mathematician-physicist Carl Anton Bjerknes. Both men placed extraordinarily great hopes in following the lead of a firm scientific conviction—the conviction that if only the mechanical properties of fluids could be understood at the foundational level then the simple harmonious patterns shown would clarify the whole of physical nature. In a very real sense the two men invested their lives in this quest.

Bjerknes usually began the story of his scientific work with the work of his father. He did so because for him a great pain was involved, one he tried to alleviate since his youth. There was a tragic aspect of his father's life-work in science, made more tragic by an apparent lack of sympathy among his colleagues for what he was trying to do. Vilhelm believed that Carl Anton was never given at home in Norway the recognition that he deserved and was given abroad. And it was this that caused Vilhelm the pain. This aspect helps one to fathom Vilhelm Bjerknes's compelling sense of duty in his science, and how he could so greatly value support from the CIW. It was, indeed, support from outside Norway, representing foreign acknowledgement of the line of investigations started by his father.

Carl Anton had tried to show through a series of extraordinary hydromechanical experiments how all the phenomena of magnetism and electricity had close parallels in the mechanics of fluids. The lack of unifying connections between magnetism and electricity on one hand and mechanics on the other seemed, to him, to be an intolerable weakness in physics. Physics was worthy of being undivided and whole. At one point in his work he became amazed, and perhaps even dazzled, by the beauty of the harmony and simple order that unexpectedly appeared on paper when he made a slight alteration in his mathematical discussion of the experiments. With this slight alteration, what previously had been impossibly complicated and contorted quite suddenly became harmonious, simple, and clear. To him it seemed that a whole new world had opened up before his very eyes. Magnetism and electricity seemed to him fully explainable by mechanics. He felt sure that he was on the track of a fundamental orderliness that would eliminate the rift marring physics. This conviction led him to spend his life trying to reveal this hitherto undeclared shared order.

Through many convincing laboratory demonstrations and years of painstaking calculations Carl Anton Bjerknes went far towards his goal. But he never reached it. He never accomplished the kind of satisfaction that comes with seeing all the various pieces of his research snap into place so as to form, or give good grounds for, a single compelling theory that would make plain all that he most earnestly wanted to show. What Vilhelm found tragic was that Carl Anton's researches became more and more cut off from other researchers' work. The researches did not appear to belong to any particular branch of science—not really to magnetism and electricity proper, not to regular mechanics or hydromechanics, not to any nameable branch of any science. His work was scientifically "homeless" and Carl Anton became an ever lonelier figure, ever more an island unto himself in his work.

This tragedy strongly marked Vilhelm's own science. Through dwelling on his father's "fate" and possible explanations for it, Vilhelm arrived at an identification of the trouble, and through this clear grasp of the trouble he shaped most deliberately and carefully his own way of doing science so as to avoid a similar fate. Vilhelm saw that Carl Anton's visions did not promise clear solutions to specific well-formulated problems. The problems that fascinated Carl Anton were perhaps too ambitious, deep, fundamental, and untamed—too philosophical perhaps—to be soluble and so to be capable of delivering safely the measure of conveyable insight that they were meant to give. It was a formative experience for Vilhelm to discover such a lesson about the importance of proper "posing" of questions in science, and to feel a son's deep sympathy for his professor-father who seemed to fail to win respect for his researches among his colleagues at Norway's only University, Kristiania, now called Oslo. Vilhelm discovered the force of the maxim: Set your scientific problems properly so that they might permit proper scientific solution! Arriving at convincing scientific clarities and realities depended on it.

Theory was the key to success, and yet also its outcome. But theory required appropriate expressive gifts or accomplishment. As a schoolboy Vilhelm had been his father's assistant. At the Paris Electrical Exhibition of 1881 Carl Anton demonstrated the strange parallels that he and Vilhelm had produced with the strange-looking hydromechanical apparatus that they had designed and built. Leading scientists declared that these demonstrations were the most intriguing of all that could be seen in that Exhibition. But those who were most eager to understand what they saw in these demonstrations clamored to ask: "When can we expect to see the theory?" Without the theory they could be intrigued, but they could hardly know what it was that they were meant to see, or even what they were looking at. They could not be sure that they saw what was being shown. With the theory, they could know. The theory would supply the missing "seeing instructions." The great snag was this: Carl Anton had expressive gifts, but these gifts did not apparently extend to finishing scientific works for publication. In Vilhelm's words, as his father "polished and filed away at his writings to get the final finish, everything fell to pieces under his hands." This drama of a missing theory touched the son, and the experience marked his life.

Vilhelm formed a deep sympathy for his father as he saw him floundering in ever greater confused despair while he tried in growing scientific isolation to show the world that he was revealing something clear and real about the true fundamentals of physical nature, without ever managing to complete his picture. Vilhelm took responsibility for doing what he knew his father could not do for himself. He resolved to go through all the material from his father's work and from a safe distance away from him at Stockholm, so as not to be caught in the same confusion, bring all the various findings into a great single treatment that would show the whole integrity of his father's work, hopefully in a form that would declare the theory of it all. He spent many years on this work, crafting into shape a book that would provide a safe proper logical home for Carl Anton's results. The task was immensely difficult, but it was accomplished a short time before his father's death. All those years of arduous duty had become resolved in the kind of satisfaction that is only known by someone who has made great sacrifices out of duty. Vilhelm experienced this kind of satisfaction when he heard his father say that with the book he was "well pleased."

The investigations which eventually brought Bjerknes into the sight of the CIW as the "exceptional man" for meteorology stemmed from Bjerknes's preparations for his lecture course in hydrodynamics at Stockholm in 1897. As part of these preparations he began to formulate theorems of a "Physical Hydrodynamics." This was in sharp contrast to the regular classical hydrodynamics that dealt only with an idealized fluid with only ideal properties. According to classical theory, storms could neither arise nor cease. This ideal hydrodynamics therefore seemed of little promise for understanding the Earth's atmosphere's real (for classical physics, *bad*) behavior. Bjerknes believed he could bring meteorologists back to physics with his physical theory. At Stockholm Bjerknes found that meteorologists and hydrographers soon plied him with requests that he bring his theory to bear on their intractable geophysical problems. He was flattered and encouraged, and he hoped to develop a theory of fluid motion directly applicable to practical investigations of the behavior of ocean and atmosphere, two fascinatingly complex geophysical fluid media. He wanted not just another "paper theory," but one useful in a busy world.

Although he was tempted to follow the meteorologists and hydrographers he initially kept physics to the fore, out of a sense of duty. While his father was alive Vilhelm spared him disappointment due to what might appear as a weakening connection with physics. Also the redaction of his father's researches precluded assuming fundamental new commitments. At best he could only make short discreet visits to geophysical sciences, and they gave him a sense of new freedom and contact with nature. Vilhelm enjoyed this close contact with nature and started to use his new-found freedom from what was a kind of scientific bondage.

The last years of Carl Anton's life were very hard years for Vilhelm as he struggled to fulfill his filial obligation. But when that obligation was fulfilled all was different. Since 1898 he had been harboring progressively firmer thoughts that his work in hydrodynamics was well-suited to providing a basis for resolving the atmosphere's behavior into soluble problems of physics. His scientific assistant at Stockholm, J. W. Sandström, mechanical worker turned scientist, had improvised mechanical-computational methods for transforming the available meteorological data into new kinds of pictures and other graphical displays illustrating Bjerknes's new atmospheric phenomena. Singularities and features never before made visible appeared on the new diagrams and charts. Without the previous restraint, he was more free to tackle the great beckoning media: the atmosphere and the sea. He took his son out on windy hills to fly kites—for scientific reasons of course, but also to enjoy with his son the sport of playing with the wind. He sketched out ways to bring the atmosphere under the scope of physical law. He leapt for scientific joy.

Bjerknes believed he could do something for meteorology that in some respects matched what he had just tried to do for his father's work. He thought that meteorology was in a pitiable state as a science, whose practitioners were unable to remedy the ailment. He believed he possessed the means to give it a proper footing through working out solutions to well-formulated scientific problems of dynamic meteorology. Physics had always been his field. He had been Hertz's first scientific assistant at Bonn and had been entrusted by Hertz to undertake experiments that made the existence of Hertzian, or radio, waves more plausible than had been the case as long as only Hertz himself had produced them. He had been lecturer then professor at Stockholm, concentrating on mechanics. Now he saw an opportunity to focus on bringing physics and meteorology together. It was a bold vision, but the excitement of the new freedom let him maintain it.

Just how bold and exciting a vision it was may be judged by how Napier Shaw, the director of the British Meteorological Office, summed up the lamentable state of meteorology in his address at the British Association meeting in 1903 at Southport, as chairman of the sub-section for astronomy and meteorology. This statement gives rich insight into Bjerknes's excitement at the prospects for his meteorology-through-physics program and how his connection with the CIW in 1905 seemed to offer him an inspiring role in science.

Early in his address Shaw implied a lethal criticism of meteorology when he announced that he would be speaking from "the point of view of an official meteorologist whose experience is not long enough to have hardened into that most comfortable of all states of mind, a pessimistic contentment." He spoke of how meteorology had been kicked out of the universities, saying:

> There was a time when Meteorology was a recognized member of the large physical family and shared the paternal affection of all professors of Physics; but when the poor nestling began to grow up and develop some individuality electricity developed simultaneously with the speed of a young cuckoo. The professors of Physics soon realized that the nest was not large enough for both . . .

The result, as Shaw put it, was that:

> [W]ith singular unanimity the science which bears in its lefthand, if not in its right, sources of wealth beyond the dreams of avarice, was recognized as a veritable Isaac, and the science wherein the fruits of discovery must be free for all the world, and in which there is not even the most distant prospect of making a fortune—that science was ejected as an Ishmael. Electrical engineering has an abundance of academic representatives; brewing has its professorship and its corps of students, but the specialised physics of the atmosphere has ceased to share the academic hospitality. So far as I know the British universities are unanimous in dissembling their love for Meteorology as a science, and if they do not actually kick it downstairs they are at least content that it has no encouragement to go up. In none is there a professorship, a lectureship, or even a scholarship, to help to form the nucleus of that corps of students which may be regarded as the primary condition of scientific development.
>
> Having cut the knot of their difficulties in this very human but not very humane method, the universities are, I think, disposed to adopt a method of justification which is not unusual in such cases; indications are not wanting which disclose an opinion that Meteorology is, after all, not a science.

Shaw reviewed certain promising developments in meteorology during the previous decades but added:

> Yet with all these achievements it must be confessed that the progress made with the problems of general or dynamical Meteorology in the last thirty years has been disappointing. When we compare the position of the subject with that of other branches of Physics it must be allowed that it still lacks what astronomy found in Newton . . . light in Young or Fresnel . . . electricity in Faraday and Maxwell. Above all, it lacks its Kepler. Let me make this clear. Kepler's contribution to physical astronomy was to formulate laws of nature which no heavenly body actually obeys, but which enabled Newton to deduce the law of gravitation. The first great step in the development of any physical science is to substitute for the indescribably complex reality of nature an ideal system that is an effective equivalent for purposes of theoretical computation . . . [M]eteorology has not yet found a satisfactory ideal problem to substitute for the complexity of nature . . .
>
> The most direct efforts to create a general theory of atmospheric circulation are those which attempt to apply Newtonian dynamics, with its more recent developments on the lines of hydrodynamics and thermodynamics . . . The "problem" is not sufficiently formulated [Shaw, 1903].

This shows very well how bold and exciting a vision it was that Bjerknes had when he formally stated his original meteorological program of 1904. He was already at the leading edge of the latest developments in hydrodynamics, in connection with his development of what he called physical hydrodynamics of real natural fluid media, as an alternative to the classical hydrodynamics which he considered as being blind in principle and in fact to all the interesting properties of air or sea. It is not at all fanciful to consider that Bjerknes, encouraged by Shaw's remarks and similar remarks made by others at the time, entertained the thought that perhaps—just perhaps—he could become the awaited Kepler of meteorology. Perhaps he might proceed even further to become also its Newton. He knew of no one better equipped. It was a large vision. Given the state of meteorology at that time, and given the promising successes of his own work in the required branch of physics, we might ask: Why not? The thought that meteorology was looking for and badly needed someone to give a new sense of direction left him excited and impatient—that much we do know.

His impatience, and his sense that time was running away, is obvious in the remarks he made in a letter written on March 14, 1904—a landmark-day for his meteorological initiatives. He was lifted out of his depression by the writing of a nine-page paper "which contains the quintessence of it all" [UOM, Bjerknes to Honoria Bjerknes, March 14, 1904]

A year had passed since the death of his father when Bjerknes wrote this quintessence and his joy shows that he was turning towards meteorology. It was becoming his science. Perhaps the most open expression of the change of heart that Bjerknes had towards this change in research interest is found in his letter to Fridtjof Nansen, September 2, 1904. Bjerknes wrote hoping to enlist support for obtaining suitable conditions of work for Sandström, whose special talents unfortunately were in danger of being wasted through not being adequately recognized by his new employer.

> I want to solve the problem of predicting the future state of the atmosphere and sea. That this really was the goal, was something to which I had previously shut my eyes—blinded, I may say, by the largeness of the problem, fearful of wanting to do too much. But it is clear that in this task all other lesser tasks are embraced, and this one and this one alone must be set as the "distant goal". . . . But, of course, from the formulation of the program to its final fulfillment is a tremendously long way, clearly much more than the work of one man. But all the same, since I have had this goal to steer towards, my meteorological-hydrographical investigations have become at least as dear to me as my old work-inheritance from my father, and it is my wish to do all I can to accomplish something whole here.[UOM, Bjerknes to Nansen, September 2, 1904]

The more Bjerknes became familiar with the way meteorologists had dealt with the atmosphere, the more he thought that meteorology's troubles as a science could not end until and unless it was incorporated into physics. He felt doubly drawn to meteorological problems. As a physicist he saw them as a provoking challenge to the honor of science. And yet he also drew a kind of essential sustenance from these problems for his own ideas in physics—a kind of sustenance for his physics that, in the circumstances around 1900 and after, he could not get from physics itself. His meteorology-through-physics plan was also a physics-through-meteorology plan. For both father and son, their great "work of life" was to try to reach an insight into the master-pattern of all physical happenings, and to express this master-pattern in a general theory of how physical reality behaves. Their personal self-esteem and sense of duty and honor were bound up with this aim. As scientists, this is surely what they lived for. A clear difference between the two was that Vilhelm was *doubly* committed to produce something clear and real through his science. That his father had not quite managed to fulfill his aim gave him the loyal son's job of fulfilling it for *both* of them. He had to produce something most clear and most real, the outcome not only of his own work, but also of his father's. And even if this something "clear and real" might belong to meteorology, it would belong also to fundamental physics at the same time.

How did the CIW enter the picture? The basic facts concerning the original intentions behind the institution were straightforwardly expressed by its president, Daniel C. Gilman, in a letter to the editor of *Nature*, September 9, 1902. He emphasized the careful arrangements that had been designed for the wise administration of the fund.

> The control of this fund is invested in a board of twenty-seven trustees, selected from widely separated parts of the United States, and including many men who have won distinction and confidence by the service they have rendered in public life. It is not a board made up of specialists, but rather . . . of men accustomed to the administration of large affairs, political, financial, philanthropic and educational. Absolute power is given to this board to devise such methods and form such plans as may seem to them wise in order to carry out the purposes of Mr. Carnegie.

Five of the six points specifically emphasized as aims by the donor were points that show how Bjerknes could be seen to fit the requirements of the Institution. These five were: the promotion of original research; the discovery and encouragement of exceptional men; the increase of facilities for higher education; the assistance of those now engaged upon research; and the publication of scientific memoirs. Consultations with other authorities and scientific bodies, at home and abroad, began the process towards laying down policies for decision-making on how and who to support. Gilman specifically remarked that:

> Attention was given to the conditions which have helped or retarded the progress of eminent men during the last generation—Darwin, Pasteur, Helmholtz, Abel and Virchov for example in Europe—Henry, Agassiz,

Dana and Rowland in America—and also the encouragements and discouragements which are encountered by the men of to-day. Time will be required for the digestion of this material in order to discover the methods which are most efficacious in the advancement of knowledge. [Gilman, 1902]

Two figures of American science played key roles in the decision of the CIW to support Bjerknes as a meteorological researcher of quality who the world might heed. One of them was Cleveland Abbe, who was alert to the need to approach the atmosphere's behavior as a problem for mechanics. Another was R. S. Woodward of the Department of Physics at Columbia University, who subsequently became president of the CIW.

Abbe was appointed in March 1902 as adviser to the Executive Committee in relation to meteorology, with the request that he submit a report of the state of research in the science. His subsequent report, of July 1902, entitled "Proposed Research in Meteorology," seems almost prophetic in regard to the subsequent choice of Bjerknes. Abbe wrote, just as Shaw expressed himself the following year at Southport, that the state of meteorology in the past had something to do with its comparative neglect by the universities. He pointed out that in consequence "the older and prominent professional meteorologists are those who originally made a special study of chemistry, or astronomy, or physics, or navigation, or engineering; this gives to each meteorologist a tendency to prosecute meteorological studies along certain pre-determined lines of thought." He suggested that the most important discoveries of the future must be expected from those who approached the subject from the point of view of mathematics and analytical mechanics. Meteorology needed a new leading figure.

Meteorology has attained the status analogous to that of astronomy in the days between Newton and La Place. It is ready to receive a new leader and is looking for him. An hundred experimentalists and thousands of observers are perfecting the data of observations,—but the crying need is for one who shall elucidate our complex phenomena to the satisfaction of the students of mechanics.

In this search for men and the pre-eminent right man, meteorologists welcome the assistance of the Carnegie Institution, and your committee would respectfully submit the following general recommendations . . .

1. Meteorology should be treated by you as a very broad subject always embracing the atmosphere as a whole. The Institution may leave it to local observers to investigate the climatology of their respective localities, embracing only one-tenth of the surface of the whole globe; the remaining nine-tenths, including the Polar regions and the high seas, are open to our investigation without exciting international jealousies or questions of propriety.

Abbe mentioned Bjerknes:

A few years ago meteorologists were encouraged to find that von Helmholtz had turned his attention in our direction. But his death in the prime of life crushed our hopes. At the present time there are several prominent workers on the mechanics of the atmosphere, such as Bigelow in America, Bjerknes and Ekholm in Sweden . . . [Abbe to Walcott, CIW, July 14, 1902]

The caution against the risks of initiatives that would excite international jealousies or questions of propriety was appropriate. At the start of formal international arrangements in meteorology in the 1850s and 1860s, the emphasis was on marine meteorology, so as to avoid trespassing on national prides. This issue was still delicate when the CIW made its plans, since it knew it had to avoid territories of agencies, universities, journals, and other bodies that might feel threatened by its initiatives. It had to avoid giving any hint of trespassing on the preserves of others. A nation's science was likely to be regarded as its badge to cultural identity and sovereignty, and that went for its meteorological services too. After his move into meteorology, Bjerknes discovered that he too had to fight against such national rivalries while trying to persuade meteorologists to accept his system. Abbe's choice of the sea and polar regions makes their force here. Similarly, in the 1920s Bjerknes and his Bergen School proposed their new system under the name "The *Polar* Front." It partly explains too, why the new meteorological code they proposed in the 1920s for replacing line-telegraphy with radio transmission was based on a code originally prepared for ships at sea. In the 1920s the problem of national pride still thwarted the easy introduction of uniform meteorological arrangements. Bjerknes and the Bergen School knew it well.

R. S. Woodward resigned from Columbia University to become CIW President in 1904. But he took steps to ensure that graduate work in physics at Columbia would be compensated in the form of special provisions in mathematical physics. It was decided to invite distinguished foreign workers in the field to offer specialized lectures. On January 26, together with two colleagues, Woodward submitted proposals to the University's president, recommending in order of preference V. Bjerknes, H. A. Lorentz, Max Planck, and J. Larmor.

Bjerknes accepted an invitation and came to New York in

late 1905. Partly through Abbe's mediation he gave a lecture in Washington on his vision of the needs of meteorology, if it were to be improved as a science. He later called this lecture the most important of his life. As a direct result of this lecture he was invited to prepare an application to the CIW for funds to cover the cost of employing his assistant in Stockholm, the "rough diamond" ex-manual worker, now priceless help, Sandström. Bjerknes and Sandström were excited by this encouragement for their work. Bjerknes formally declared his plan of producing a treatise meant to provide what meteorology sorely needed: a great leading idea. The grant was awarded, and it was renewed subsequently without break until after the start of World War II.

For several years, after quickly completing and publishing volume one, *Statics*, of this treatise, Bjerknes struggled in the clutches of a difficulty that hindered him as he tried to complete later volumes. The observations that were available from international meteorology were not the kind he required to advance his theoretical work. But—and here was the rub—his theoretical work, published, was to be his device for persuading meteorologists to make the right kind of observations. In 1910 he said in despair: I cannot publish the book because I have not the observations, and I cannot get the observations before by the publication of the book I have proved their necessity.

Volume two, *Kinematics*, was published at the end of 1911. The third volume, *Dynamics*, proved the great stumbling block. The rest of Bjerknes's life is a story of his struggle to do justice to the themes that belong to this third volume. It was never completed as the third volume of the originally conceived treatise. When he accepted a call from Germany and became the first head of a newly established Geophysics Institute at the University of Leipzig, beginning in January 1913, he believed that in such a position at a world-renowned university he would find ideal conditions for his work. But war broke out the following year and his conditions in Leipzig became difficult to bear. Consequently, in 1917 he and a little nucleus of his research team moved to Bergen [Good, 1991, pp. 304 and 315]

In that new setting on the coast of Western Norway, and in the emergency conditions of a world still at war Bjerknes found a new mood and way ahead for his work. He became caught up in a mood of bold improvisations, prompted by the necessity of developing new kinds of weather forecasts to aid Norwegian agriculture and fisheries while the nation faced food shortages. This was a far cry from the steady, strict work of producing polished scientific studies of the kind he had been encouraging his team to produce until then.

In this new setting of a different mood and new ways of working, the completion of the third volume, and the unfinished treatise, eventually seemed no longer feasible or even desirable. From the vantage point of the fresh work carried out in Bergen the unfinished treatise seemed to go somewhat stale, and difficult to save merely by the expedient of grafting on a later volume or two. Fundamentals had changed. From 1918 his young assistants in Bergen, who became known as the Bergen School, showed that there was a surprisingly different way to bring the atmosphere under scientific management [Friedman, 1989]. Bjerknes saw it as a more "bourgeois" way ahead—less aristocratically proud in its demands for theoretical purity, but so much more in touch with the real surprises delivered by the weather. Even the fishermen up and down the coast came to trust the Bergen forecasts more than their own judgment, and this meant a lot to Bjerknes when he recalled that the great Helmholtz had admitted the superiority of fishermen over physicists in outguessing the actual developments of the weather. He eventually came to terms with this new "Bergen School" path together with its changed mentality, and increasingly came to look upon his original path—that of his unfinished "Carnegie Treatise"—as a thing of his stricter past.

The developments within Bjerknes's meteorological initiatives between 1911 and 1925 are complex and fascinating. I shall illustrate how these initiatives affected his sense of duty and honor in his meteorological investigations through three occasions when he mentioned them. The first is a letter to the CIW in 1925; the second is a letter to the Swedish physicist Oseen in 1929; and finally a most expressive letter to the Faculty of Mathematics and Natural Sciences of the University of Oslo in 1931. This will show how he perceived it as a duty and a matter of honor to deliver a complete scientific victory in return for the support that he had received from the CIW.

In the 1925 letter to Merriam, Bjerknes used the image of battle and victory in science to introduce the theme of how he wanted to make an honorable exit from meteorology and return to physics. To Bjerknes, resolving one's work into proper order before one left it—into a theory—was a matter of a scientist's honor.

> When eighteen years ago I began that meteorological work which has been so generously granted by the Carnegie Institution, I did not underestimate whether [sic] its extent nor its difficulty. But I believed to be able to come through in limiting myself strictly to the working out of *formal methods*, leaving to the meteorologists to make the practical use of them.
>
> But as the work proceeded, I realized on the one side how much a deeper empirical knowledge of atmospheric conditions would help my work, and on the other side

how little the great meteorological Institutes were inclined to adopt any change in their methods or habits.

This led me to take up empirical work parallel to the theoretical investigations. From the year 1918 it was made in a very efficient way. I organized then for the Norwegian Government a Weather Service for Western Norway. My previous Carnegie-Assistants, who took up the problem from the new point of view and using the new methods were the acting forecasters. Through their daily work new empirical results came out which were both of immediate practical value, and of high importance for my continued theoretical work: the atmospheric discontinuities, which had been considered more or less as exceptional phenomena, were seen actually to govern the weather.

Before this result was found I had seen only one way of attacking the meteorological phenomena mathematically, namely by the laborious method of "mechanical integration". But now I recognized that the problem of the disturbance of the equilibrium of an atmospheric surface of discontinuity could be reduced upon *linear* differential equations, and therefore be accessible by ordinary analytical methods. . . . But the full solution of the problem will require a great amount of mathematical work, which, to be fruitful, has to be guided in close touch with empirical conditions.

My wish is, before I conclude my meteorological work, to make a serious attack upon this problem. I feel confident that its solution will prove to be of similar importance for meteorology as that of the motion of two attracting bodies has been for astronomy. [CIW, Bjerknes to Merriam, January 4, 1925]

President Merriam and those who advised him were greatly impressed by Bjerknes's work and his further plans. Bjerknes was asked to confirm his plans later that year so that a threefold-increase of his Carnegie grant could be considered. At the end of September Bjerknes sent these plans, which he hoped would bring his meteorological work "to a satisfactory end." To do this would require: a mathematical solution of the problem of cyclones, completion of volume three on dynamics, and new editions of volumes one and two [CIW, Bjerknes to Merriam, September 30, 1925]. We see in this way of formulating the plans, that the treatise was suffering from severe compositional stress. Its original design was becoming outmoded, so that later parts could only be produced on the basis of altering the design of the already published parts. He concluded the detailed statement of his plans by admitting one of his reasons for wishing to get the work finished: "To accomplish it while I am still myself at full force,—nobody knows when one's forces will begin to go down."

It was in the 1929 letter to Oseen that Bjerknes uttered the defiant words of the ballad mentioned at the beginning. When writing this letter Bjerknes was at a kind of cross-roads in his life-work. Here the track of the new kind of meteorology that the Bergen School had been developing since 1918 joined the track of his fundamental hopes as a physicist. At this juncture thoughts about empirical discoveries and thoughts about theory came together, and here he tried to come to terms with the limits of what, as a scientist he had accomplished and what accomplishment still escaped him. He had seen his meteorological work as a "physics-investment"—an investment of physics for the ultimate benefit of physics. Now, in 1929, he was eager and impatient to draw the physics from all this meteorological work. Now he wanted to see the beauty and power of this nearly expressed, nearly revealed, fundamental physics.

This attempt to draw forth the physics of it all from the rich experiences connected with the meteorological work that had been largely inspired by him, involved two aims. Considered as part of his meteorological initiatives, he aimed to encourage his more mathematical assistants to produce a fully fledged mathematical treatment of the wave-character of cyclones which the Bergen School discoveries appeared to show and require. The other aim was more directly connected with his attempt to contribute to the teaching of physics in Norway, by preparing a textbook of theoretical physics for the benefit of Norwegian physics students who, he believed, were not being educated well enough at Norway's sole university.

The first part of this textbook came out in 1929, and Oseen at Uppsala had read it. Bjerknes was not exactly proud of this work—a work that he had written almost out of a sense of despair, in an emergency. It was extremely difficult to obtain Norwegian recruits for his Bergen School who were sufficiently competent in physics. He considered this a critical matter, and he felt compelled to remedy it quickly and in the right way. The textbook was this attempt. He felt obliged to explain to Oseen his immediate purpose in writing it, and said how much he wished the second part of the book was finished too. He alluded to the classical image of the thread of Ariadne, and spoke of the Ariadne thread running towards the end of the maze of the work. In a most expressive admission he spoke of Halvor Solberg, one of the young Bergen School pioneers who was in the forefront of the attempt to treat the Bergen School findings theoretically and show the mathematical forms required for the new wave-theory of cyclones:

He is the only one who now has this Ariadne thread, and the strength to follow it. I too, of course, have the thread, but I no longer have the strength to follow it. From 1916 my strength has gone fully in the search for one end of thread, i.e. to win the clarity that the problem that has to be *solved* could also be *posed* . . . My work now, for as long as I have Solberg (sooner or later he will surely go abroad) is to assemble and get published everything that we have of solutions for "equations for perturbations", (for my part from 1916 onwards) i.e. for solved partial problems around the great unsolved cyclone problem . . . It is my hope that this work, if my powers last out, may be the banner that stands when the man has to fall [UOM, Bjerknes to Oseen, July 16, 1929].

This gives some of the essentials of what was involved for Bjerknes, as he reviewed his life work in 1929. Theory was the banner that would stand, even when he fell. Theory would maintain all that he so defiantly and firmly stood for in science. Just as the banner had been a rallying point in the battle of Stiklestad, so the polar-front theory, made firm by a mathematically developed wave theory for cyclones, would remain the rallying point for meteorologists of the Bergen School. This theory would be the means by which they could easily recognize each other and declare and develop their common allegiance to their common cause.

The battle image is entirely apt since although Bjerknes was a great man of peace he relished the metaphorical force of battle imagery, when it came to science. Indeed even without the distance of metaphor he relished polemical confrontation and battle as a means of sharpening one's science. He had a habit of speaking in terms of "attacking problems," or "skirmishing around the outer defenses of a great problem" in the preparatory stages of work, of "mustering up one's forces," of "holding out for the complete victory." The atmospheric fronts themselves, of course, are a perfect expression of battle and confrontation, since the fronts are the battle-zones where all the exciting rivalry and action is played out among the competing masses of air which make up the atmosphere as a whole dynamic natural system. The Bergen School's meteorology is the meteorology to find, show, and deploy the best possible image of the system of this vast natural struggle aloft. It was almost as if being such a man of peace gave him the freedom so openly to use the imagery of conflict in his science.

My story ends with the thoughts that Bjerknes expressed in a letter he wrote in 1931 to the Faculty of Mathematics, University of Oslo, the third of the three occasions mentioned above. It shows us that in Bjerknes's view the matter of his honor was also a matter of Norway's honor. After reviewing his work in theoretical physics and in meteorology, clarifying how they were connected yet distinct, he explained how the discoveries made in Bergen thoroughly transformed his prospects for winning a complete victory in science.

On the mathematical side for many years I had only managed to arrive at a DISCUSSION of the atmospheric motions. In spite of the most strenuous efforts every attempt to reach the formulation of CLEAR FUNDAMENTAL PROBLEMS which should be given the form of equations and conveyed to mathematicians for EXPLICIT SOLUTION failed. Under these circumstances I could feel the truth of Helmholtz's words: "Among all of Nature's phenomena, the ones of meteorology are the most mysteriously changeable, the most fluid, the most difficult to grasp. They escape all our efforts to bring them into the enclosure of law." But then the wonderful thing happened, that the empirical discoveries in Bergen gave the insight: I FOUND THE FORM OF THE PROBLEM WHICH HAD TO BE SOLVED. Through the construction of what I had called the "equations of atmospheric perturbations" the problem of atmospheric disturbances, in my view, has gained, once and for all, a mathematical form. But with the construction of these equations, and with the accomplishment of a qualitative discussion of the type of solution they will give in the decisive cases, the limit has been reached for what I can do. The years have done their inevitable work. I am no longer young enough for the task that remains to be done, to achieve the exact mathematical solution of these equations. My fault has been that I have taken on a job of work that was beyond my capacity.

Had this been a defeat which only affected me there would be nothing more to say about it. Everyone must be prepared to fail in the matter he takes up. But as things have turned out, it no longer concerns only me. It has achieved a national aspect. Out in the world at large, one speaks quite generally about the "Bergen Methods", and about "The Norwegians" or "the Norwegian School" in meteorology. The name is connected with a breakthrough which is nevertheless only half way through. But only in the case of a COMPLETE breakthrough will it stand in history as a name of honor which is not to be displaced by others. This COMPLETE victory will not be won unless it is possible definitely to bring meteorology in under Applied Mathematics by the sufficiently comprehensive integrations of the equations for atmospheric perturbations.

And the matter has still another aspect which concerns our national honor. That it has been possible to come as far as we have is due not least to the support I have

far as we have is due not least to the support I have received from the Carnegie Institution of Washington since 1906. Nearly 200,000 kroner has been channeled to our science through this grant. This money has been almost exclusively used to aid young Norwegian scientists. . . . It is thus not only a matter which concerns me personally, but it concerns also the honor of Norwegian science as a whole, whether or not the Carnegie Institution of Washington as a compensation for its generosity to Norwegian science shall be able to point to a thoroughly completed scientific work. [UOM, Bjerknes to Faculty of Mathematics and Natural Sciences, University of Oslo, September 8, 1931]

But Bjerknes's sense of duty and honor went even deeper than this. He wanted to live a life in which he could "matter in science" in the very way that his father sadly had failed. Vilhelm Bjerknes's commitment was deeper than a commitment to meteorology, deeper even than one to physics. It was a commitment to science itself, as an expression of a duty-filled life—a life lived fully in the power of a sense of duty that one's life should make a difference, and that one's life should matter. Bjerknes saw science as the way of potentially mattering most, in this new research age. Science for him was a way of living so that consequences of the most unsurveyably far-reaching kind might potentially connect one's own life with the lives of those who have lived before, those who live today, and those who will live in the future. Given these potential connections, he saw science as a most responsible undertaking. For Bjerknes, science was the ultimate in the recognition and acceptance of responsibility in one's life. That gave him the sense of duty that I wanted to speak about, and the sense of honor.

In 1929 Bjerknes wondered if there would be an emblem to remain standing, bearing the impress of what he stood for in science. Today, whenever we see a weather map showing weather fronts, we can say: Yes, it still stands!

ARCHIVAL SOURCES

Materials in the Bjerknes Collection at the University Library, Oslo, Department of Manuscripts are abbreviated as UOM. Materials in the Bjerknes files at the CIW administrative building are cited as simply CIW.

REFERENCES

Friedman, R. M. *Appropriating the weather: Vilhelm Bjerknes and the construction of a modern meteorology*, 251 pp., Cornell Univ. Press, Ithaca, NY, 1989.

Gilman, D. C., Letter to the Editor, *Nature*, *66*, 548-549, 1902

Good, G. A., The Rockefeller Foundation, the Leipzig Geophysical Institute, and National Socialism in the 1930s, *Hist. Stud. Phys. Biol. Sci.*, *21*, 299-316, 1991.

Shaw, W. N., Methods of meteorological investigation, *Nature*, *68*, 468-472, 1903.

Sivle, P., Tord Folesen (poem), in *Skrifter*, 2nd edn., *1*, pp. 16-17, Gyldendalske Bokhandel, Oslo, 1925 (1909-10).

Climate and History: Raphael Pumpelly's Geoarcheological Expeditions to Turkestan

PEGGY CHAMPLIN
2169 Linda Flora Drive
Los Angeles, CA 90077

When the Carnegie Institution of Washington (CIW) announced its intention of funding scientific research, Raphael Pumpelly was one of the first to apply for a grant. With CIW support he led two expeditions to Central Asia in 1903 and 1904 that pioneered in the application of scientific methods to the prehistoric archeology of that region, in the use of specialists from several disciplines to analyze the remains, and in the study of the geological and climatic factors that had affected an archeological site.

After a long career as a mining geologist, explorer, and director of government scientific bureaus in the United States, Pumpelly, at sixty-five, wanted nothing less than to search for the homeland of the Aryans. These were the people many nineteenth-century scholars believed spoke a language from which the Indo-European group of languages evolved. Pumpelly's hypothesis was that this proto-Indo-European language had evolved among a people who developed agriculture, domestication, and metal working. Thomas Huxley, among others, thought the Aryan homeland had been in northern Europe, perhaps unintentionally lending support to the growing belief in German superiority. Pumpelly, however, like many other students of the subject, was convinced that the homeland had been in Central Asia, an inner-continental region once filled with a large post-glacial sea, which he thought must have supported a large population in Neolithic times. According to his theory, increasing aridity would have caused the shrinking of the sea and forced the migrations of those people westward, where they introduced their language, agriculture, and other skills to the Stone Age people of Europe [Pumpelly, 1905, p. 3; 1918, pp. 698-751].

In his proposal to CIW secretary Charles Walcott, Pumpelly outlined a study to correlate the physical geography and geology of Central Asia with the archeological remains of prehistoric cultures, hoping to find evidence to support his hypothesis. His proposal might have sounded far-fetched coming from anyone else, but Pumpelly was one of those "exceptional men" Andrew Carnegie had in mind when he established the CIW. He was known to CIW president Daniel C. Gilman, secretary Walcott, and others on the CIW Executive Committee and Board of Trustees, and his long career had given him a reputation for getting things done. As John Shaw Billings remarked at a Trustees meeting in December, 1903, the buried cities of the Transcaspian region were to be explored by "a mining engineer capable of doing that kind of work."

Born in Owego, NY in 1837, Pumpelly received his training as a mining geologist in Germany at the Freiberg Mining Academy, where many Americans studied until this country established its own mining schools. He was hired in 1860 to develop a silver mine in Arizona. His courageous conduct of the business of identifying the ores and extracting the silver while dodging Apache arrows led to an assignment as technical adviser to Japan's infant mining industry. He then explored parts of China and Mongolia, where enormous fertile loess deposits aroused his interest in the relation between climate, soil, and history. He returned home to take positions on the Michigan and Missouri geological surveys and to direct the collection of mineral statistics for the U.S. Tenth Census of 1880. From 1881 to 1883 he organized an ambitious survey of the Northwest for the Northern Pacific Railroad, and from 1884 to 1892 he headed the Archean Division of the USGS, doing a highly-regarded structural study of the Green Mountains in Massachusetts, which some thought had solved America's long-standing "Taconic controversy" [Champlin, in press].

Pumpelly had often discussed his climatic theory with his friends, including historians Henry and Brooks Adams, and had earlier considered forming a society with Brooks Adams and Henry S. Pritchett to raise money for an expedition to Central Asia to test his hypothesis. Brooks Adams enthusiastically supported Pumpelly's application for a CIW grant, even visiting Walcott to promote it. Adams suggested in his *The New Empire* (1902) that the trade routes influencing the development of civilizations depended on geography and possession of minerals. Pumpelly's proposed expedition could provide clues to the geographic changes that had caused shifts in the ancient trade routes.

The Earth, the Heavens and the Carnegie Institution of Washington
History of Geophysics Volume 5

Other geologists shared Pumpelly's interest in the effect of the environment on human societies. Nathaniel Shaler's *Nature and Man in America* (1891), for example, was a study of the influence of geography on American history, although it had overtones of racial superiority that Pumpelly's writings did not have. The ideas of Charles Van Hise and T. C. Chamberlin, Frederick Jackson Turner's mentors at the University of Wisconsin, are thought to have contributed to the development of Turner's frontier thesis [Livingstone, 1987, pp. 158-191; Coleman, 1966, pp. 28-29].

In its early days the CIW gave few grants for geographic exploration or archeology, but on December 16, 1902 the Executive Committee granted Pumpelly $5,000 to make a preliminary examination of the Transcaspian region and to collect information needed for organizing further investigations. On learning that Pumpelly had received the grant, Adams congratulated Walcott and incidentally revealed his antipathy to academia by saying

> I take it as an indication that you have obtained full control of the Carnegie Trust. With you to direct I have little doubt that the Institution will do great work. Were it left in academic hands it would be pure waste [CIW, Pumpelly File, Adams to Walcott, Dec. 21, 1902].

Pumpelly began to prepare for his expedition at once, one of his first moves being to persuade William Morris Davis to go with him to take charge of the "physico-geographic" and climatological part of the work. "He is as enthusiastic over the whole scope and importance of the work as I am," Pumpelly told Walcott [CIW, Pumpelly File, Pumpelly to Walcott, Jan. 21, 1903]. Davis had once been Pumpelly's student at Harvard and had worked on the Northern Pacific survey where he had first conceived the idea of the cycle of erosion. In 1903 he was professor of geology and geography at Harvard and the recognized leader in physical geography, which he shortened to "physiography" [Chorley et al., 1973]. Davis was developing his own ideas about environmental influences, explaining to president Gilman that he had coined the term "ontography" to include

> all that part of geography which is concerned with the response of organisms to their environment; while physiography includes all that part concerned with the physical environment of organisms [CIW, Pumpelly File, Davis to Gilman, May 28, 1903].

On the Pumpelly expedition Davis had many opportunities to observe the interrelation of physiography and ontography. As his assistant, Davis took Ellsworth Huntington, a promising graduate student, who had worked for the Turkish government, had fallen in love with Asia, and wanted to return. While with Pumpelly, Huntington began to develop his theories about the effect of climate on history, resulting in *The Pulse of Asia* (1907) and other works that made him a well-known proponent of environmental determinism [Martin, 1973].

The archeologist with the expedition of 1903 was Richard Norton, director of the American School of Archaeology in Rome. Norton's primary interest was the classical period, but he realized that little was known about the prehistoric archeology of Central Asia and that it needed to be investigated. Norton did not agree with Pumpelly that climate had been a factor in the abandonment of cities, however. He thought inhabitants abandoned old towns "to avoid the ghosts and traditions of their predecessors and to get the first draw of the river water for their new gardens" [Norton, 1905]. Norton did not return in 1904.

On March 18, 1903, Pumpelly sailed for Europe with his son, Raphael Welles Pumpelly, a Harvard junior studying geology with Shaler. After three weeks in St. Petersburg trying to get permission to travel and work in Russian Turkestan, Pumpelly received "final assurance of authority from the Minister of War who controls Turkestan very jealously," he wrote Gilman. Pumpelly's gentlemanly manner, his wide experience of travel, and, not least, the prestige of the Carnegie name undoubtedly helped him in his cause. He was given letters of introduction to provincial governors that assured his hospitable reception, even by Kirghiz tribal chieftains in the mountains, and the use of a private railroad car that could be pulled off at a siding whenever desired [CIW, Pumpelly File, Pumpelly to Gilman, May 14, 1903].

Pumpelly and Davis wrote a number of letters to president Gilman informing him of their progress. Pumpelly told Gilman that, although the Russians had done excellent work in the geology and physiography of Turkestan, they had done almost nothing in archeology. "They do not seem to have grasped the idea of the possibly extremely remote antiquity of the civilizations," he wrote, or of the "relation between these and their changing environment."

> The nearer I approach to the region and the more I learn about it, the more the problem appears vast, fascinating and complex. From the Caspian to Eastern Mongolia there seems to be a continuous cemetery of cities" [ibid.].

On this expedition Pumpelly emphasized the preliminary physiographic work. His list of "topographic and geologic work to be done" in the first of his Transcaspian notebooks indicates the ambitious scope of his plans. He wanted to prepare a profile of the watersheds and geology of the

Caspian-Aral region; study the distribution, elevation and fossil content of terraces; determine the extent of former mountain glaciers; and establish the relation of glaciers to past climate and rainfall in the Transcaspian basin. He hoped that loess deposits would provide clues to the processes of progressive aridity, such as changes in soil fertility and the encroachment of sand onto cultivated areas [HL-M, Pumpelly Collection, Transcaspian Notes, 1902].

Davis and Huntington met Pumpelly at Baku, and spent a week examining old shorelines along the Caspian Sea, trying to establish former sea levels. The study somewhat resembled Grove Karl Gilbert's classic study of Lake Bonneville in which he had tried to demonstrate the correlation between glacial cycles in the mountains and water levels in the former sea. Davis and Huntington found some well-defined terraces and beach gravels at various levels, which Pumpelly hoped to correlate with the elevation of ancient settlements.

From the Caspian the party traveled eastward by the Transcaspian Railroad. The belt of alluvial fans at the base of the Kopet Dagh, the mountains separating southern Russia from Iran, especially interested Pumpelly because on the fans, or "delta oases" as he called them, were weathered mounds and abandoned irrigation ditches indicating formerly inhabited sites. He realized that the fans were constantly being aggraded by the debris washed down from the mountains during wet periods, and that a geological study of the fans might provide clues to past climatic change. At Ashkhabad, the capital of Turkmenistan, his party took a five-day side trip by horseback into the Kopet Dagh looking for signs of former glaciers but finding none. He also visited ruins at the ancient trade centers at Old Merv, Bokhara, and Samarkand, and consulted with local archeologists. But the much older weathered tumuli interested him most, as apparent remains of a much older culture [Pumpelly, 1905].

At the end of the railroad at Andizhan the party split into two groups, Davis and Huntington ascending by pack train into the Tien Shan range, Pumpelly, his son, and Richard Norton into the Pamirs. Even at sixty-five, Pumpelly found the journey exhilarating, riding twenty-five miles or more a day without difficulty. This was the first opportunity Americans had had to investigate the existence of successive glacial epochs in high Asia. Pumpelly and his son thought they saw evidence of three glacial epochs, Davis and Huntington at least two. Pumpelly, who had written a paper in 1879 on the origin of loess, paid special attention to the great loess deposits that mantle the mountain sides in many places. He theorized that the fine material derived in part from the breakdown of rocks by the extremes of cold and then passed through a desert stage where it was "spread out on the fans and plains, impregnated by capillary attraction in moist periods with the desert salts, and thence carried by the winds to be deposited on grassy surfaces beyond the desert as loess" [Pumpelly, 1918, pp. 707-712; HL-M, Pumpelly Collection, Transcaspian Notes, 1902].

Pumpelly and Davis both knew that their physiographic results were inconclusive and recommended further work the following year, but the Executive Committee, although approving a second grant of $18,000, requested that Pumpelly concentrate on archeology. Trustee John Shaw Billings had told Pumpelly that he found his reports very interesting but did not think that physiographic work in Central Asia was a good use of CIW funds [CIW, Pumpelly File, Pumpelly to Walcott, Nov. 12, 1903]. Davis did not return with the 1904 expedition, although Ellsworth Huntington was included as assistant to the archeologist. A factor influencing the committee's decision to emphasize archeology may have been the knowledge that a large French expedition under Jacques De Morgan was excavating at the ancient Chaldean site of Susa in Persia. Considerable prestige could come to the Institution by sponsoring similar excavations.

In December 1904 Pumpelly sailed again for Europe, this time accompanied by his wife Eliza, and at the Museum of Ethnology in Berlin he found the archeologist who helped to assure the success of his 1904 expedition. Hubert Schmidt, the museum's ceramics curator, had worked with the archeologist who had introduced the latest excavation techniques at Troy. Schmidt's methods at Anau proved to be meticulous and thorough in a region where most digging had been done indiscriminately in a search for treasure. In February Pumpelly returned to St. Petersburg to obtain permission from the Archeological Commission to excavate in Russian Turkestan. On a visit to Moscow he and Eliza saw soldiers leaving for the war zone, for on February 6 the Russo-Japanese War had begun. As Americans the Pumpellys found a warm welcome wherever they stopped. At Tbilisi (then Tiflis), the capitol of Georgia, they were entertained at the home of a royal prince and princess, where toasts were made to harmonious Russian-American relations. The Georgians performed traditional songs and dances for the Americans and in return Langdon Warner, a college friend of young Raphael's, danced the cakewalk to loud applause [HL-M, Pumpelly Collection, Eliza Pumpelly diary, 1903-1904].

The site Pumpelly selected for excavation was close to the abandoned city of Anau, near Ashkhabad. In 1903 he had examined two mounds there 50 to 60 feet (12 to 15.2 m) high, kurgans he called them, composed of disintegrated sun-dried brick. Several years earlier a Russian general had dug a trench in one of the mounds, exposing fragments of broken pottery, charcoal, and bones. Well-defined zones containing different types of pottery indicated to Pumpelly that several different cultures had occupied the site over a period of perhaps thousands of years (Figure 1).

On March 24 excavation began on the older north kurgan and after two weeks shifted to the south kurgan. Schmidt directed the excavation of a series of terraces and shafts that allowed access to the middle and lower levels. He and Pumpelly established a datum point at the level of the plain and labeled all items with their horizontal position and distance above or below the datum point. Schmidt classified the pottery, which Pumpelly likened to the fossils that characterize geological horizons, and with the different pottery types as the principal criteria, but referring to other finds as well, such as objects of stone and metal, hearths, cooking utensils, and the remains of the skeletons of children buried near the hearths, he divided the kurgan cultures into four main periods. Anau I, at the bottom of the north kurgan, dated from about 3000 B.C., according to Schmidt's estimate. Pumpelly always insisted that it

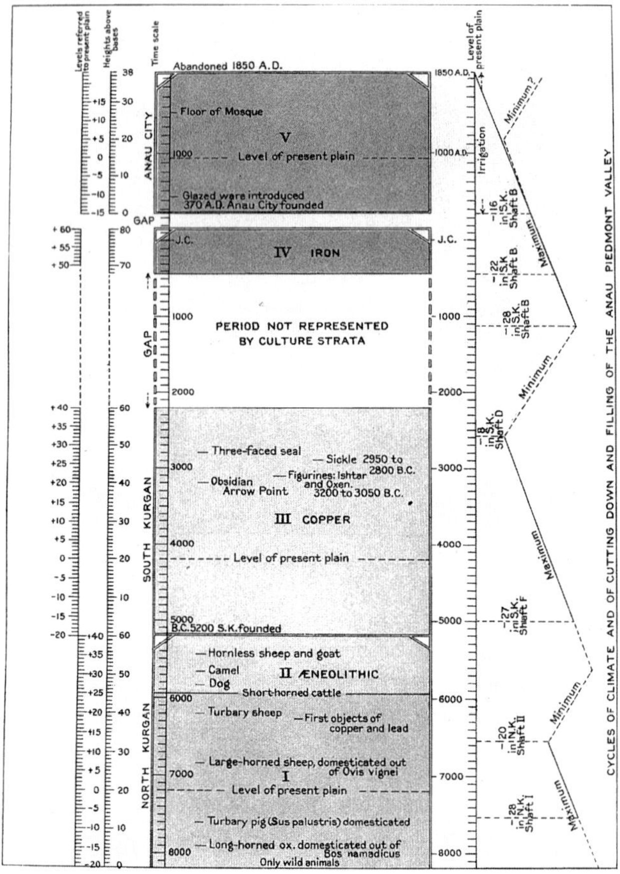

Fig. 1. Pumpelly's illustration of sites near Anau, from his Explorations in Turkestan, 1904, vol. 1.

belonged in the Neolithic period. Above that was Anau II, a stone and copper period known as the Chalcolithic. After this the site had apparently been abandoned and a new site developed by a later copper culture constituting Anau III, at the bottom of the south kurgan. A culture gap indicated that the site had then been unoccupied for an indefinite period. An Iron Age culture, designated Anau IV, beginning about 500 B.C. appeared at the top of the south kurgan. Pumpelly added Anau V, representing the culture which built the city of Anau, abandoned about 1850, containing the ruins of a blue-tiled Mohammedan mosque [Pumpelly, 1908, vol. 1, pp. 83-210].

What set the work at Anau apart, making it something other than just another archeological dig, even though an unusual one for its time, was Pumpelly's attention to the geological setting and to the changes in past climate that his geological work revealed. He knew that the fans where the kurgans were located had undergone several periods of build-up and erosion. According to his "geological reasoning," the growth of alluvial strata during wet periods paralleled the growth of the culture strata, and a decline in the growth on the fan during dry periods coincided with a time when the kurgans had been abandoned. To confirm this hypothesis Pumpelly and his son dug twenty-four shafts on and near the kurgans and carefully measured the natural and irrigation sediments and layers of cultural debris that had been washed down into irrigation ditches. He found "evidence of three separate growths of alluvial strata between the founding of the north kurgan and the beginning of irrigation, and of two intervening degradations." A complex diagram prepared for his final report shows the interrelations between depth of shafts, time, cultural development, and climate change. The diagram indicates the major dry periods when the plains were dissected and the kurgans were abandoned, as well as three wetter periods when silt was deposited and cultures flourished. Pumpelly's geological work also indicated an average ratio of one foot (0.3 m) of natural growth to two and one-half feet (0.76 m) of cultural strata growth, which determined his tentative chronology for the Anau cultures [Pumpelly, 1908, vol. 1, pp. 50-55 and plate 5].

Pumpelly asked specialists from several disciplines to analyze materials from the kurgans. A paleobotanist found fragments of cultivated wheat and barley in pottery from the lowest level, indicating early agricultural practice. Ulrich Duerst, a comparative anatomist from the University of Bonn, who made a lengthy study of the many animal bones, determined that those from the lowest level of the older mound were from wild species of sheep, ox, and pig, while those from the upper part of Anau I were of domesticated species. Pumpelly's estimates set the date for Anau I at 8,000 B.C. (a figure he later revised to around

6,000 B.C., about the same as today's estimates), leading him to claim that in Anau I "we have the oldest instance of an organized agricultural civilization at present known," and that these people had learned to domesticate animals. The absence of arrowheads or other weapons indicated to him that this was a peaceful community [ibid., pp. 67-68, 435-436].

In May, an infestation of grasshoppers brought an end to the work at Anau, but Pumpelly hoped to return the following year to complete the work and to dig elsewhere. The CIW approved a new grant of $26,000, but, because of Bolshevik activity in Russia and recurring strikes protesting the war with Japan, Pumpelly had to postpone the next expedition indefinitely. To his regret he never returned. He spent the next four years traveling and consulting with experts in Egypt and Europe and preparing his final report, published in 1908. The two-volume report is a remarkable collection of interdisciplinary work, containing 97 plates and reports by specialists on the plants, animal bones, human skulls and skeletons, stone and metal implements, and pottery, as well as on the physical geography and geology of the deserts and oases.

Pumpelly's introductory chapter, "Ancient Anau and the Oasis World," reveals his Humboldtian perception of the Earth as an organic whole and summarizes his ideas about climate and the evolution of the oasis culture. He speaks of the inner-continental region as a "geographic organism," every part affected by developments in every other part. At first the grass of the loess plains nourished the "teeming herds" that supported the earliest inhabitants, but progressive aridity finally restricted human life "to the mountains and to the borders of the few remaining streams and the deltas." Pumpelly believed that the increasing aridity had stimulated the inventiveness of these early oasis dwellers, who survived by developing irrigation agriculture and raising herds. It was among these people that the primitive forms of Aryan speech originated, he thought, because the development of an organized society and new industries must have been accompanied by progress in vocabulary and grammatical construction [ibid., pp. 11-13].

Pumpelly wrote a chapter on Indo-European origins entitled "The Migrations and the Aryan Problem," but he did not include it in his 1908 report because he thought it "a hazardous venture for one not already an authority on the subject." In the unpublished manuscript in the Huntington Library Pumpelly suggested that as the population grew beyond the ability of the environment to support it, probably during one of the periods when the kurgans had been abandoned, the oasis dwellers began to move eastward and westward in a "slow-moving tidal displacement," introducing their skills and their language to other, more primitive cultures. Pumpelly was no doubt right not to publish the chapter as much of it was hypothetical, but it still holds interest because the question of the origin of the Indo-Europeans remains a lively one, today surrounded by controversy over whether they were warlike, mounted nomads of the Russian steppes or peaceful agriculturalists, as Pumpelly proposed [Mallory, 1989].

Pumpelly's reports on his two expeditions received favorable comment from reviewers in both the scientific and popular press, who praised Pumpelly's scientific treatment of archeology and physical geography even when they did not agree with all of his conclusions. His friend, publisher Henry Holt, thought Pumpelly's reputation as an archeologist was now on a level with his achievements as a geologist and that he had "destroyed at a stroke the megalomaniac claim to be the cradle of civilization that Germany had lately been cooking up" [Holt, 1923, p. 229]. A writer for the *Nation* thought Pumpelly's 1903 expedition dealt with "an historical mystery," the solution of which would be just as important as "the investigation of double stars or cryptograms." The *Nation* had criticized the CIW the year before for emphasizing the study of physical phenomena and neglecting history and the arts ["The Carnegie Institution," 1904; "A Year of the Carnegie Institution," 1903].

For many years Anau was considered a major Middle Eastern archeological site, on a level with those in Mesopotamia and elsewhere. Today dozens of sites have been excavated east of the Caspian, along the base of the Kopet Dagh, and on the western flanks of the Pamirs and the Tien Shan, confirming Pumpelly's conviction that a populous, complex civilization emerged in southern Central Asia in prehistoric times. The Djeitun culture, located north of Ashkhabad, is now a major Neolithic site with which the lowest level at Anau is compared, its date estimated to be about 6500 B.C. The Namazga culture, a large site west of Anau, first excavated in 1916, is the type site for a copper culture comparable to Anau II and III. Archeologists today generally agree that agriculture began somewhere in the Middle East in Neolithic times [Gupta, 1979, vol. 1, pp. 56-68].

Pumpelly's attention to the total environment of the Anau culture was an unusual interdisciplinary approach to archeology for his time. He is considered a pioneer in archeological geology, which has become a well established specialty, attracting geologists, archeologists, and anthropologists, although Pumpelly's work did not attract followers for several decades. [Rapp and Gifford, 1985, pp. 9-10]. Perhaps he accomplished so much because, like other geologists of his generation, Pumpelly had investigated a wide range of subjects during his career and understood their relationships. He well appreciated the value of specialists, however, and knew when to utilize their expertise. As an individual with no academic affiliations he could transcend the boundaries then forming in graduate departments of geology, anthropology, and archeology, where scholars were defending their territory rather than cooperating with those from other disciplines. Apart from private subscription, the CIW was one of the few means of support for making such an expedition possible, an experience which Pumpelly called the most interesting part of his adventurous life.

ARCHIVAL SOURCES

The Pumpelly File at the CIW includes Pumpelly's grant proposal, documents relating to his grants, and letters from Pumpelly and William Morris Davis to Daniel Coit Gilman. The Pumpelly Collection in the Manuscripts Department of the Henry E. Huntington Library, San Marino, CA (HL-M) contains the papers of Pumpelly and his wife Eliza Shepard Pumpelly. These include Pumpelly's field notebooks for the 1903 and 1904 expeditions and the diaries his wife kept during the expedition of 1904. The manuscript of Pumpelly's unpublished chapter "The Migrations and the Aryan Problem," is in the Bailey Willis Collection at the Huntington Library.

REFERENCES

Champlin, P., *Raphael Pumpelly: Gentleman Geologist of the Gilded Age*, Univ. of Alabama Press, Tuscaloosa, in press (1994).

The Carnegie Institution, *Nation*, *78*, 26, Jan. 14, 1904; A year of the Carnegie Institution, *Nation*, *76*, 187, Mar. 5, 1903.

Chorley, R. J., R. P. Beckinsale, and A. J. Dunn, *The Life and Work of William Morris Davis*, 896 pp., Methuen, London, 1973.

Coleman, W., Science and symbol in the Turner frontier hypothesis, *Am. Hist. Rev.*, *72*, 22-49, 1966.

Gupta, S. P., *Archaeology of Soviet Central Asia, and the Indian Borderlands*, 2 vols., 577 pp., B. R. Publishing, Delhi,1979.

Holt, H., *Garrulities of an Octogenarian Editor*, 460 pp., Houghton Mifflin, New York, 1923.

Livingstone, D. N., *Nathaniel Southgate Shaler and the Culture of American Science*, 395 pp., Univ. of Alabama Press, Tuscaloosa, 1987.

Mallory, J. P., *In Search of the Indo-Europeans*, 288 pp., Thames and Hudson, New York, 1989.

Martin, G. J., *Ellsworth Huntington, His Life and Thought*, 315 pp., Archon Books, Hamden, Conn., 1973.

Norton, R., Report on archaeological remains in Turkestan, *Supplementary Papers of the American School of Classical Studies in Rome*, *1*, 196-216, 1905.

Pumpelly, R., *My Reminiscences*, 2 vols., 844 pp., Holt, New York, 1918.

Pumpelly, R. (Ed.), *Explorations in Turkestan, With an Account of the Basin of Eastern Persia and Sistan, Expedition of 1903*, 324 pp., CIW, Washington, 1905.

Pumpelly, R. (Ed.), *Explorations in Turkestan; Expedition of 1904. Prehistoric Civilizations of Anau; Origins, Growth and Influence of Environment*, 2 vols., 494 pp., CIW, Washington, 1908.

Rapp, G., Jr. and J. A. Gifford, *Archaeological Geology*, 435 pp., Yale Univ. Press, New Haven, 1985.

Weighing the Earth from a Submarine: The Gravity Measuring Cruise of the U.S.S. S-21

NAOMI ORESKES
Department of Earth Sciences, Dartmouth College
Hanover, NH 03755-3571

A central topic in current history of science is the emergence of research traditions. Andrew Carnegie believed in the historical driving force of the "exceptional man," but contemporary historiography sees the solitary scientific genius as a rare and poetic exception. Most scientific work is done in communities of men and women, researchers and technicians, leaders and followers, who cluster around particular forms of scientific practice [Latour and Woolgar, 1979; Rudwick, 1985; Latour, 1987; Pickering, 1992]. Research groups may amalgamate around a common methodological approach, such as an observational technique, a form of experimentation, or a specific piece of instrumentation. Or the unifying agent may be a research focus: a disease, an animal, a geographic locale.

Several recent historical studies have discussed the emergence and evolution of research traditions in American science [Galison, 1985; Warner, 1986; Pauly 1987; Galison and Assmus, 1989; Servos, 1990; Maienschein, 1991]. Maienschein [1991] has emphasized that new research traditions are commonly formulated by the gradual evolution and subtle transformation of an existing tradition. Such a "birth" of a new tradition would be impossible to "date." However, new research traditions may also emerge that are linked to identifiable historical events, while a root tradition lives on as a complementary or competing approach. Marine geophysics in America is an example of the latter. With roots in continental geodesy, marine geophysics emerged from an existing institutional and theoretical framework in the early 20th century. By mid-century, it had become a widely recognized, independent research tradition, in the process of developing its own institutional and theoretical framework. In the 1960s, the data produced from this tradition had a transforming impact on all of the Earth sciences. How did this new research tradition emerge? The purpose of this paper is to answer this question by examining the first American attempt to obtain marine geophysical data: the gravity-measuring cruise of the U.S.S. S-21 submarine, a joint initiative between the Carnegie Institution of Washington (CIW), the U.S. Coast and Geodetic Survey (USCGS), and the U.S. Navy. The S-21 expedition marks the birth of marine geophysics in America. Its history illustrates a confluence between theory and utility that characterized much of American Earth science in the late 19th and early 20th century.

THE S-21 SUBMARINE EXPEDITION

In September 1928, the Dutch geodesist F. A. Vening Meinesz (1887-1966) arrived in the U.S. to measure gravity aboard a U.S. Naval Submarine. The goal of the S-21 submarine expedition was to measure the acceleration of gravity in the Gulf of Mexico and the Caribbean, or, as William Bowie (1872-1940), Chief of the Geodesy Division of the USCGS, put it, to "weigh the Earth from a submarine" [Bowie, 1929d]. Vening Meinesz arrived in Washington on September 25, calibrated his apparatus at USCGS headquarters, and installed the device on board the submarine. On October 2, the U.S.S. S-21 sailed from the Naval Yard in Hampton Roads, Virginia, accompanied by two Eagle boats in case of emergency, and headed for the Gulf of Mexico (Figure 1). To assist with the scientific work and learn the technique of gravity measurement at sea, two American scientists accompanied Veining-Meinesz on board: Fred E. Wright, a petrologist at the Geophysical Laboratory (GL) of the CIW, and Elmer B. Collins, principal scientist of the Naval Hydrographic Office. The outward trip took the scientists down the Atlantic Coast to Key West and across the Sigsbee Deep to Galveston, Texas. At Galveston, they turned around, and returned across the Mississippi Delta, along the coast of Cuba, across the Bartlett and Nares Deeps, and into Guantanamo Bay (Figure 2). The final leg brought them home on November 27, 1928 [Lamson, 1930; Wright, 1929; Bowie, 1930; US-NA RG24 18W4: Logbook of the USS S-21, p. 605-613; CIW GL Misc File 1908-1935, Curtis Wilbur to John Merriam, June 19, 1928].

The cruise was a tremendous success. The submarine covered a distance of 7000 miles in just under two months, measuring gravity at forty-five stations at sea. Five

The Earth, the Heavens and the Carnegie Institution of Washington
History of Geophysics Volume 5

Fig. 1. Photos from the S-21 expedition. Upper left: The U.S.S. S-21 at sea. Upper right: The S-21 docked at Hampton Roads, prior to leaving on the gravity-measuring cruise. Bottom photos: Vening Meinesz on board the S-21, greeting officials of the U.S. Navy. Navy Secretary Curtis Wilbur and several other officers attended the send-off. The man shaking hands with Vening Meinesz may be the ship's captain, Lt. J. L. Fisher. Pictures taken by F. E. Wright.

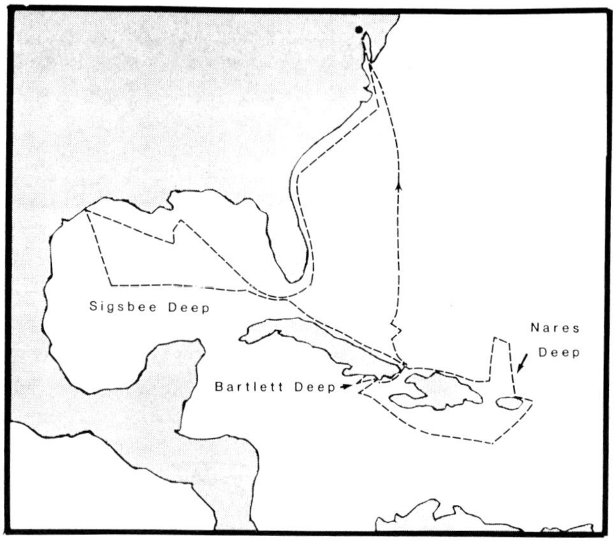

Fig. 2. The route of the S-21 expedition. After Vening Meinesz and Wright, 1930.

additional measurements were made in harbor at Hampton Roads, Key West, Galveston, Guantanamo, and St. Thomas. Of these, only Galveston had previously been the site of gravity measurements, for a total of 49 new gravity stations [Vening Meinesz, 1929]. At each station, depths were recorded at five-minute intervals using a sonic depth finder to allow for accurate topographic corrections. Additional soundings throughout the cruise assisted in topographic and geological interpretation of the results. In calm seas, measuring stations were located at periscope depth [42-45 feet]. In rough water, the submarine submerged to depths of 60-90 feet to minimize the effects of surface waves. Throughout the cruise, measurements were taken of longitude and latitude for accurate positioning, temperature, humidity, and barometric pressure for corrections to the pendulum, and the direction and velocity of the ocean current to correct for rotational velocity in computing centrifugal acceleration. The typical dive time was 35-60 minutes, of which about half an hour was spent in measurements, but in one case, measurement was completed in only 18 minutes on a total dive of 22 minutes.

The cruise was considered a success in part because of the tremendous efficiency of the operation. Efficiency had long been a preoccupation of American geodesists for both practical and intellectual reasons. In the nineteenth century, geodetic measurement in the young country had been a notoriously slow and difficult process. Hours or days were spent moving equipment by hand or horse; men were frequently injured climbing rocks and beating back brush; insects, sickness, and exhaustion were frequent companions; and gravity pendula were big, heavy, and required a solid base [Manning, 1988]. In forested or uneven terrain, survey teams had to build wooden towers to obtain levelling data; in the early 20th century it had been suggested that the USCGS should abandon triangulation in such cases and resort to simple traverses. Bowie refused such compromises with their resulting loss of accuracy and precision, and instead promoted the use of trucks (instead of horses) and the development of the 'Bilby' portable steel levelling tower [Bowie, 1932]. Nevertheless, in the 1920s it still commonly took more than a week to complete a single gravity station [Fleming and Piggot, 1956, p. 328]. In his annual reports, Bowie described the increasing pressure to improve the Survey's productivity to satisfy growing demand for geodetic data [NA-US RG23 USCGS, Series 33, Annual Office Reports]. The prospect of a radically more efficient operation was thus enticing both for reasons of personal safety and in response to the financial pressures endemic in this federal agency.

Productivity pressed the Coast Survey for intellectual reasons as well, because a large quantity of data was essential for an accurate model of the figure of the Earth. In geodesy, more data meant better science. Towards this end, Bowie's predecessor, John Hayford (1868-1925), had pioneered the application of the theory of isostasy in the interpretation of geodetic data. Hayford's innovation was to use the concept of Pratt isostasy, with its simplifying assumption of a uniform depth of isostatic compensation, to streamline the laborious calculations required to reduce raw field data. Hayford's strategy dramatically increased the efficiency of calculation, and resulted in a new model for the figure of the Earth: the "Hayford spheroid," adopted in 1924 as the standard of the International Union of Geodesy and Geophysics [Hayford, 1909; Hayford and Bowie, 1912; Bowie 1922; Burger, 1931]. The S-21 expedition was considered a fulfillment of Hayford's legacy, measured in part by the *quantity* of data produced: about 1/6 as many gravity measurements as had *ever* been collected in the terrestrial U.S. [Wright, 1929].

ORIGINS OF THE S-21 EXPEDITION

The Hayford model for the figure of the Earth was considered a foremost accomplishment of American Earth science. In a discussion of a paper by Bowie presented at the Royal Geographical Society in 1923, British geophysicist Harold Jeffreys called the Coast Survey work "one of the outstanding scientific achievements of our time" [Bowie, 1924, p. 44; also see Burger, 1931]. Colonel Sir Sidney Burrand, retired Superintendent of the Trigonometrical Survey of India, recalled that "When Hayford's method had been introduced into India we realized that it would mark an epoch in the history of geodesy" [Bowie, 1924; p. 36]. Burrand said he had first heard of Hayford's work from the great German geodesist Friedrich Helmert (1843-1917), "the foremost geodesist of the age," and recalled Helmert's realization, "with sadness, that Hayford had hit on the solution which he had been vainly seeking" [Bowie, 1924, p. 36].

Hayford's innovation was two-fold. First, using the Pratt assumption of a uniform depth of isostatic compensation, he streamlined the calculations of the deflection of the vertical used in trigonometrical surveys, vastly increasing the rate at which these calculations could be done [Hayford, 1909; also see Bowie, 1922; 1924; 1927; and 1929e]. Then, working with Bowie, he developed a new method of calculating isostatic anomalies based on variations in the acceleration of gravity [Hayford and Bowie, 1912]. This work suggested that the major features of the Earth's crust were isostatically compensated. It appeared that the theory of isostasy held true, and therefore gravity measurements could proxy for traditional geodetic data [see Bowie, US-NA RG 23 USCGS, Entry 30, General

Correspondence, "Terrestrial Magnetism" 1908-1914].

As suggested above, Hayford was primarily motivated by issues of efficiency. A civil engineer by training, his choice of the Pratt model was not based on geological considerations, but on facility of calculation. In his words, "The assumption [of Pratt isostasy] was adopted as a working hypothesis, because it happens to be that one of the reasonable assumptions which lends itself most readily to computation" [Hayford, 1909, p. 147; also see Vening Meinesz and Wright, 1930, p.11 and Reingold, 1970, p. 188-9]. Bowie's contribution was to use this computational strategy as the basis of a theoretical interpretation of the Earth's structure and origin. Bowie argued that if all the major physiographic features of the Earth were isostatically compensated then they must be very old. The Earth must be essentially stable. In his words, "If the earth's crust is in isostatic equilibrium . . . then we are justified in assuming that the isostatic condition has obtained since earliest geological times" [Bowie, 1924]. For this reason, Bowie became an adherent of the Darwin-Fisher hypothesis of fissiparturition, which placed the origin of the continents and oceans during break-up of a proto-crust when the moon separated from the Earth early in planetary history [Bowie, 1929a; Bowie, 1935b; Yale University Archives, Charles Schuchert Papers: Bowie to Schuchert June 17, 1927, Box 19 Folder 166; and October 11, 1928, Box 21 Folder 181]. As a corollary, it followed that few if any major stresses were being sustained in the crust at present. This was one reason why Bowie rejected the idea of continental drift: there simply weren't sufficient stresses to move continents.

By the mid 1920s, the practical success of the Pratt-Hayford model and Bowie's promotion of its geological results had increased acceptance of the underlying geological theory: most American geologists accepted isostasy theory as a general statement of crustal dynamics. In 1925, William Bowie declared in *The New York Times* that isostasy theory had been "proved" [Bowie, 1925]. Shortly, he claimed this in more scholarly journals as well [Bowie 1927; Bowie, 1929a; Bowie 1929c]. However, despite his optimistic public pronouncements, Bowie was well aware of a major lacuna in the empirical data base: a complete absence of data from the oceans basins. Furthermore, the available land-based data were almost entirely concentrated in Europe, Asia, and North America. This raised questions both about the geodetic models produced and any geological interpretations placed upon them. Thus there were two motivations to go to sea. One was to continue to improve estimates for the figure of the Earth. To sustain the Hayford spheroid, or replace it with something better, more geographically widespread data from hitherto unexplored regions were needed. Second, to apply isostasy theory broadly applied to problems of crustal dynamics, data from the entire Earth, not just the continents, were essential. Did isostatic equilibrium obtain over the oceans basins? To answer this question, one needed to measure gravity at sea. As Fred Wright put it, "The late Dr. Hayford told me years ago that if a gravity apparatus for use at sea were available, a single ship could, in the course of a year, contribute more important data bearing on the figure of the Earth and the theory of isostasy than have been collected in the last generation on land." [CIW-GL Misc File 1908-1938 #2: Wright to Merriam, December 27, 1928].

The obstacle to Hayford's ambition was technical. Gravity measurements were based on the well-known relation between the acceleration of gravity and the period of a pendulum, $g = 4\pi^2 L / T^2$, where T is the period of the pendulum and L is the length. If the bulk Earth were a regular spheroid with topography superimposed on it, then one could predict the value of gravity at any location based on its elevation. Differences between measured and predicted values would reflect either divergences from the calculated figure of the Earth or uncompensated isostatic anomalies [Bowie, US-NA RG23 USCGS Entry 30, General Correspondence, "Terrestrial Magnetism," 1908-1914]. However, this conclusion presupposed no external disturbances. The only forces acting upon the pendulum should be the acceleration of gravity and the centrifugal acceleration caused by the Earth's rotation. Random accelerations encountered on board a ship would render measurement impossible. Attempts to measure gravity at sea had failed; the isostatic condition of the oceans remained unknown [Vening Meinesz and Wright, 1930; Laudan, 1980]. However, in the early 1920s, the situation changed.

Vening Meinesz was a civil engineer by training, employed by the Geodetic Commission of Holland. While American geodesists labored to overcome black flies and the Ozark Mountains, Vening Meinesz and his Dutch colleagues struggled with ground vibrations induced by storms and waves in coastal Holland. In the early 1920s, Vening Meinesz designed a gravimeter that would work in unstable conditions. Two pendula of nearly the same vibrational period swung in the same vertical plane, and the difference in their angles of elongation was photographically recorded. If the two pendula are equally affected by extraneous horizontal acceleration, then the difference between them is due to the acceleration of gravity at that location. That is, the difference between the two pendula is equivalent to a single "virtual" pendulum free of horizontal disturbance. In application, the device actually used three pendula in the vertical plane, with the outer two set in motion to create two pairs of pendula swinging in opposite phase. In addition, Vening Meinesz's gravimeter contained three "dummy" pendula to record temperature and humidity inside the apparatus, and to record the motion of the entire device. The Meinesz apparatus overcame the effect of horizontal acceleration that otherwise disturbed the motion of a single pendulum at sea [Wright, 1929; Vening Meinesz and Wright, 1930; Laudan, 1980].

In 1923, Vening Meinesz tested his device on the Netherlands K-II submarine expedition to Indonesia. A second cruise to Indonesia in the opposite direction—the K XIII, which travelled westward from Holland via the Panama Canal—resulted in a complete circumnavigation of the globe. On the latter expedition, Vening Meinesz completed a detailed survey of the enigmatic Java Trench. Over 200 measurements of gravity-at-sea were obtained—a staggering level of productivity [Bowie, 1929d; Veining Meinesz and Wright 1930, p. 1; Vening Meinesz, 1932]. The success of these expeditions caught Bowie's attention, and he began to discuss the results with Arthur Day, director of the CIW Geophysical Lab.

Bowie and Day saw the significance of Vening Meinesz's work. On a practical level, this was Hayford's ambition realized, with the potential to solidify—or radically transform—existing geodetic practice. On a theoretical level, it was an opportunity to test the theory of isostasy across the globe. In the spring of 1928, Bowie and Day approached C. S. Freeman, Superintendent of the U.S. Naval Observatory, with a plan to invite Vening Meinesz to measure gravity aboard a U.S. Naval submarine. The Navy would supply the submarine, the Carnegie Institution would invite Vening Meinesz and pay for his trip, and the Coast Survey would provide the base station for calibrating the gravimeter and the personnel and computational expertise to perform the data reductions [Vening Meinesz and Wright, 1930, p. 1-2; CIW GL Misc File 1, 1908-1935]. CIW would also supply Fred Wright to assist with the investigations. In the early 1920s, Wright had developed a torsion gravimeter to use density differences to investigate volcanic processes, and he was anxious to learn the details of Vening Meinesz's technique [Fleming and Piggott, 1956; Wright, 1916; Wright and England, 1938; Wright 1941; and CIW GL Misc. File, 1908-1935 #1]. In June 1928, Navy Secretary Curtis Wilbur agreed to the proposal, with the stipulation that CIW would be sure to pay for all of Vening Meinesz's personal expenses, including meals on board the submarine [CIW GL Misc. File 1908-1935 #1, Wilbur to Merriam June 19, 1928].

Although some geologists had questioned the conclusion of complete and local equilibrium [e.g. Barrell 1919, 1927] Bowie plowed ahead in working through the geological implications of the Hayford-Bowie result. If isostatic disturbances were to be found, they should be in areas of recent loading, such as river deltas. Positive gravity anomalies were therefore expected in coastal regions where large quantities of sediment were accumulating, but not over the deeper portions of the ocean which were thought to be ancient, fully equilibrated, features. Indeed, following the earlier work of Dutton, Bowie believed that sedimentation was the *only* cause of isostatic imbalance [Bowie, 1925; 1929b; 1929d; 1931a; 1931b; 1935b]. But Vening-Meinesz's cruises had suggested the opposite result. Measurements over the Nile Delta revealed no gravitational anomaly despite the huge volume of sediment known to be accumulating there [Bowie, 1927; 1929a]. Furthermore, unexpected negative anomalies had been found over the Java Trench. Vening Meinesz's preliminary interpretation of the Java Trench was that it was a recent or even active downwarping in the crust that was not yet adequately compensated. This suggested that, contrary to Bowie's views, there *were* active stresses present in the crust [see, for example, discussion by Evans in Bowie, 1924, p. 43-44]. Bowie's initial impulse was to explain away the missing anomaly over the Nile Delta by suggesting that isostatic adjustment was even more rapid than had hitherto been supposed. Quoting from the geologist and Arctic explorer Fritjof Nansen, in reference to the isostatic rebound of Fennoscandia, Bowie suggested that perhaps "the earth's crust . . . approaches its level of equilibrium much more closely than even the most extreme advocates (like Hayford) of perfect isostasy have considered to be possible" [Bowie 1929b, p. 589]. Yet the negative anomalies over the Java Trench remained to be explained.

GOALS OF THE S-21 EXPEDITION

The itinerary of the S-21 expedition was designed to take on the theoretical issues raised by Vening Meinesz's preliminary results. As Eleanor Lamson, Head of the Computing Section that analyzed the results, put it, the goal was "to include as many stations as possible which would assist in solving the geophysical and geological problems in and near the West Indies" [Lamson, 1929]. The overall theoretical framework was explained in detail by Wright in his preliminary report of the cruise, sent to John Merriam, CIW president, in February 1929:

> Gravity measurements over the land areas of the Earth prove that the earth's crust is everywhere in a state approaching equilibrium. Wherever large departures do occur, they indicate excess or deficiency of load and these in turn produce stresses in the Earth's crust. It is an axiom in geology, as in other matters, that extremes are temporary in character; high mountain masses are not eternal but are soon worn down and effaced. If they are actually extra loads on the Earth's crust, they give

rise to abnormally high gravity values. [Thus g]ravity anomalies above a certain value serve to locate unstable portions of the Earth's crust where movements are taking place so rapidly that compensation has not kept pace with the disturbance and where therefore earthquakes due to shearing stresses are likely to occur. A knowledge of these factors, especially of the order of magnitude of the stresses active in mountain building, earthquakes, and other crustal movements, is fundamental geologic theory and can best be ascertained by gravity measurements over areas not yet compensated. [Therefore i]n planning the itinerary of the cruise of the S-21, the effort was made to select areas that promised results of interest, such as the continental shelf, the Sigsbee Deep and the Mississippi Delta of the Gulf of Mexico, the Bartlett Deep of the Caribbean Sea, the Nares deep north of Porto Rico [sic] and the normal value of gravity over the deep portions of the Caribbean Sea and of the Atlantic Ocean. [Wright 1929]

Thus the plan was to cross the Mississippi Delta to check for an expected positive anomaly, and the Nares, Bartlett, and Sigsbee Deeps, to check for negative anomalies. The Deeps were especially close to the hearts of Coast Survey scientists, as the discovery of the Sigsbee Deep by their colleague Charles Sigsbee was considered one of the major accomplishments of 19th century hydrography. As historian Thomas Manning has put it, the discovery of great depths was a passion of 19th-century hydrographers and Sigsbee's discovery was one of the deepest [Manning, 1988, pp. 37-38]. Sigsbee's work was followed by that of hydrographer John Bartlett. The Sigsbee and Bartlett deeps perplexed American geologists [Anonymous,1888]. Why should there be differences in elevation in the ocean basins greater than those known on land? What geological forces formed them? Existing tectonic theories gave no account of any significant heterogeneities in the ocean floor. Until the late 19th century, they had never needed to. However, in the early 20th century, preliminary geodetic work in Puerto Rico revealed deflections of the vertical in excess to that attributable to the mass of the island, or the mass deficiency of the surrounding ocean, suggesting that isostatic anomalies might be associated with these Deeps [Bowie, 1926a] .

Earthquakes were also relevant to the itinerary. Some geologists argued that off-shore earthquakes were proof of large-scale crustal movement in the ocean basins. This idea could be tested by looking for isostatic disturbances:

If, for instance, the crust is subject to tangential stress [horizontal compression], its position may [change] and this gives rise to an excess of mass that reveals itself in positive isostatic anomalies. . . . If, therefore, we succeed in determining the true isostatic anomaly for a certain part of the earth's crust, we may obtain data on the trend and the magnitude of the tectonic stresses existing in that region. [Vening Meinesz and Wright, 1930]

In other words, isostatic anomalies could not only reveal the location of active stresses in the crust, but also should be proportional to the magnitude and direction of those stresses, thus providing a test of the idea of continental drift.

The problem of drift was prominent in the minds of several American geologists in the late 1920s, including Merriam and Wright. In October 1928, while Wright and Vening Meinesz were on board the S-21, Merriam arranged for the Dutch scientist to receive a copy of Alex Du Toit's new book, *A Geological Comparison of South Africa with South America* [1927; CIW General Files, Meinesz folder, Memorandum, October 1, 1928]. Du Toit's study had been funded by the CIW to test the evidence of geological similarities between the two continents attributed to continental drift. Wright was directly connected with this work. He and Reginald Daly of Harvard University had proposed Du Toit's project to the CIW, after travelling with the South African geologist on a CIW-sponsored field trip in 1922. In South Africa, Wright and Daly had also met G. A. F. Molengraaf, Holland's leading expert on the geology of the East Indies, and a proponent of some form of continental drift because of its apparent applicability to the geology of the East Indies [CIW General files, Wright folder; Molengraaf, 1928; van der Gracht, 1928; Laudan, 1986]. Upon his return from South Africa, Wright became Du Toit's principal contact at the CIW, and Merriam turned to Wright to review Du Toit's preliminary manuscript before accepting it for publication.

Wegener's theory of drift suggested definite predictions about the isostatic condition of the oceans. These predictions might be tested by the gravity-measuring expedition. Vening Meinesz and Wright explained:

In areas subject to orogenic movement, [large] stresses . . . are to be expected; gravity measurements in such areas are especially important and serve to increase knowledge of the factors that tend to disturb the shape of the Earth's crust. [Therefore m]any of the measurements made on the voyage of the S-21 were over areas in which crustal movements have recently taken place. [Vening Meinesz and Wright, 1930, p. 11]

Gravity anomalies might provide a test of continental drift.

RESULTS OF THE S-21 EXPEDITION

The results of the S-21 expedition confirmed the suggestion that isostatic compensation was considerably more complicated than Hayford and Bowie had thought. Like the Nile Delta, the Mississippi Delta seemed to be almost entirely compensated, "in spite of the fact that each year a load [of] nearly 12 billion tons [of sediment] is being laid down" [Wright, 1929, p. 9]. This suggested that the oceanic crust was staggeringly weak, and that equilibration was virtually instantaneous, "proceeding concomitantly with the deposition of the load." [Vening Meinesz and Wright, 1930, p. 77]. On the other hand, the Nares Deep was indeed the site of a large negative isostatic anomaly, suggesting that it was a recent geological feature "in which shearing stresses of large magnitude are present" [Wright, 1929, p. 9]. Furthermore, the negative anomaly extended beyond the length of the Deep proving that the anomaly was not caused by the mass deficit of the Deep, but rather that both anomaly and Deep were caused by movement of the Earth's crust in geologically recent time. The Bartlett and Sigsbee Deeps were partly compensated, suggesting that they were older features.

But the most unexpected result of the cruise was the discovery of a systematic positive isostatic anomaly throughout much of the Gulf of Mexico, in an area of relatively uniform depth that they labelled the "Gulf of Mexico Plate." This positive anomaly increased abruptly by an order of magnitude at the edge of the continental shelf. These findings utterly contradicted the expectation of general isostatic equilibrium over the table portions of the ocean basins. A major isostatic disturbance had been revealed that had little or no topographic expression. Furthermore, the overall results seemed to give contradictory indications about the strength of the Earth's crust. On the one hand, the results from the Delta suggested a very weak crust, responding almost instantaneously to sedimentary load, consistent with the traditional Hayford-Bowie view. On the other hand, the positive anomalies over the Gulf of Mexico "plate" suggested a strong crust supporting regionally extensive excess loads. Vening Meinesz and Wright were at a loss to explain these results, and the possibility of a new explanatory framework began to develop in their minds:

> The theory of isostasy has been so well established that it is not easy to understand actual excess loads on the Earth's crust over areas of such vast extent. If stresses, active in the crust, are responsible for them, then the engineering difficulty arises of explaining their maintenance over such great areas. [Vening Meinesz and Wright, 1930, p. 76]

It appeared that the crust needed to be extremely strong and extremely weak at the same time.

The authors also discussed their data as a test of the hypothesis of continental drift. Vening Meinesz noted that some positive anomalies had been detected on the Pacific side of the American continent during his earlier expeditions, which might be taken as evidence in support of drift: westward migration would cause buckling of the crust, leading to excess mass and positive anomalies. However, the S-21 expedition had revealed positive anomalies on the *Atlantic* side: the opposite of that predicted on the theory of westward drift.

> The fact that . . . pronounced positive anomalies are found at the foot of the continental slopes bordering the Pacific Ocean might seem to point toward a westward drift of the American continent in accord with the theory of Wegener—because they might be explained—in connection with the time element required to re-establish equilibrium—by the effect of the pressure exerted by the continent on its front side; but in this case one would expect a negative anomaly on the lee side, i.e. the east coast. [Vening Meinesz and Wright, 1930, p. 76]

On the face of it, the North American continent appeared to be moving in both directions simultaneously. Interpreted in terms of Wegener's theory of *westward* drift, "the evidence . . . so far as it goes, is not in favor of this hypothesis" [Vening Meinesz and Wright, 1930, p. 76]. But neither did the evidence confirm existing views.

A NEW FRAMEWORK FOR RESEARCH

The results of the S-21 expedition led Vening Meinesz and Wright to two conclusions at odds with the mainstream of American geodetic thinking. First, they concluded that major regional stresses *were* present in the oceanic crust. The ocean basins were neither a passive substrate for floating continental rafts, nor were they a fully compensated region of higher than average crustal density. They were geologically active provinces sustaining regionally extensive tectonic stresses. Second, they concluded that if the crust did sustain significant stress, then it must not be so weak as the Hayford-Bowie school supposed. It must contain "some residual strength." These conclusions implied a re-thinking of isostatic processes and their relation to geological change. A change in Wright's perspective is clearly apparent in one passage, which was significantly revised between his preliminary report to the CIW and the final report published with Vening Meinesz one year later. Wright's axiom, quoted above, that "in geology, as in other matters . . . extremes are temporary in character" was modified in his final report, and placed in a quite different context. It was no longer his own opinion, but a truism found in the geological literature, which was perhaps being interpreted in faulty or misleading ways:

> In geological literature emphasis is placed on the fact that extremes in topography, such as lofty mountain ranges and great ocean deeps or troughs, are geologically young features and essentially evanescent in character. They are the culminating centralized effects of the action of mountain-making or orogenic forces; there we may expect to find departures from normal gravity equilibrium or balance. In view of the fact that the magnitude of orogenic forces is quite beyond direct study in the laboratory, it is necessary, if we would evaluate them, to study their effects in the field where they are now active. We know from a study of the rocks themselves what changes [these stresses] produce and how large are the masses they can move. But the mechanical relations are so complex and the quantities involved so prodigious that we can not, by any direct method, measure the order of magnitude of the forces themselves. Gravity measurements afford the only available approach to this problem which is fundamental to geological theory [Vening Meinesz and Wright, 1930, p. 53].

This statement was a direct indictment of colleagues who presumed to understand the mechanics of the Earth's crust based on grossly scaled-down experimental studies or theoretical models rife with unverified assumptions. It was clear to these two scientists that the Earth's crust simply was not understood. Measuring gravity at sea demonstrated the inadequacy of existing knowledge and concepts. The causes of large-scale deformation over geological time-frames needed to be studied via their observable and measurable effects. Thus, Vening Meinesz and Wright concluded their report by prescribing an explicit research program:

> [T]he contribution of the S-21 cruise to the theory of isostasy and to fundamental geological theory is of lasting value and should form the starting point for other measurements of similar nature over more extended areas. The regions surveyed in a reconnaissance way should be explored in detail with both the gravity apparatus and the sonic depth finder in order to obtain accurate information on the physiography and the distribution of the gravity anomalies and of the stresses present in the Earth's crust. It is believed that much of the detailed gravity and sonic depth-finding work in the Caribbean area can be done on a submarine temporarily assigned for a few weeks to the task. [Vening Meinesz and Wright, 1930, p. 4]

The proposed program began immediately, as Vening Meinesz embarked in 1929 on another Dutch cruise to the East Indies, which confirmed and expanded his earlier results [Vening Meinesz, 1932; Vening Meinesz et al., 1934; Vening Meinesz, 1941]. Meanwhile, the USCGS organized an immediate land-based expedition to Haiti, Cuba, and Puerto Rico, and Bowie made arrangements for future submarine expeditions. Bowie was confident that the S-21 was "only the beginning of the use of American submarines on gravity surveys" [Bowie, 1929d, p. 220]. And in April 1929, the Hydrographic Office announced plans for additional submarine-based investigations [Freeman, 1929]. These plans came to fruition in the *International Expedition to the West Indies*, led by Richard Field of Princeton University in the early 1930s, with junior colleagues Harry Hess and Maurice Ewing. A series of further expeditions followed [Field et al., 1933; Field, 1937; Ewing, 1937; Hess 1937; Worzel, 1965], and Hess and Ewing subsequently became leaders in marine geophysics and the development of plate tectonics [Bullard, 1975; Bates et al., 1982; Menard, 1986; Allwardt, 1990]. The S-21 expedition spawned a new research tradition, beginning with measuring gravity at sea, soon expanding to include geological magnetics and seismic profiling (C. Drake, pers. comm., 1993; Bates et al., 1982]. The result was a new *kind* of scientific data—marine geophysical data—which ultimately led not only to a new explanatory framework for isostasy, but for all of the Earth sciences.

A RESEARCH TRADITION THAT WASN'T

From the start of the S-21 project, John Merriam and Arthur Day planned to continue gravity studies on the ship *Carnegie* after Vening Meinesz had gone home. This was the principal reason why Fred Wright was sent on the cruise: his prior experience with gravity apparatus suggested that he could facilitate a technology transfer [F.E. Wright, U.S. Patent no. 1,579,273, 1926; CIW GL: Patents folder: F.E. Wright]. Recall that the rationale for using submarines was to avoid the effect of surface disturbances by submerging below wave base; the S-21 measurements demonstrated that rolls up to 6° could in fact be tolerated [Vening Meinesz and Wright, 1930]. Day was not convinced that the *Carnegie* would prove this stable, and expressed hesitation about spending large amounts of CIW funds on instrumentation that might not function. However, he agreed to purchase the equipment given assurances from the Navy Secretary it would be "no imposition to . . . install such an apparatus upon appropriate submarine cruises in the future" if the work on the *Carnegie* did not succeed [CIW-GL Misc. File #2

1908-1938, Day to Merriam, January 9, 1929]. With this back-up, the CIW Executive Committee approved the purchase of a Meinesz gravimeter from its Dutch manufacturer, to be installed on the *Carnegie* when it next docked in San Francisco.

The gravimeter arrived in Washington in August 1929. Wright calibrated it and travelled with it to San Francisco. At this point the program was taken over by Scott Forbush of the Department of Terrestrial Magnetism [CIW GL Misc File 2 1908-1938; Bowie to Merriam, August 30, 1929; Merriam to Bowie, Sept. 8, 1929]. Gravity measurements during the seventh cruise of the *Carnegie* between September 3 and November 24, 1929 yielded some preliminary data indicating a large major negative anomaly over the Tonga Deep bordered on both sides by positive anomalies [CIW-DTM General Files 1900-1935, Forbush File; Forbush, 1946]. However, as Day had feared, the ship's instability limited the number and reproducibility of these results. Forbush was confident that the problems could be resolved, but before he had a chance to try, on November 29, 1929, the ship *Carnegie* tragically burned [Forbush, 1946; Harland, 1967]. The gravity program on the *Carnegie* was over. Future work would be done on submarines.

The Stabilization of a New Research Tradition

The S-21 gravity-measuring cruise marked the start of a new research tradition. This raises the question of what begets a successful research program. How was a 56-day cruise transformed into a decades-long research program? Here, we can identify three agents that helped to define and unify the emergent tradition: a set of theoretical questions, an instrument, and a geographic locale.

A set of theoretical questions: Gravity measurement at sea grew out of continental geodesy, a field rooted in cartography and driven by unapologetically utilitarian concerns. But the theoretical implications of geodetic work and the significance of isostatic compensation for models of Earth processes had been widely recognized in the late 19th and early 20th centuries. The S-21 cruise was a nearly seamless extension of the land-based geodetic tradition, both in terms of the measurements being made and the hopes for their significance. However, the observational results obtained challenged some of the fundamental beliefs of that tradition, and thereby brought into sharp focus a set of theoretical questions that related problems in isostasy to larger geological issues. The most important of these questions was the relation between isostatic disturbances and the forces that generate large-scale geological features in the crust. In articulating this issue, Vening Meinesz and Wright created an intellectual justification and framework for continued work centered on geological, rather than geodetic, questions. The recognition of this new context was facilitated by Wright's broad geological background.

An instrument: The tradition of measuring gravity at sea was also the direct result of instrumental development by an individual motivated by a practical problem. Vening-Meinesz's technical innovation permitted accurate measurements to be made that were previously impossible. In the S-21 expedition, technology drove science. Without this technology, the research tradition could not have developed no matter how interesting or important the problem to be solved. Furthermore, as Robert Kohler has suggested in a different context, instruments may help to facilitate cooperation among scientists from diverse disciplinary (or institutional) backgrounds by permitting them to share a common technique; this occurred in the 1930s, when large numbers of physicists joined geologists to create marine geophysics (Kohler, 1991, pp. 358-64, esp. 360). But what of the use of submarines? Was this an essential element, or merely a romantic but coincidental detail? As discussed above, the reason for submarine-based work was to minimize random accelerations at sea by submerging below wave-base. But the goals of the CIW, and the results on the S-21, suggested that the work might be done on a conventional ship. Had the *Carnegie* not burned, the initial difficulties might well have been worked out, and the research program could have continued under civilian auspices. But note that Day's willingness to purchase the gravity apparatus for the *Carnegie* was *contingent* upon the military back-up. Submarines may not have been essential, but the logistical, financial, and psychological support of the Navy was.

A geographic locale: Geologists commonly organize their research programs around a specific locale or a physiographic province, but geophysicists often reject the intellectual premises of "localism" [Le Grand, 1986; 1988, esp. pp 80-89, and refs. cit therein]. Indeed, some scholars have attempted to *define* geophysics in contradistinction to geology as founded in a concern with the structure of the globe as a whole, or with processes that transcend local manifestations [Wood, 1985]. Elements of this view may be true, perhaps for historical reasons, but the S-21 expedition suggests that such a distinction is an oversimplification. In the geophysical work described in this paper, the geographic locale of the ocean basins was a unifying intellectual theme: the *raison d'etre* of this work was to explore and understand an uncharted physical province. Before long, the theoretical questions began to change, and so did the instruments, but the focus on the oceans remained [Laudan, 1980].

Three unifying factors—theoretical, instrumental, and geographical—gave conceptual focus to the emergent research tradition. Institutional back-up from the U.S. Navy gave civilian scientists the confidence to embark on an extended and expensive research program. But what created the successful collaboration in the first place? Why did a private philanthropic institution become involved in a joint venture with a government agency and the military? How did a group of scientists convince the Navy to allow them to occupy a submarine for the better part of two months? The answers to these questions are essential to understanding the extensive Navy-civilian cooperation that so radically affected the Earth sciences in the middle and later parts of the 20th century—cooperation which began well before the start of the second World War. (On earlier links between geology and the U.S. military, see Goetzmann, 1959, 1966; on contemporary links, see Mukerji, 1989). One key element in the S-21 collaboration was the facilitating role of scientific and government administrators: William Bowie at the Coast Survey, John Merriam at the Carnegie Institution, and Curtis Wilbur, Secretary of the U.S. Navy. Each of these men had reasons for supporting the S-21 expedition; institutional and personal ties between the Navy, the USCGS, and the CIW were critical in creating an expedition that required the resources of all three.

THE ROOTS OF COLLABORATIVE RESEARCH

Two important factors in the initiation and success of the S-21 collaboration were the organizational prowess of William Bowie and the institutional resources to which he had access. Bowie's work at the Survey had given him decades of experience organizing field parties for scientific and technical purposes, and from this field-oriented and team-driven tradition sprung a drive towards collaborative scientific research initiatives [Fleming, 1951; Whitten, 1992]. In 1939, Bowie became the first recipient of an American Geophysical Union medal bearing his name, and dedicated in his honor to rewarding "unselfish cooperative research" [Fleming, 1951]. But Bowie's position at the Coast Survey not only inspired him towards collaborative efforts, but also permitted him to supply the specific computational expertise and personnel required for the S-21 project.

The Coast Survey was a uniformed service, and a long institutional tie with the U.S. Navy—a tradition dating back to Alexander Dallas Bache—would have suggested the possibility of access to a submarine [see Dupree, 1985, esp. pp. 133-34; Reingold, 1991, p. 112]. However, Bowie's work at the Coast Survey did not actually place him in contact with the Navy Secretary. Rather, it was Merriam who provided Bowie's first letter of introduction to Curtis Wilbur [LC JCM, Merriam to Wilbur, April 11 1925, 9507—Box 134]. Merriam, president of the National Research Council (NRC) prior to joining the CIW, was highly connected in Washington political and scientific circles [Stock, 1951], and Merriam and Bowie became acquainted through their mutual work on the NRC. They also served together on the National Academy of Sciences Committee on Oceanography formed in the late 1920s [LC JCM Box 22 Bowie folder, Letters through 1932]. Also linked to this "Washington network" for geology [see Hevly, this volume] was Arthur Day, chair of the NRC section on "geophysical chemistry."

Merriam was highly receptive to the concept of collaborative research, although for different reasons than Bowie. Merriam believed that his institution would better serve Andrew Carnegie's aim of improving the condition of mankind if it worked in concert with others. In a report written for Merriam in 1933, Fred Wright argued that the CIW could not only contribute more scientifically through cooperative efforts with other researchers, but that the institution should be a role model in this respect: "[T]he major contribution that can be made by the Carnegie Institution of Washington is that of teaching, by example, the value of cooperation in the attack on scientific problems" [LC JCM Box 185 Wright folder; Letters through 1934; Report by F. E. Wright for Merriam, 25 October 1933, p. 3.] Indeed, Wright suggested it was peculiar to think otherwise: "The Army and the Navy realize the need of pulling together according to carefully prepared plan. Team work in sports is axiomatic . . . " . Furthermore, in the short history of the institution it had already become apparent that "exceptional men" were exceptionally rare. Fortunately, there was an alternative which might achieve comparable results: groups of good men acting in concert. Cooperative research could also facilitate knowledge and technology transfer, multiplying the impact and effect of the "exceptional man" when he was found [Ibid., p. 4].

Merriam's commitment to the cooperative and interdisciplinary approach outlined in Wright's report stemmed at least in part from his personal experience in paleontology, a discipline which had been rife with priority and interpretive disputes. It was Merriam's opinion that many of these problems stemmed from the excessively specialized nature of paleontology which

> tends to narrow itself and through this development to become relatively ineffective. It would be good for paleontology to have physics, chemistry, biology and all the other sciences working on the problem of life just as it would be good for geology, and possibly in a few instances physics, to know the outcome of paleontology. [LC JCM Box 185; Wright folder, Merriam to Wright, March 11, 1932]

Thus Merriam supported interdisciplinary projects as diverse as the *Carnegie Committee on the Surface Features of the Moon* and *The Committee for the Study of the Color of the Water of Crater Lake* [LC JCM; also see Yoder, 1989; Doel, this volume; and Hetherington, this volume]. The S-21 expedition appeared a particularly suitable venture in several respects: it was large-scale and easily publicized, thus helping to fulfill the goal of setting an example for the rest of the scientific community; it was organized around an exceptional man whose expertise would be shared and ultimately transferred; and it involved cooperation with the Coast Survey and Navy, organizations with experience in collaborative work. But what of this third component in the S-21 expedition: the U.S. Navy? Why did the Navy agree to tie up one of its submarines in an apparently esoteric scientific enterprise? Navy Secretary Wilbur actively supported not only the original expedition, but also the concept of a long-term collaborative venture. Why?

Part of the Navy's enthusiasm arose from Wilbur's technophilic philosophy. Wilbur was driven by a fundamental belief that the history of the Navy was the history of technological advance, with technology on the side of the victors. Any Navy that did not keep up with technological innovation would soon become the victim of one that did. In stump speeches given on behalf of presidential candidate Herbert Hoover, Wilbur focused specifically on Hoover's technical background as his greatest attribute and qualification for the job of commander-in-chief. In various other public addresses, Wilbur emphasized the controlling impact of science and technology in the modern world in general and the military in particular. In the Navy, technology had been and would continue to be critical in the obvious areas of power and navigation, but Wilbur argued that the impact was *pervasive*, from the diesel engines that replaced the boiler room to "the electric potato peelers in the modern mess." [LC CDW, Box 2, Speeches 1924-27]. Whilst the contribution of potato peelers to victory might not be measurable, other innovations had turned out to be decisive, often in unexpected ways. But given that the Navy had no research labs of its own, support for civilian scientists had to be considered part of the Navy's mandate. Thus Wilbur declared the Navy's arms "outstretched to the inventor and the scientist":

> What does it mean to have a radio-controlled, electronically operated Navy? It means constant experimentation, constant research, constant development. What does it mean to have aircraft incorporated into the Naval forces, directing its gunfire by radio from the air, assailing its enemies in the air, and on the water and beneath the water? It means invention, research, improved planes, improved engines. . . . [LC CDW Box 2, Speeches 1924-1927, "Recent accomplishments in the Navy," ca. 1927, p 2]

Wilbur furthermore accepted the stereotype that creativity and invention were a particular strength of American culture, and should therefore be a strength of the American military. He recounted the following apocryphal story:

> It is said that when the German Kaiser was considering the question of war with America he said he had no fear of the American army or the American Navy or the wealth and resources of the American people, but he did fear the American's inventive genius . . . [Therefore] it is essential and indeed vital that the American genius for invention should be constantly stimulated along lines looking toward success in war, as well as in peace [LC CDW Box 3, Press releases, 1924-1929: Sept. 5, 1924; p. 1-5].

Since genius was to be found in diverse places, the task of the military was to find it, support it, and remind it of military concerns. One particular concern of Wilbur was U.S. dependence on imported Mexican oil, and the risk of a wartime "energy crisis." Therefore, he took a strong interest in the work of geologists, and advocated the development of a strategic petroleum reserve:

> Five years ago, the Federal Oil Commission reported that according to the best judgment of the oil industry the Nation's supply of crude oil would be exhausted in ten years, that is in 1934. This would mean that the Navy, which is almost wholly oil burning, would be helpless, unless new oil was discovered or new equipment provided. . . ." [LC CDW, Box 2, Speeches 1924-1927, "Progress in the Navy," ca. 1929, p. 5]

He also wrote:

> That an ample supply [of oil] should be available within our own continental borders in time of war needs no argument. The commodity so absolutely vital to the very life of the nation must not be dependent upon foreign supply and overseas transportation, else disaster and defeat will be inevitable, for even if there were overseas sources upon which we could draw freely without international complications as to violations of neutrality, etc., it would be necessary to call so heavily on the fleet for protection of the lines of communication that the armed forces in the actual combatant areas must needs be seriously depleted. What we need and must have are huge ground reserves of oil capable of being drawn

upon without delay to supply the petroleum products needs by the armed forces and industries in the event of war, and a sufficient amount of overground storage to take care of our week to week wants [LC CDW Box 2, "Speeches, 1929 and undated, "Petroleum and National Defense"].

A strategic petroleum reserve would prevent short-term supply cut-off during hostilities, but Wilbur wanted to be proactive to ensure the long-term supply by funding research into alternative energy sources:

> [W]e must remember that petroleum, like other mineral raw materials, is a wasting asset. No other mineral among those essential to the well being of the country has shown so rapid an increase in production and also in the rate of depletion. The question is how long can we keep up the pace we are going and still have any reserve stored in the natural reservoirs. . . . The authoritative estimates by geologists is that the original supply of petroleum in the United States is forty per cent gone . . .
>
> [Therefore] the Navy has been planning, experimenting and testing a new plan of supplying fuel. The oil shale reserves of the Navy and the nation are tremendous. If the Navy can develop a method of producing oil from oil shale, in commercial quantities and with reasonable economy, the problem of national defense, so far as fuel for aircraft and for ships, will be solved . . . [LC CDW "Petroleum and National Defense," Box 2, "Speeches 1929, and undated" and Speeches, 1924-1927, "Progress in the Navy," p. 5].

While promoting research on alternative fuel sources, Wilbur also continued to advocate technological innovation in the conventional arenas of power and navigation. In July 1924, Wilbur organized a meeting of "representatives of the executive departments and scientific establishments of the Government of the U.S. Navy" to discuss the importance of further conventional exploration of the oceans and the need for technological improvements in this area. He reminded his colleagues that the oceans were still far from fully explored, and that there remained many "unsounded depths and undiscovered secrets" beneath the waves [LC CDW, Box 3: Press releases, 1928 " July 1, 1924 [sic]]

The S-21 expedition was consistent with Wilbur's goals on a variety of levels. Gravity measurement at sea was related clearly to the overall physiography of the ocean basins. Wilbur's general sympathy to innovation would have predisposed him to look favorably on an expedition to test a new technological device. His preoccupation with energy resources would have made him sympathetic to having geologists on board. The type of data being produced—improved information about the shape of the Earth—was directly relevant to the latitude corrections required for gyroscopic measurement of true north and, indeed, for all navigational problems [LC PCDW, Box 3 Press releases, "Extracts," p. 5]. Wilbur's professional interests for the Navy thus coincided on several levels with the research needs of the scientists of the S-21 expedition. Wilbur fostered a technocratic milieu in which scientific research was perceived as beneficial to the long-term interests of the Navy, and in which the Navy was on the lookout for research of potential practical relevance. The Navy would support science so that scientists would be on board [in this case, literally] when the Navy needed them. The Navy would know when potentially important innovations arose, and scientists would keep the Navy in mind when they did their work [cf. Mukerji, 1989]. The result was a symbiosis between a military establishment on the look-out for relevant technological innovation and a scientific establishment seeking logistical and material support to expand its domain of research.

BASIC OR APPLIED RESEARCH?

Nathan Reingold has eloquently argued that historians have been hindered in their understanding of American science by adherence to the dichotomous categories of "basic" and "applied" research, which perhaps exist more for contemporary historians than they did for past American scientists [Reingold and Reingold, 1981, pp. 1-6; Reingold, 1991, pp. 60-61]. The S-21 expedition is a case in point. William Bowie and Felix Vening Meinesz were geodesists employed to provide their countries with better geodetic control, but this required a fundamental concern with the shape, the structure, and even the history of the Earth. Consequently, these men thought about, talked about, and published articles on issues that we would label "theoretical" geology. They called themselves geodesists, but they published papers in *Nature*, *Science*, and the *Bulletin of the Geological Society of America*, as well as in the publications of their sponsoring institutions.

When Bowie referred to the S-21 expedition, he repeatedly wrote of "the *purpose* and scientific *significance* of the determination of gravity at sea." [e.g. Bowie, 1929d; emphasis added]. Bowie viewed his work as simultaneously practical and theoretically significant. Utility and knowledge arose congruently from the same activity. There is no evidence that Bowie and his colleagues saw these two goals as mutually exclusive, competing, or that one was an excuse for the other. Nor is their any evidence that the pressure to do "applied" research had a "debilitating effect" on their theoretical work [see Reingold, 1991a p. 59]. Rather, practical and theoretical goals arose congruently from the nature of geodetic work. Bowie, Vening Meinesz, and Wright were all exceedingly productive scientists, and part of the pleasure of their work seems to have been that it *was* both purposeful and significant.

Likewise Fred Wright aimed for "a more accurate determination both of the figure of the Earth and of the state of equilibrium of its surface features" [Vening Meinesz and Wright, 1930, p. 1]. The Earth's figure was primarily a practical problem, the state of equilibrium was primarily a theoretical one. Two goals were accomplished through one activity. Wright never suggested that one or the other was his "real" goal. This might seem surprising from a Carnegie man, employed full-time at an institution dedicated to advances in fundamental knowledge and conceived of as a "sheltered enclave for basic research" [Reingold and Reingold, 1981, p. 3; also see Servos, 1984; Reingold, 1991; Kohler, 1991]. But the concept of science as service was not alien at the Carnegie. During the first World War, Wright had researched and supervised the production of optical quality glass; this work resulted in a number of scientific papers during and after the war [Schairer, 1954]. Nor did Wright's colleagues dismiss these practical applications: Wright's memorial in *American Mineralogist* emphasized the importance of his work on the ternary system CaO-Al_2O3-SiO_2 in the development of Portland cement.

Vening Meinesz and Wright's science and invention developed concurrently. Wright developed his torsion gravimeter because of its relevance to geological problems; Vening Meinesz was well enough aware of geological issues to put his geodetic invention quickly to work on them. Among Wright's other inventions he counted a device for distinguishing between cultured and natural pearls and a method to decrease the "scratching proclivities" of scouring powder. There is no evidence that these useful inventions offended Wright's patrons in any way; on the contrary, Merriam approved and sponsored Wright's patent application for the gravimeter [CIW General Files: F. E. Wright folder]. Perhaps this only seems surprising in retrospect, given current notions of what it means to do "pure" science. But recall that the CIW founders were steeped in the traditions of American federal science, and hence of federal *service* [Reingold, 1991, p. 218-219; Yochelson, this volume]. The founders of the CIW had done science *in* the federal government, and in the 1920s the scientists there continued to do science *with* the federal government.

The notion of science as service was developed explicitly by Wright in later years when he wrote a report for Merriam on the "Public Relations Problem of the Institution" [Wright, 1937]. Wright emphasized that the purpose of the CIW was fundamentally the same as Andrew

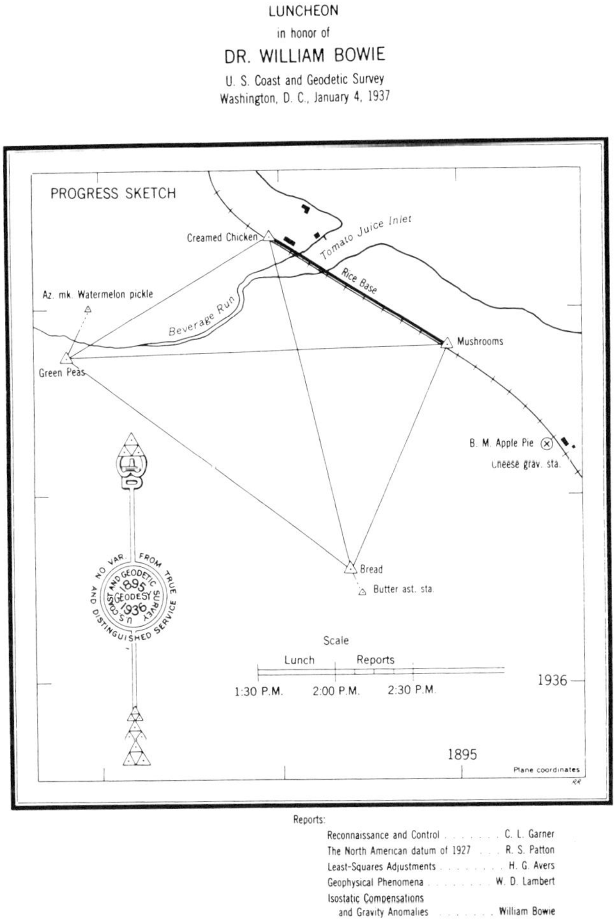

Fig. 3. The luncheon program for the retirement of [Captain] William Bowie from the U.S. Coast and Geodetic Survey, 1937. Note the north arrow, which points in only one direction, thus visually encapsulating the ethos of the Coast Survey: "true and distinguished service."

Carnegie's other endowments: all were dedicated to "progress of the human race." Thus, closely parallelling Bowie's view of the government scientist, Wright saw the task of the CIW scientist as two-faceted: "a] to contribute to progress in science through research of the highest quality; and b] to interpret the results of research in terms of their value to mankind" [Wright, 1937, p. 1]. Part of the task of interpretation included an obligation both to fellow citizens and scientists to explain and disseminate one's results. Scientists would benefit by being forced to confront the meaning of their endeavors. The two aspects of inquiry and interpretation—research and application—were seen as synergistic, not competing. In Wright's words, "these two objects are intimately linked and actually represent different phases of the same problem" [Wright,1937, p. 2]. Ultimately, CIW would suffer if it failed to maintain this essential link, for the life of this "semi-public" institution was inescapably linked to public support of it [Wright 1937, p. 3-4].

Thus although one might expect a different sense of purpose at the CIW and the USCGS, in fact a similar ethos and sense of mission facilitated cooperation. In the same year that Wright was expressing this verbally at the CIW, the ethos of science as service was encapsulated visually by the program for a luncheon in honor of William Bowie on the occasion of his retirement. Bowie's career was illustrated metaphorically by a map representing the various courses of the meal, but in place of the usual multiple arrows for true, magnetic, and grid north, this map displays only one arrow, pointed firmly in the direction of "true and distinguished service" (Figure 3).

CONCLUSION

The gravity-measuring cruise of the U.S. Submarine S-21 was simultaneously an endeavor in basic and applied science, an outgrowth of a research tradition in which utility and knowledge were, if not precisely identical, at least intimately related functions [Reingold, 1978, p. 171]. Reingold [1978, p. 167] has cited the U.S. Coast and Geodetic Survey as embodying "classic American strategies for combining culture and service, theory and practice, mass and elite," but this strategic confluence was promoted and practiced by colleagues at the Carnegie Institution of Washington as well. Among the American Earth scientists discussed in this paper, this combination fostered a mindset in which civilian-military collaboration was a natural outgrowth of shared interests and overlapping realms of concern.

A belief in the essential interconnectedness of theory and practice led American Earth scientists to the application of instrumental traditions of geodesy into theoretical realms of geology. The result was the emergence of a new research tradition: marine geophysics. This tradition crystallized around a conceptual question, an instrument with which to answer that question, and a particular place in which to use that instrument. A military establishment committed to technological advance provided material and psychological support without which the emergent tradition might well have foundered.

ARCHIVAL SOURCES

The principal archival sources used in this work are the CIW General Files, the Records of the U.S. Coast and Geodetic Survey and the John C. Merriam papers at the U.S. Library of Congress. Most of the CIW materials on the S-21 expedition are found in the Files of the Geophysical Lab [CIW-GL], or in the Director's Files. The CIW internal reports cited are as follows:

- Wright [1926]: CIW General Files; Patents: Wright

- Wright [1929]: CIW-GL Director's file #2, 1905-1935.

- Forbush [undated] "Gravity measurements on the Carnegie" and "The Meinesz gravity apparatus on the Carnegie" in CIW General Files 1900-1935, DTM: Scott Forbush folder.

The Records of the USCGS are in the US National Archive, Record Group 23 and are cited here as [US-NA USCGS]. They were inventoried by Nathan Reingold. The most useful Coast Survey materials are Series 33, Annual Office Reports, especially Box 774, 1920-1925, and 775-776, 1926-1933. Also useful are the published "Report of the Director of the U.S. Coast and Geodetic Survey", Government Documents. These give a good summary of the objectives and working conditions of the Agency.

There is a tremendous amount of material in the Merriam papers, AC 9507 at the Library of Congress, here cited as [LC JCM]. These include extensive correspondence on numerous aspects of American geology and geophysics in the early 20th century. These papers desperately need a finding guide!

Additional sources consulted include the Curtis D. Wilbur papers (Library of Congress, AC 3172), cited as [LC CDW]. These papers give an interesting snapshot into the mindset of the Navy Secretary through his speeches and other public addresses. There is no personal correspondence. The logbook of the S-21 expedition is in US-NA, Record Group 24 [US-NA RG24 LS21 Area 18W4 pp 605-613]. It confirms the accounts given in the published reports. There was essential nothing in the Records of the U.S. Naval Observatory [US-NA, RG 78, Entry 14 General Correspondence, 1925-1929] and the inventories are very poor. Lastly, there are the Collected papers of William Bowie, National Oceanic and Atmospheric Administration. These two volumes include unpublished speeches, published papers, and miscellaneous materials such as invitations to speeches, luncheon programs, etc. These give a very good sense of the range of William Bowie's scientific and professional activities. There is no personal correspondence.

ACKNOWLEDGEMENTS

I am grateful to the following persons for help in accessing historical materials: Grace Sollers, U.S. National Oceanic and Atmospheric Administration, Charles Whitten, retired, U.S. Coast and Geodetic Survey; Susan Vasquez, Carnegie Institution of Washington; Richard von Dornhoff, U.S. National Archives and Record Administration; Barbara de Felice, Dartmouth College Libraries, and Helen Wright

Greuter and Finley Wright, daughter and son of the late F. E. Wright, who provided the photographs of the S-21. Ron Doel, Charles Drake, and Greg Good graciously provided comments on an earlier draft of this paper. I am indebted to the police and librarians at the U.S. Library of Congress for not arresting me when I accidentally put several original letters in my pile of photocopies; fortunately, the materials I had were "worthless." Financial support from Dartmouth College, Burke Research Initiation Fund, is gratefully acknowledged.

REFERENCES

Alessio, A, Doubts and suggestions on terrestrial isostasy, *Geogr. J.*, *13*, 33-45, 1924.

Allwardt, A. S., The Roles of Arthur Holmes and Harry Hess in the Development of Modern Global Tectonics, PhD diss., Univ. California, Santa Cruz, 1990.

Anonymous Review, *Am. J. Sci.*, *135*, 495-497, 1888.

Barrell, J., The nature and bearings of isostasy, *Am. J. Sci.*, *198*, 281-290, 1919a.

Barrell, J., The status of the theory of isostasy, *Am. J. Sci.*, *198*, 291-338, 1919b.

Barrell, J., On continental fragmentation and the geologic bearing of the Earth's surface features, *Am. J. Sci.*, *213* [Fifth series XIII], 283-314, 1927 [published posthumously].

Bates, C. C., Gaskell, T. F., and Rice, R. B., *Geophysics in the affairs of man*, 492 pp., Pergamon Press, Oxford, 1982.

Bowie, W., The Earth's crust and isostasy, *Geogr. Rev.*, *12*, 613-627. 1922.

Bowie, W., Abnormal densities in the Earth's crust disclosed by analysis of geodetic data, *Geogr. J.*, *13*, 26-33, discussion on pp. 35-45, 1924.

Bowie, W., Scientists to weight the floating Earth's crust, *The New York Times*, Sunday, September 20, 1925.

Bowie, W., The deflection of the vertical in Porto Rico, *J. Wash. Acad. Sci.*, *15*, 29-31, 1926a.

Bowie, W., Proposed theory, in harmony with isostasy, to account for major changes in the elevation of the Earth's surface, *Gerlands Beiträge zur Geophysik*, *15*, 103-115, 1926b.

Bowie, W., The effects of the shape of the geoid on values of gravity at sea, *Am. J. Sci.*, 214, 222-227, 1927.

Bowie, W., *Isostasy*, 275 pp, E.P. Dutton, Co. New York, 1927.

Bowie, W., Possible origins of ocean and continents, *Gerlands Beiträge zur Geophysik*, *21*, 178-182, 1929a.

Bowie, W., Zones of weakness in the Earth's crust, *Science*, *70*, 589-592, 1929b.

Bowie, W., Isostasy and geological thought, *Sci. Monthly*, *28*, 385-392, 1929c.

Bowie, W., Weighing the Earth from a submarine, *Sci. Am.*, 218-221, March 1929d.

Bowie, W., The figure of the Earth derived by triangulation methods, *Am. J. Sci.*, *218*, 53-64, 1929e.

Bowie, W., International cooperation in geographical work, *Science*, *71*, 425-429, 1930.

Bowie, W., The Earth as an engineering structure, *Sci. Monthly*, *32*, 457-460, 1931a.

Bowie, W., Shaping the Earth, *J. Wash. Acad. Sci.*, *21*, 103-125, 1931b.

Bowie, W., Significance of gravity anomalies at stations in the West Indies, *Bull. Geol. Soc. Am.*, *46*, 869-878, 1935a.

Bowie, W., The origins of continents and oceans, *Sci. Monthly*, *41*, 444-449, 1935.

Bullard, Sir E., William Maurice Ewing, 1906-1974, *Biogr. Mem. Fellows Roy. Soc.*, *21*, 268-311, 1975.

Burger, W. H., John Fillmore Hayford, *Biogr. Mem. Nat. Acad. Sci.*, *16*, 159-252, 1931.

Chamberlin, T. C., Isostasy in light of the planetesimal theory, *Am. J. Sci.*, *192* (Series 4, no 42), 371, 1916.

Dupree, A. H., *Science in the federal government*, 460 pp., The Johns Hopkins Univ. Press, Baltimore, Md., 1986.

Ewing, M., Marine gravimetric methods and surveys, *Proc. Amer. Phil. Soc.*, *79*, 47-70, 1937.

Field, R. M., Brown, T. T., Collins, E. B., and Hess, H. H., *The Navy-Princeton gravity expedition to the West indies in 1932*, U.S. G.P.O., Washington, D.C., 1933.

Field, R. M., The importance of geophysics to submarine geology, *Proc. Amer. Phil. Soc.*, *79*, 1-8, 1937.

Fleming, J. A., William Bowie, *Biogr. Mem. Nat. Acad. Sci.*, *26*, 61-97, 1951.

Fleming, J. A. and Piggott, C. S., Frederick Eugene Wright 1877-1953, *Biogr. Mem. Nat. Acad. Sci.*, *29*, 1956.

Forbush, S., *Gravity Measurements on the Carnegie*, Unpublished and undated report, CIW General Files 1900-1935, DTM Scott Forbush folder.

Forbush, S., *The Meinesz gravity apparatus on the "Carnegie"*, Unpublished and undated report, CIW General Files 1900-1935, DTM Scott Forbush folder.

Forbush, S., Gravity determinations on the "Carnegie" in *Scientific Results of Cruise VII of the "Carnegie," 1928-1929 under command of Captain J.P. Ault*, CIW Publication No. 571, 1946.

Galison, P. L., Bubble chambers and the experimental workplace, in *Observation, Experiment, and Hypothesis in Modern Physical Science*, edited by Peter Achinstein and Owen Hannaway, pp. 309-373, MIT Press, Cambridge, Massachusetts, 1985.

Galison, P. L. and Assmus, A., Artificial clouds, real particles, in *The uses of experiment*, edited by D. Gooding, T. Pinch, and S. Schaffer, pp. 225-274, Cambridge Univ. Press, Cambridge, England, 1989.

Goetzmann, W. H., *Army Exploration in the American West*, 509 pp., Yale Univ. Press, New Haven, 1959.

Goetzmann, W. H., *Exploration and Empire*, 656 pp., Alfred A. Knopf, New York, 1966.

Harland, J. P., The last cruise of the Carnegie, in *The Search for Understanding, Selected Writings of Scientists of the Carnegie Institution*, edited by C. P. Haskins, pp. 281-298, CIW, Washington D.C., 1967.

Hayford, J. F., *The figure of the Earth and isostasy from measurements in the United States*, U.S. G.P.O., Washington D.C., 1909.

Hayford J. F. and Bowie, W., The effect of topography and isostatic compensation upon the intensity of gravity, *U.S. Coast and Geodetic Survey Special Publ. 10, Geodesy*, Washington D.C., 1912.

Hess, H. H., Gravity anomalies and island arc structure with particular reference to the West Indies, *Proc. Amer. Phil. Soc.*, *79*, 71-96, 1937.

Kohler, R. E., *Partners in Science: Foundations and Natural Scientists 1900-1945*, 415 pp., Univ. of Chicago Press, Chicago, 1991.

Lamson, E. A. Gravity expedition of the United States Navy in 1928, in William Bowie, ed., *Geodetic Operations in the United States*, U.S. Coast and Geodetic Survey, Special Pub. 166, 1930.

Laudan, R., Oceanography and geophysical theory in the first half of the twentieth century: The Dutch school, in *Oceanography: The Past*, edited by M. Sears and D. Merriman, pp. 656-666, Springer Verlag, New York, 1980.

Laudan, R., Redefinitions of a discipline: Histories of geology and geological history, in *Functions and uses of disciplinary history*, edited by L. Graham, W. Lepenies. and P. Weingart, pp. 79-104, on 93, Reidel, Dordrecht, 1983.

Laudan, R., Frank Bursey Taylor's theory of continental drift, *Earth Sci. Hist.*, *4*, 118-121, 1986.

Le Grand, H. E., Specialities, problems and localism: the reception of continental drift in Australia, 1920-1940, *Earth Sci. Hist.*, *5*, 84-95, 1986.

Le Grand, H. E., *Drifting continents and shifting theories*, 313 pp., Cambridge Univ. Press, Cambridge, England, 1988.

Maienschein, J., *Transforming traditions in American biology*, 1880-1915, 366 pp., Johns Hopkins Univ. Press, Baltimore, 1991.

Manning, T. G., *U.S. Coast Survey vs. Naval Hydrographic Office: A 19th century rivalry in science and politics*, Univ. of Alabama Press,

Tuscaloosa, 1988.
Menard, H. W., *The Ocean of truth: A personal history of global tectonics*, 353pp, Princeton Univ. Press, Princeton, 1986.
Miller, H. S,. *Dollars for research: Science and its patrons in nineteenth century America*, 258 pp., Univ. of Washington Press, Seattle, 1970.
Molengraaf, G. A. F., Wegener's continental drift, in *Theory of Continental Drift*, edited by W. A. van Waterschoot van der Gracht, Amer. Assoc. of Petroleum Geologists, Tulsa, 1928.
Mukerji, C., *A Fragile Power: Scientists and the State*, 253 pp., Princeton Univ. Press, Princeton, 1989.
Pauly, P. J. *Controlling life: Jacques Loeb and the engineering ideal in biology*, 251 pp., Oxford Univ. Press, New York, 1987.
Reingold, N., National style in the sciences: The United States case, in *Human Implications of Scientific Advance, Proc. XVth Int. Congr. Hist. Sci.*, E.G. Forbes, ed., 1978.
Reingold, N. and Reingold, I. H. , eds, *Science in America: A documentary history, 1900-1939*, 490 pp, Univ. of Chicago Press, Chicago, 1981.
Reingold, N., Alexander Dallas Bache: Science and technology in the American idiom, pp. 110-126, in *Science American Style*, Rutgers Univ. Press, New Brunswick, 1991a.
Reingold, N., American indifference to basic research reappraised, pp. 54-75, in *Science American Style*, Rutgers Univ. Press, New Brunswick, 1991b.
Reingold, N., National science policy in a private foundation: The Carnegie Institution of Washington, pp. 190-223 in *Science American Style*, Rutgers Univ. Press, New Brunswick, 1991c.
Schairer, J. F., Memorial of Frederick Eugene Wright, *Am. Mineralogist*, *39*, 284-292, 1954.
Servos, J. W., To explore the borderland: The foundation of the Geophysical Laboratory of the Carnegie Institution of Washington, *Hist. Stud. Phys. Sci.*, *14*, 147-186, 1984.
Stock, C., John Campbell Merriam, *Biogr. Mem. Nat. Acad. Sci.*, *26*, 209-232, 1951.
van der Gracht, W.A. van Waterschoot, The problem of continental drift, in *Theory of Continental Drift*, edited by W.A. van Waterschoot van der Gracht, Amer. Assoc. Petroleum Geologists, Tulsa, 1928.
Vening Meinesz, F. A., Gravity expedition of the U.S. Navy, *Nature*, 123, 473-475, 1929.
Vening Meinesz, F. A., *Gravity expeditions at sea, 1923-30, vol. I*, Netherlands Geodetic Commission, Delft, 1932.
Vening Meinesz, F. A. with Umbrove, J. H. F. and Kuenen, Ph. H., *Gravity expeditions at sea, 1923-32, vol. II*, Netherlands Geodetic Commission, Delft, 1934.
Vening Meinesz, F. A., *Gravity expeditions at sea, 1934-1939, vol. III*, Netherlands Geodetic Commission, Delft, 1941.
Vening Meinesz, F. A. and F. E. Wright, The gravity measuring cruise of the U.S. submarine S-21, with an appendix on computational procedure by Miss Eleanor A. Lamson, *Publications of the U.S. Naval Observatory*, Second series, *13*, appendix 1, 94 pp., 1930.
Warner, J. H., *The therapeutic perspective: Medical practice, knowledge, and identity in America, 1820-1885*, 367 pp., Harvard Univ. Press, Cambridge, MA, 1986.
Whitten, C., William Bowie: Engineer, administrator, diplomat, *EOS*, *73*, 113-125, 1992.
Wood, R. M., *The dark side of the Earth*, 246 pp., Allen and Unwin, London, 1985.
Worzel, J. L., *Pendulum gravity measurements at sea, 1936-1959*, 422 pp., John Wiley and Sons, New York, 1965.
Wright, F. E., Apparatus for the measurement of the variations in the force of gravity, U.S. Patent Office patent #1,579,273, April 6, 1926.
Wright, F. E., Gravity measuring cruise of the submarine U.S.S. S-21, *Preliminary report to the CIW*, 1929.
Wright, F. E., The public relations problem of the Carnegie Institution of Washington, *CIW Internal Report*, January 29, 1937.
Wright, F. E., Gravity Measurements in Guatemala, *Trans. AGU*, *22*, 512-515, 1941.
Wright, F. E. and J. L. England, An improved torsion gravity meter, *Am. J. Sci.*, *35A*, 373-383, 1938.
Yoder, H. S., Scientific highlights of the Geophysical Laboratory, 1905-1989, *CIW Annual Report of the Director of the Geophysical Lab*, p. 143- 203, 1989.

Amundsen and Edmonds: Entrepreneurial and Institutional Exploration

S. M. SILVERMAN
Physics Department, Boston College
Chestnut Hill, MA 02167

MARION EDMONDS SMITH
Department of Neurology, Stanford University
Stanford, CA 94305

In August 1910 Roald Amundsen and his companions left Norway on what was ostensibly the beginning of an expedition to drift across the Arctic Ocean. In fact he intended to travel south to the Antarctic for an attempt to be the first to reach the south geographic pole. A few years earlier he had been the first to complete the Northwest Passage. That voyage had been officially dedicated to an accurate location of the North Magnetic Pole. On completion of the voyage he took the magnetic instruments for recalibration to the U.S. Coast and Geodetic Survey (USCGS) magnetic observatory at Sitka, Alaska, then in charge of Harry Edmonds. Amundsen was apparently impressed by Edmonds qualifications and, in planning for the Arctic drift, invited him to be the magnetician onboard ship, with instruments and expenses to be provided and paid for by the Carnegie Institution of Washington (CIW). In the event, when Amundsen sailed for the South Pole instead of the Arctic, Edmonds was not on board, and no magnetic measurements were carried out.

This paper details these events, and examines the reasons why the magnetic measurements, so central in the Northwest Passage expedition, were planned for and then not carried out in the South Pole attempt. The incident illustrates two modes of exploration prevalent around 1900 and still existing today: the entrepreneurial explorer, typified by Amundsen, and the institutional explorer, exemplified by Edmonds. The event also demonstrates the role of the CIW's Department of Terrestrial Magnetism (DTM), under Louis Bauer, in fostering international cooperation in carrying out scientific measurements.

The Earth, the Heavens and the Carnegie Institution of Washington
History of Geophysics Volume 5

ROALD AMUNDSEN (1872-1928)

"Whatever I have accomplished in exploration has been the result of lifelong planning, painstaking preparation, and the hardest kind of conscientious work."

Roald Amundsen, *My Life as an Explorer*, 1928.

Roald Amundsen (Figure 1) was born near Oslo, Norway on July 16, 1872. From the age of 15, influenced by the writings of Sir John Franklin, the British Arctic explorer, Amundsen determined also to be an explorer [Amundsen, 1928, p. 1]. At the insistence of his mother, however, he enrolled in medical school, but withdrew after her death. A year later he, together with a single companion, attempted to cross a Norwegian plateau on skis in the midst of winter, an adventure which almost cost him his life. No one had accomplished this, and, in view of his subsequent career, the wish to be first may have influenced the attempt. They had been, unknowingly, within a hundred yards of their goal when they turned and recrossed the plateau. Amundsen commented that the adventure involved as many hardships and dangers as anything he later encountered [ibid., pp. 6-18].

Following his army training he signed on as a sailor during the summers of 1894-1896 in preparation for obtaining a master's license. He had read all the books on Arctic exploration and recognized the need for a single commander on an expedition, not splitting functions between the commander of the expedition and the captain of the ship. In 1897 he was first mate on the *Belgica* expedition, the first to winter over in the Antarctic. One of his colleagues on this voyage, Frederick Cook, later claimed to have discovered the North Pole. Amundsen developed a high regard for Cook and credited him with a significant role in the safe return of the *Belgica*.

Fig. 1. Roald Amundsen. (Courtesy of Robert Eather.)

In 1900 Amundsen obtained his master's license and planned his first expedition. He visited Fridjof Nansen who endorsed his plans and provided introductions to people who might help him. Nansen, in the 1890s, had built a boat, the *Fram*, which could be frozen in the ice to drift with it across the Arctic Ocean. During 1893-1896, the *Fram* reached a latitude of almost 86°. Nansen, with one companion, had left the *Fram* at a latitude of 84° 4' in March 1895 in an attempt for the North Pole. He reached 86° 14' before turning south. Nansen's plan later became the basis for Amundsen's own plan for reaching the North Pole which was aborted on news of Robert Peary's success and Cook's reported success.

Encouraged by his interview with Nansen, Amundsen took up magnetic studies. His motivation was, in his own words: "My expedition must have a scientific purpose as well as the purpose of exploration. Otherwise I should not be taken seriously and would not get backing." Here we find a requisite of the entrepreneurial explorer—the need for financial support, and the justification for this in purported scientific accomplishment. In line with his need to learn magnetics he first approached the director of the Kew Observatory, England, who refused his request. He then obtained a letter of introduction to the director of the Deutsche Seewarte in Hamburg, George von Neumayer. He explained to him that he wished to become an explorer on his own account, "and that I must learn how to take magnetic observations so that I might acquire scientific data to justify my adventure." On being further questioned by von Neumayer he disclosed his ambition to be first to complete the Northwest Passage, and to make conclusive observations of the true location of the North Magnetic Pole. At that von Neumayer reacted with enthusiasm: "Young man, if you do that, you will be the benefactor of mankind for ages to come. *That* is the great adventure." Amundsen had found his justification.

For several months Amundsen studied magnetic observation in Hamburg, and also visited the observatories at Wilhelmshaven and Potsdam. In 1900 he bought a small fishing smack, the *Gjoa*. The following summer he made oceanographic observations in the North Atlantic between Norway and Greenland, presenting the results to Nansen. Finally in 1902-1903 he prepared for the Northwest Passage. His problems again were financial: "I besieged every possible source of funds—the learned societies and the private patrons of science." But: "Despair almost overcame me at times, because, in spite of everything, sufficient funds were not forthcoming." His creditors were becoming demanding. One demanded payment within 24 hours and threatened to seize the vessel and arrest him for fraud. In desperation Amundsen and his companions secretly set sail at midnight of June 16, 1903. Three years later, having successfully transited the Northwest Passage, he returned to Norway. How his truculent creditor managed during those three years is not recorded.

Amundsen and his companions wintered for two years near the North Magnetic Pole [Good, 1991]. His measurements (observatory, Figure 2) served as the basis for its location for several decades thereafter. Following the successful transit of the Northwest Passage Amundsen spent 1906 and 1907 lecturing in Europe and the U.S., returning to Norway with sufficient funds to repay his creditors.

On arriving in Alaska Amundsen had taken the magnetic instruments to the USCGS magnetic observatory in Sitka for recalibration. Here he met Harry Edmonds.

HARRY MARCUS WESTON EDMONDS (1862-1945)

Harry Edmonds (Figure 3) was born in Oshkosh, Wisconsin on June 25, 1862. Both his parents were New Englanders, the mother from Maine, the father from Vermont. Like many New Englanders they emigrated first to Wisconsin, and later to California. The father went first to California and then sent for his family. They sailed from New York in 1867, crossed the Isthmus of Panama, and arrived in San Francisco [Mary Elizabeth Edmonds, Journal]. The father was an attorney and one-term judge. Edmonds' mother, Mary Elizabeth, was among the first women admitted to the California Academy of Science in 1878, and among the first women (possibly the first) to graduate (in her late 40s) from medical school in San Francisco.

Harry Edmonds graduated from the University of California at Berkeley in 1882 and went to Germany for post-graduate studies until 1885. His motivation and his area of study are not known, but they did not result in any degree. In 1887 he began studying at San Francisco's Hahnemann College, where his mother had obtained her medical degree. His interests, however, were clearly on travel and exploration. From 1889-1891 he served on the Alaska Arctic Boundary Survey as assistant astronomer and observer for the USCGS. He went east to attend medical school, graduating from Hahnemann Medical College in 1892. In 1893 he joined his mother in her office [M. E. Edmonds Journal, November 29, 1892]. A few months later, however, he was again off to Alaska for the summer on boundary work (Figure 4). He rejoined his mother in her office for the winter and again spent the 1894 summer in Alaska. He repeated this seasonal pattern several more times in the late 1890s, until his mother died in 1899.

The Alaskan work of 1899 involved a reconnaissance of the mouth of the Yukon river. This was carried out by Edmonds, a cook, a rodman, and a canoeman [USCGS, Report, 1901, p. 191]. The shoreline, previous to Edmond's work, had been erroneously placed at least 10

Fig. 2. The observatory for absolute magnetic measurements, Gjoahavn, 1904-1905. (Amundsen, *The Northwest Passage*, volume 1, p. 108.)

Fig. 3. Harry Edmonds, after his retirement (From Edmonds family).

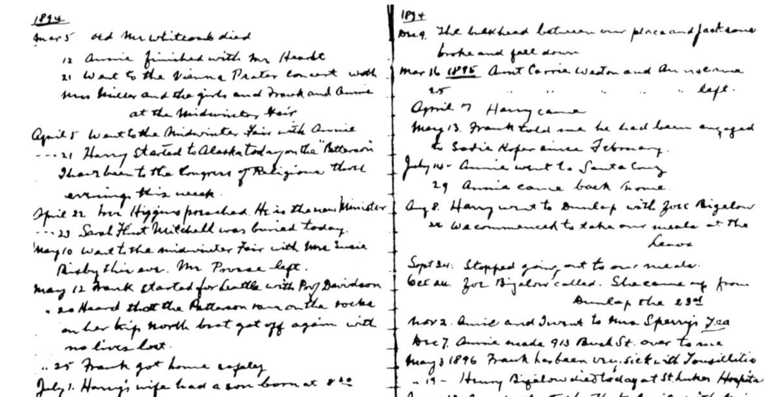

1894
Mar 5 Old Mr Whitcomb died
12 Annie finished with Mr Haadt
21 Went to the Vienna Prater concert with Mrs Miller and the girls and Frank and Annie at the Midwinter Fair
April 5 Went to the Midwinter Fair with Annie
... 21 Harry started to Alaska today on the "Patterson"
I have been to the Congress of Religions three evenings this week.
April 22 Mr Higgins preached. He is the new Minister
... 23 Sarah Hunt Mitchell was buried today.
May 10 Went to the Midwinter Fair with Mrs Susie Risby this eve. Mr Prosse left.
May 12 Frank started for Seattle with Prof Davidson
.. 20 Heard that the Patterson ran on the rocks on her trip north but got off again with no lives lost.
.. 25 Frank got home safely
July 1. Harry's wife had a son born at 8 o'c this morning.
... 5. Annie went to Lake Co for a rest at Wambolds - Blue Lake
... 21 Annie came home
... 24 F. R. Whitcomb came to the house sick
Aug 6 - " " left.
Annie commenced teaching again
Sept 6 - Harry came back from Alaska today
25 Harry left for Sampsons Flat

1894
Dec 9. The bulkhead between our place and Jacksons broke and fell down
Mar 16 1895 Aunt Carrie Weston and Annie came
25 " " " " left.
April 7 Harry came
May 13. Frank told me he had been engaged to Sadie Hofer since February.
July 14 - Annie went to Santa Cruz
29 Annie came back home.
Aug 8. Harry went to Dunlap with Jose Bigelow
26 We commenced to take our meals at the Lenox
Sept 24. Stopped going out to our meals.
Oct 26. Jose Bigelow called. She came up from Dunlap the 23rd
Nov 2. Annie and I went to Mrs Sperry's Tea
Dec 7. Annie made 913 Bush St. over to me
May 3 1896 Frank has been very sick with Tonsillitis
.. 19 - Henry Bigelow died today at St Lukes Hospital
June 12. Annie started for the Yosemite with Miss Lombard
... 23. Mr Whitcomb sent up a St Bernard dog today ... full blooded. Frank named him Duke Waldo
July 4. Annie got home from the Yosemite Valley
... 7 Mr & Mrs Higgins left for Santa Barbara
... 24 Harry had a severe attack of Sciatica
Aug 8. Frank went to Windsor to stay 2 weeks

Fig. 4. Pages from the journal of Mary Elizabeth Edmonds (unpublished)

miles to the eastward. The Superintendent's report, evaluating the effort, states:

> The many difficulties overcome, discomforts experienced, and valuable geographic knowledge acquired can not be adequately described in a short narrative. The task, to use Assistant Putnam's words, required the greatest endurance, perseverance, and courage, and credit is also due to the men who accompanied Dr. Edmonds [USCGS, Report, 1901, p. 191].

After his mother's death in 1899, Edmonds worked full time for the Survey. Amundsen had similarly taken his chosen path only after his mother's death. In 1900 Edmonds was appointed Surgeon at a salary of $125/month [NA-US-USCGS, Record of Employees, 1895-1910, 3. Disbursing Office]. In 1901 he was transferred and appointed as Magnetic Observer, at a lower salary of $100/month [ibid.]. After training in Washington and at the Cheltenham, MD magnetic observatory, then under the direction of Bauer, he was sent to take charge of the magnetic observatory at Sitka [USCGS, Report, 1903] (Figure 5), where he remained until 1909 [Harradon, 1945]. It was here that Amundsen and Edmonds paths crossed.

AMUNDSEN RECALIBRATES HIS MAGNETIC INSTRUMENTS. During the final phases of the Northwest Passage Amundsen made plans for the recalibration of his magnetic instruments. Beset by ice on September 2, 1905 the expedition was forced to winter over at King Point, Canada, a few miles from Herschel Island, a wintering location for whaling ships. Nearby was the beached whaler *Bonanza*, whose captain was anxious to travel overland to a point where he could continue on to San Francisco. Amundsen, eager to telegraph the news of his success, joined him, though apprehensively. Again, financial considerations played a part. Amundsen notes: "First of all, he had the money and I had none, so he would be in command of our little party. In the nature of things, I was sure that he was far less competent for this kind of an expedition than I." [Amundsen, 1928, p. 54] They left King Point on October 24 and arrived in Fort Yukon on November 20. Since Fort Yukon had no telegraph, they continued on to Eagle City, 200 miles south. There were expensive road houses along the way spaced at intervals of about twenty miles. Arriving at Eagle City on December 5 Amundsen could finally send his telegraph messages [Amundsen, 1908, pp. 212ff]. Amundsen stayed in Eagle City until February 1906.

The Norwegian consul, Henry Lund, in San Francisco, advised Amundsen that Sitka was the best place to check his instruments after leaving Bering Straits [NL-Oslo, Amundsen to Aksel Steen, January 19, 1906, Ms. fol. 3666 F:II]. He then contacted Edmonds by telegram, who then notified the Superintendent of the Survey by telegram [NA-US, RG 23, telegram, Edmonds to Superintendent, January 21, 1906] and Edmonds was ordered to offer Amundsen every facility [NA-US, RG 23, Edmonds, Annual Report to Superintendent, July 10, 1906]. Edmonds then asked Amundsen for details as to the work to be done at the Sitka Observatory [NL-Oslo, Amundsen to Edmonds, January 24, 1906]. Amundsen provided some of these details and noted: "None of us have a scientific education (I refer to the crew of the Gjoa) and I shall be very much pleased if I can have the assistance of some competent magnetic observer, in making the final observations." He speculated that the Gjoa would arrive at Sitka the middle of the following September, and asked that preliminary preparations be completed, "as time is very valuable to me." [ibid.] He then wrote to Aksel Steen in Oslo asking him to give Edmonds detailed information on the necessary preparations for the re-standardization [NL-Oslo, Steen to Bauer, undated copy; Ms. fol. 3666 F:II]. Steen wrote to Bauer, still in charge of USCGS magnetic work asking him to give the necessary instructions, as "the head authority of the U.S. magnetical service," since he, Steen, was not familiar with the equipment and arrangement of the Sitka observatory [ibid.]. Steen also suggested that the instruments, originally tested in Potsdam, should be returned there for re-testing.

Conditions at the Sitka Observatory were not always comfortable. In January 1907, for example, Edmonds noted: "It is rather uncomfortable at present in the office. Everything is frozen up and all the water I use is from melted snow. . . . There is no coal in town, so I am burning scrap wood." [NA-US-USCGS, Edmonds to Superintendent, January 18, 1907]. Shortly before receiving Amundsen's request, in December 1905, the ceiling of the variation building gave way because of dry rot in the sawdust packing and the contiguous wood, requiring extensive renovation [NA-US-USCGS, Edmonds, Annual Report, July 10, 1906]. As a result of Amundsen's request it was then decided to put up an auxiliary structure for the variation instruments while the original building was repaired. The auxiliary building was also to serve as a place for comparison of instruments generally even after variation instruments had been returned to their proper place [ibid.]. The construction and repairs of the observatories were carried out between April and July [ibid., abstract of data] when the instruments were removed from the auxiliary observatory back to the main building [NA-US-USCGS, Edmonds, semi-annual report to Superintendent, January

Fig. 5. Sitka in 1902. (Sheldon Jackson, *12th Annual Report on Introduction of Domestic Reindeer into Alaska*, Plate 22, G.P.O., Washington, D.C., 1903.)

Fig. 6. The magnetic observatory at Sitka, Alaska, where Amundsen's magnetic instruments were standardized after the completion of the Northwest Passage. (L. A. Bauer, Hunting the Magnetic Pole, *Van Norden Magazine*, vol. 2, p. 64, November 1907.)

Fig. 7. Amundsen and Edmonds, Sitka, Alaska, 1906.

10, 1907]. By August everything was ready (Figure 6) and awaiting Amundsen's arrival.

On September 7 Amundsen left with the magnetic instruments for Sitka [Amundsen, 1907], arriving there on the 20th, and leaving on the 27th (Figure 7). The comparison began on September 21 and continued into October, with determination of azimuth lines at the station [NA-US-USCGS, Edmonds, semi-annual report, January 10, 1907]. Most of the instruments were left behind by Amundsen in order to get temperature and other data (NA-US-USCGS, telegram, Edmonds to Superintendent, September 22, 1906; Edmonds to Superintendent, September 29, 1906]. The recording instruments were not set up and were taken to Seattle. The Seemann magnetometer was lost track of at Nome and had not been traced [NA-US-USCGS, Edmonds to Superintendent, September 29, 1906]. The Tschau magnetometer was the same one used by Nansen on one of his expeditions [ibid.]. Edmonds telegraphed for descriptions of the instruments and the principles of their operation, "in any language," since the instruments were to be left there for a while and perhaps indefinitely [ibid.]. By May 30, 1907 the Amundsen observations were being completed [NA-US-USCGS, Edmonds to Superintendent, May 30, 1907].

Bauer was responsible for cooperation with Amundsen while at the Coast and Geodetic Survey. He was also in charge of the new Department of Terrestrial Magnetism at the CIW. In 1906 he was required to make a choice of one or the other. He chose the CIW. As he noted to O. H. Tittman, USCGS superintendent:

> From our various conversations you already know that it is not merely a question of the considerable increase in compensation which has finally led me to accept this offer, but rather the fact of the greatly enlarged opportunities. The work of this Department as you are aware has already grown to such an extent that it is impossible for me much longer to perform the duties and have the responsibilities of its directorship, while at the same time holding the position I have the honor to occupy in your Bureau." [NA-US-USCGS, Bauer to Tittman, March 30, 1906]

Bauer's resignation from the Survey became effective August 31, 1906 [NA-US-USCGS, Record of Employees, 1895-1910, 3. Disbursing Office]. Bauer, in April 1908, forwarded a query from Amundsen as to the observations to the superintendent of the Survey with the hope that the reduction of the observations would soon be completed (NL-Oslo, Bauer to Amundsen, April 15, 1908; Ms. fol. 3666 F:II). The observations were then sent on to Aksel Steen in Norway from the Survey (NA-US-USCGS, Amundsen to Superintendent, May 12, 1908]. Since Amundsen continued to rely on Bauer in his planning for magnetic measurements the focus of their cooperation then shifted to the DTM.

Cooperation between the USCGS and the DTM also continued after Bauer's resignation from the Survey. The *Galilee*, a yacht chartered by the Carnegie Institution, equipped to make magnetic measurements on the oceans, stopped at Sitka for intercomparisons with the magnetic instruments there in July and August 1907 [NA-US-USCGS, Edmonds, Semi-Annual Report to Superintendent, December 31, 1907]. Bauer commended Edmond's work in a letter of October 14, 1907 to Tittmann, [NA-US-USCGS]: "I have already informed you that Dr. Edmonds, of the Sitka Magnetic Observatory, likewise did everything in his power to make the intercomparisons of instruments successful at that Observatory as also in facilitating the necessary work of the 'Galilee' at Sitka."

PLANNING FOR THE NORTH POLE.

After the completion of the Northwest Passage Amundsen resolved to attain the north geographic pole. His plan was essentially to repeat Nansen's drift on the Arctic Ocean currents [Amundsen, 1928]. The plan was to enter the Arctic Ocean through Bering Strait, where he believed the drift would take him over or near the pole The *Fram*, after all, had drifted to almost 86°. For this purpose he obtained the *Fram*, well tested in Arctic ice. The voyage had been portrayed as a purely scientific expedition, rather than as one to set a new record for furthest north [Amundsen, 1912, p. 42]. In accordance with the purported scientific focus, magnetic measurements were to be included in the program. Amundsen had clearly been impressed with Edmonds' capabilities in this regard. He had sent him a courteous thank you note for his recalibration of the magnetic instruments after the Northwest Passage voyage [Edmonds papers, Amundsen to Edmonds, November 15, 1906]. He now invited him, in a letter dated August 31, 1909 [Edmonds papers] to act primarily as magnetician and secondarily as doctor of the expedition. Amundsen, as usual, had to arrange for funding from a variety of sources. The *Fram* had been lent for the expedition through the Norwegian parliament, which provided funds for repairs and necessary alterations [Amundsen, 1912, p. 54]. Private contributors supplied more funds. And Amundsen was in debt for supplies. The magnetic work, then, had also to be funded from outside sources. Bauer offered to do this. Amundsen noted, in his letter of August 31, 1909 to Edmonds:

> I regret my resources are not important but I have had an offer last year from Dr Bauer to charge the Carnegie

Institution with the costs of the man that I might select for magnetic observations. I am therefore writing Dr Bauer by the same post proposing him to communicate immediately with you and ask you to see him first in Washington & after arrangement fixed, then if possible proceed to Christiania [now Oslo] to close the matter with me. I think this will be the only way of doing, my time being rather close to the departure from here in January next. I hope very much indeed that you will be able to arrange a meeting with Dr Bauer and I have asked him to send me a cablegram as soon as possible if it can be done [Edmonds family].

Bauer evidently agreed to the proposal and Edmonds, having resigned from the USCGS, joined the CIW staff. Edmonds then spent from January to June at the DTM preparing for the expedition, and doing some regular departmental work [CIW *Year Book* for 1910, p. 203].

The date of Amundsen's letter to Edmonds, August 31, 1909, is significant. The next day a telegram was received from Frederick Cook claiming that he had reached the pole. A few days later, on September 6, a message arrived from Peary with the same claim. The resulting controversy over these claims is relevant here only in its impact on Amundsen's plans. An essential basis for Amundsen's plan to drift across the Arctic Ocean, the first attainment of the pole, had now vanished. In Amundsen's words: "This was a blow indeed! If I was to maintain my prestige as an explorer, I must quickly achieve a sensational success of some sort." [Amundsen, 1928, p. 64]. Amundsen immediately reversed direction and decided to head for the South Pole, the last great problem [Amundsen, 1912, p. 42]. Financial considerations were in the forefront: ". . . in view of the altered circumstances, and the small prospect I now had of obtaining funds for my original plan, I considered it neither mean nor unfair to my supporters to strike a blow that would at once put the whole enterprise on its feet, retrieve the heavy expenses that the expedition had already incurred, and save the contributions from being wasted." [ibid., p. 43].

Amundsen now resolved on secrecy about his change of plan, which was to impact on Edmonds and the magnetic measurements. He stated three reasons, not entirely convincing, soon after the event. First, he was merely delaying the Arctic drift for a year or two, and that the contributors to the drift plan, "all men of position, and above discussing the application of the sums they had dedicated to the enterprise . . . [would] know that when the time came their contributions would be used for the purpose for which they were given" [Amundsen, 1912, p. 43-44]. Second, he would inform Scott, whose plans for the Antarctic were well publicized at the time, in sufficient time so as not to interfere with Scott's own plans. In any event, "The British expedition was designed entirely for scientific research. The Pole was only a side-issue, whereas in my extended plan it was the main object" [ibid., p. 44]. This disingenuous statement of Scott's plan was subsequently contradicted by Amundsen with the statement that after he was safely at sea, he "disclosed fully my intention to compete with Captain Scott for the South Pole" [Amundsen, 1928, p. 66]. An additional expedition to the Antarctic was planned by the Japanese, but they apparently did not intend to try for the pole. Amundsen gave a third reason, that had he made his intention public there would have been a great deal of newspaper discussion, possibly ending the project at its birth [Amundsen, 1912, p. 45]. The truth appears likely to be that he had received many contributions for an Arctic drift, and his contributors might have felt that he should continue with the plan, especially since it had been publicized as a strictly scientific undertaking. Convincing his benefactors and his creditors on short notice of a major change of direction and focus might well have destroyed the whole project. The expedition was always on the edge of financial disaster [Amundsen, 1912, p. 61]. On the other hand, if he succeeded all would be forgiven, and the additional moneys from books and lectures would cover the expenses and insure funding for future ventures.

Thus without divulging his plans, except to his brother and the first officer of the *Fram*, Amundsen set sail from Norway in August 1910. In the Madeira Islands, the last landfall before the Antarctic, Amundsen told his crew and gave them the option of returning home or continuing. All opted to continue. Safely at sea, his plans were made public.

No provision had been made for Edmonds and the magnetic instruments to accompany them. We can only speculate on the reasons. First was the secrecy, and having Edmonds meet them either in Norway or the Madeira Islands would have raised questions. Second was probably a time question. Edmonds was still in Washington in June 1910 preparing for the Arctic voyage. At this time the *Fram* was being loaded for the voyage. Possibly additional material for the Antarctic expedition was needed, reducing the space available for an extra man and his instruments. A large hut was carried on the *Fram* to be used as an Antarctic shore base, something not needed on an Arctic drift. The magnetic measurements also would have seemed less important. Amundsen thought in terms of location: the geographic pole, the magnetic pole, rather than in terms of the magnetic variations in quiet and disturbed times. The South Magnetic Pole had already been located by Shackleton's expedition in 1908-1909, and thus represented nothing new [Amundsen, 1912, p. 40]. Furthermore, the

South Magnetic Pole was further north than the Bay of Whales, where Amundsen had determined to set up his base, and the additional distance would have mattered in going to the pole. Thus the magnetic measurements must have seemed secondary at the time, and were included only to provide scientific verisimilitude to the enterprise. But the main reason was that the science was now clearly cast aside. Amundsen had to do something spectacular to maintain his hopes of future funding. There may also have been a nationalistic tinge. Amundsen constantly expressed his gratitude to the King. The Norwegian parliament had financially supported his endeavor (see also, e.g., the farewell scene in Norway where the crew unites in a "God preserve the King and Fatherland"). This would have been diluted if he had a foreigner along. Thus Edmonds was abandoned, and no magnetic measurements were made on the expedition. Edmonds' secondary role as doctor on the expedition also lapsed; no doctor was taken on the expedition, and Amundsen took on this responsibility himself [Amundsen, 1912, p. 71].

CONCLUSIONS

Amundsen and Edmonds both needed to explore. Amundsen's way was entrepreneurial—he needed to be in command. To carry out his plans he needed funds. To obtain these funds he felt he had to carry out spectacular, high-visibility operations. He needed to be first: first to complete the Northwest Passage, first to the South Pole, first to get to the North Pole by air, first to have been at both poles. He sought both public and private funding. To obtain public funds he used the nationalistic feelings of Norway, which achieved its independence from Sweden in 1905. Glory by association attracted private funds. Amundsen also needed to obtain funds through the publication of books, rights to newspaper articles, and lectures. For this reason both the completion of the Northwest Passage and the attainment of the South Pole were followed within a year by books detailing these accomplishments. The need for funds was a constant concern.

Amundsen's chronic need for funds or for contributions of supplies and equipment is well documented throughout his life, with perhaps a short respite for a few years after 1917. During the First World War Norway was neutral, and Norwegian shipping in demand. Amundsen invested in shipping and made enough profits to outfit a new vessel, the *Maud*, for the Arctic drift. In 1918 the *Maud* transited the Northeast Passage. By 1922 the continuance of the *Maud* Arctic expedition was again in doubt because of postwar decline in the value of the currency, and Amundsen was again beset by financial problems. The Arctic drift of the *Maud* was continued in 1923 by Wisting, in cooperation with the Carnegie Institution. By this time Amundsen had become obsessed with a vision of the airplane as the future of Arctic exploration, but this too required financing. He had put his trust in a fellow countryman, who committed him to expenses which bankrupted him. By 1924 he had again to try to recoup his fortunes by lecturing and writing. When Lincoln Ellsworth approached him offering to provide funding and join him in an attempt to reach the North Pole by plane, he gladly agreed. The attempt was unsuccessful, but a later flight in a dirigible designed by the Italian Nobile was successful. He had difficulties with Nobile's need to be center stage, and his basic incompetence. Amundsen devotes much of his book *My Life as an Explorer* to an attack on Nobile. It is ironic that Amundsen in the end lost his life in an attempt to rescue Nobile, marooned on an ice floe after a failed Arctic venture.

Edmonds, on the other hand, exemplified what might be called the institutional explorer. He was associated throughout his career with institutions concerned with measurements carried out on a large geographic scale. Initially with the U.S. government he saw many parts of Alaska. A broader scope was available with the Carnegie Institution's DTM. Under Louis Bauer the Department measured magnetic parameters on a worldwide scale. Edmonds, abandoned by Amundsen in the race for the South Pole, continued working at the CIW for the remainder of his career. He was first assigned to the *Carnegie*, the non-magnetic boat designed for magnetic observations on the oceans of the world. In 1913 he was detached to carry out a magnetic survey of the region west of Hudson's Bay, a region of interest because one of areas of maximum total intensity of the Earth's magnetic field is there. There he traversed approximately 2000 miles, of which 500 was along an unsheltered open coast. He then returned to the *Carnegie*. In 1917, after the declaration of war of the U.S. against Germany he was given command of the vessel on its return from Buenos Aires to Washington. In 1919 he was given the job of initiating the construction and equipping of the Huancayo Magnetic Observatory in Peru, where he remained until 1921. He was then sent to Apia, Western Samoa to carry out magnetic work in the Pacific and atmospheric electric measurements at Apia. On his return to Washington he worked on the reduction of the magnetic data obtained at Watheroo, Australia. In all of these activities the focus was on the science—being first was not the issue. Funding was more stable, at least once the CIW was committed to the project. While Edmonds published some papers this clearly was not particularly important to him, nor did he have the incentive for it. He retired from the CIW on December 31, 1930, and died in Berkeley, California on April 4, 1945.

The interaction of Amundsen and Edmonds also illustrates another aspect of science around 1900, the role of cooperative efforts. Here Bauer played a major role, setting up programs at the CIW involving scientists from many countries. His influence on world geophysics was immense through this. He also provided for joint ventures, such as the one initially planned by Amundsen for the Arctic drift, where the CIW provided the manpower and instrumentation for expeditions headed by others. Bauer, in his efforts to spread international cooperation, also founded the journal *Terrestrial Magnetism*, which became the *Journal of Geophysical Research*, now perhaps the major world journal in the field. Bauer's contribution to world geophysics was immense. His story deserves to be told in detail.

ACKNOWLEDGMENTS AND ARCHIVAL SOURCES

The material on Harry Edmonds has relied partly on documents published by or available from the Carnegie Institution of Washington, and partly on materials from the Edmonds family, one of the authors (M.E.S.) being a granddaughter of Edmonds. We have used especially the journal of Edmonds' mother, Mary Elizabeth Edmonds (unpublished).

Much material on Amundsen is available in those other books of his not specifically mentioned in the text above. A very thorough and meticulously researched book on the race for the South Pole and its background is that of Roland Huntford, *Scott and Amundsen*, 1979 [reprinted 1985, under the title, *The Last Place on Earth*]. Huntford has also edited and introduced a collection of photographs used by Amundsen in his lectures [*The Amundsen Photographs*, New York: The Atlantic Monthly Press, 1987]. The books by Amundsen and Huntford give many details about Amundsen's chronic financial worries.

We are also grateful to Mr. Oddvar Vasstveit, at the Department of Manuscripts, Universitetsbibliotek (National Library), Oslo for his help in locating and copying relevant unpublished materials in their collection. This material is cited as NL-Oslo. Manuscript material of the USCGS is in Record Group 23 at the U.S. National Archives. It is cited as NA-US-USCGS.

Finally, we thank the editor, Gregory Good, who clearly spent much time eliminating excess verbiage, and materially improving the flow of the text.

REFERENCES

Amundsen, R., To the North Magnetic Pole and through the Northwest Passage, *Annual Rept. Smithsonian Inst. for 1906*, Govt. Printing Office, Washington, D.C., 1907. pp. 249-273. Reprinted from *The Geographical Journal, London, 29*, 1907.

Amundsen, R., *The Northwest Passage*, 2 vols., E.P. Dutton, New York, 1908.

Amundsen, R., *The South Pole*, 2 vols., John Murray, London, 1912.

Amundsen, R., *My Life as an Explorer*, Doubleday, Doran, Garden City, N.Y., 1928.

Good, G. A., Follow the Needle: Seeking the Magnetic Poles, *Earth Sci. Hist., 10*, 154-167, 1991.

Harradon, H. D., Harry Marcus Weston Edmonds, 1862-1945, *Terr. Mag. Atm. Electr., 50*, 145-146, 1945.

U.S. Coast and Geodetic Survey, *Report of the Superintendent for July 1, 1899 to June 30, 1900*, Government Printing Office, Washington, D.C., 1901.

U.S. Coast and Geodetic Survey, *Report of the Superintendent for July 1, 1901 to June 30, 1902*, Government Printing Office, Washington, D.C., 1903.

Expeditions and the CIW: Comments and Contentions

RONALD E. DOEL
Center for History of Physics, American Institute of Physics
American Center of Physics, One Physics Ellipse
College Park MD 20740

EXPEDITIONARY SCIENCE IN HISTORICAL PERSPECTIVE

Between 1902 and World War II, the Carnegie Institution of Washington (CIW) emerged as a formidable patron of expeditionary science. Grants from the CIW funded a wide range of field studies, among them archeological explorations in Central Asia and geological surveys of China. They also permitted Carnegie scientists themselves to form and join expeditions. During the 1920s alone, CIW scientists studied geomagnetism near the North Pole, examined evidence of continental drift in South Africa, and made gravity measurements from a submarine crossing the Gulf of Mexico. Symbolic of the CIW's commitment to expeditionary science was its three-masted schooner, the *Carnegie*, designed to record geomagnetic data from the far corners of the globe. Expeditionary science rivalled and supplemented studies at its permanent geoscience branches, the Department of Terrestrial Magnetism (DTM) and the Geophysical Laboratory (GL).

The CIW of course was not alone in sponsoring major scientific expeditions in America during the early twentieth century. Until about 1890 the federal government had remained the primary patron able to mount major scientific and exploring surveys. The King, Wheeler, and Powell expeditions to the American West were symbolic of this post-Civil War era [Dupree, 1957; Goetzmann, 1972]. By 1900, however, this situation had greatly changed. The growth of individual fortunes and adherence to an ideal of *noblesse oblige* led to the creation of major metropolitan museums, foundations, and scientific institutions. Many of these organizations possessed luxurious endowments. Major research universities also benefitted from the wealth of industrial capitalists by the twentieth century [Miller, 1971; Geiger, 1986; and Kohler, 1990]. These developments allowed universities and museums to join the federal government as sponsors of scientific expeditions. By 1910 the American Museum of Natural History of New York helped finance Robert E. Peary's first expeditions to Greenland, Franz Boas's anthropological investigations in Alaska and Siberia, and the botanical expedition of Herbert Lang and James Chapin to the Congo rain forest, all major explorations for their day [Perkins, 1981]. Aided by a grant from the General Education Board, Princeton University launched its own excavation of Corinth [Anonymous, 1932]. Less than two miles from the CIW's P Street headquarters, the Smithsonian Institution, energetically led by Charles Greeley Abbot, oversaw about a dozen expeditions annually by the late 1920s [True, 1928]. Expeditions were also supported by private organizations, most notably the National Geographic Society, founded in 1890 [Abramson, 1987].

What distinguished the CIW from this diverse field was its enormous resources. As Robert Kohler has noted, the CIW's extramural grants program alone invested over $100,000 per year between 1903 and 1920, a sum that exceeded the total of all other American grant funds put together [Kohler, 1991, pp. 15] Much of this sum supported expeditions. Although other agencies financed larger and more extensive expeditions than the CIW, few had the freedom to organize them with the spontaneity and energy that the CIW's resources allowed. Despite its own generous endowment, for example, the Smithsonian typically needed to secure outside patrons to support its expeditions [True, 1928].

From 1902 through the 1930s, Carnegie administrators supported expeditionary science in two ways. The first was by providing outright grants to individuals not connected with the CIW. The second was by funding the expeditionary requirements of its own scientists. Funds of the second kind for example paid for the extensive geomagnetic survey work promoted by Louis A. Bauer, the first director of the DTM. But in the Institution's earliest days, individual grants to outside researchers consumed a large share of the CIW's resources. Designed to reward "the exceptional man," a condition Andrew Carnegie stipulated in his endowment, the grants program was initially intended to be one of the CIW's operating arms, its permanent research divisions the other.

One of the CIW's first expeditions was Raphael Pumpelly's study of civilization and climate in Turkestan, Central Asia. Pumpelly was an ideal candidate for CIW patronage. A mining engineer who had directed the New England section of the U.S. Geological Survey and studied

The Earth, the Heavens and the Carnegie Institution of Washington
History of Geophysics Volume 5

in the U.S. and Europe, Pumpelly was skilled in several languages, competent in several disciplines, and brimmed with ideas about applying scientific methods to prehistoric archeology. To the key members of the CIW's initial board of trustees, including John Shaw Billings, a leading creator of the New York Public Library system, and Charles D. Walcott, director of the U.S. Geological Survey, Pumpelly was a model of Carnegie's "exceptional man." The trustees then largely directed the CIW, and Pumpelly quickly received funds for his expedition [Champlin, this volume; Champlin, 1993; on CIW policy see Kohler, 1991, p. 17; Reingold, 1991]. His plans beat out other compelling proposals, including David Starr Jordan's campaign to explore the ichthyology of the Pacific Ocean [Anonymous, 1902].

No less importantly, Pumpelly's ambitions resonated with another mandate Carnegie had imposed on the CIW: to encourage "investigation, research, and discovery, and the application of knowledge to the improvement of mankind" [see Merriam, 1938, p. 2504]. His plan to determine "the physical basis of the human history"—in essence, discovering whether Central Asia was the cradle of Indo-European civilization—was precisely the kind of fundamental question that the CIW's presidents and trustees sought to address. Pumpelly failed to find sufficient evidence for his ideas during his expeditions of 1903 and 1904, and the Russian Revolution of 1905 prevented him from returning to Turkestan. Nor did his limited contacts with European and Asian researchers influence the pattern of science in these locales, unlike in other instances of western scientists in their countries' colonies in the early twentieth century [Amstutz, 1975; Pyenson, 1985]. Nevertheless, Pumpelly's expedition demonstrated what became one of the CIW's great strengths: the ability to promote and sustain research that straddled the traditional boundaries of scientific disciplines. Pumpelly's team included William Morris Davis, the Harvard geomorphologist and geographer, the Berlin archeologist Hubert Schmidt, and Ellsworth Huntington, a graduate student later involved in climate research who, like Davis, became a major figure in American geography [Livingstone, 1993]. Science disciplines were rapidly professionalizing in the early twentieth century, making it difficult to organize Humboldtian enterprises of this kind at universities [Furner, 1975]. Aware of this trend, CIW leaders consciously advanced interdisciplinary research, and did so long before "cooperative" research became a cornerstone of American science ideology in the 1920s [Day, 1938; Bugos, 1989; DeVorkin, this volume; Kargon, 1977; Servos, 1983].

Fewer and fewer grants were made to outside researchers after 1905, as Robert S. Woodward, second president of the CIW, successfully shifted most operating funds to the emerging departments [Kohler, 1990, pp. 21-27]. Yet CIW leaders soon funded new expeditions at the Geophysical Lab and the DTM. Most visibly it did so through continued support for the *Carnegie*, which made seven voyages, acquiring diverse oceanographic and meteorological as well as geomagnetic data, before exploding in Apia harbor in 1929 [Ault, 1942]. But the CIW also took advantage of other planned expeditions to complete its mission: in essence, hitching a ride on other expeditionary vehicles. Such an occasion occurred in 1909, when DTM director Bauer supported a proposal by the Norwegian explorer and entrepreneur, Roald Amundsen, to include the Alaska researcher Harry Edmonds on an anticipated Arctic voyage. Edmonds was to make magnetic measurements on the CIW's behalf as Amundsen's ship drifted across the Arctic Ocean. The arrangement promised benefit to all parties concerned: Bauer would obtain valuable measurements, Edmonds would gain stature and exposure, while Amundsen would obtain the coveted endorsement of the CIW needed to promote the "scientific" value of his mission to prospective patrons. Nonetheless Amundsen's plan fell through. Peary beat Amundsen to the North Pole, and Amundsen quickly set his sights on winning the race to the South Pole, abandoning Edmonds and the CIW's legitimization in his haste to recruit a new team [Silverman and Smith, this volume]. But a pattern was set. Lloyd Berkner, later prominent in the DTM, sailed with Byrd's first expedition. And in 1921 the young Harald Sverdrup, fresh from another voyage with Amundsen, temporarily joined the CIW to reduce his geomagnetic observations [Spjeldnaes, 1976]. Such expeditions served as apprenticeships or even de facto graduate schools when few distinct academic programs of geophysics were in existence, and gave participants valued experience in data-collection, instrument operation, and scientific practice [Dupree, 1957; Bruce, 1987; Good, 1988; Levere, 1988; Pyne, 1986, p. 106].

In the 1920s the CIW expanded its range of collaborative partners to include the U.S. Government. One of the more important expeditions of this kind was the S21 submarine mission to obtain gravimetric measurements at sea. This 1928 expedition involved cooperation between the Dutch geodesist Felix A. Vening Meinesz, the petrologist Fred E. Wright of the CIW's Geophysical Lab, and Elmer B. Collins, the chief scientist of the Naval Hydrographic Laboratory. As in the case of Edmonds, cooperation among these parties promised benefit for CIW: Vening Meinesz's sensitive gravimeters allowed gravity measurements over the previously uncharted expanse of the oceans, a subject of considerable theoretical and practical interest to Wright. Typical of the approach advocated by CIW leaders, the expedition did not emphasize data-gathering so much as the testing of particular hypotheses. Wright and Arthur Day, director of the Geophysical Lab, knew well that gravimetry could test the theory of isostasy. In this instance, access to transportation was less critical than access to new instrumentation [Oreskes, this volume].

What permitted collaborative expeditions with partners such as the U.S. Navy to occur—or, better put, to take place with little institutional friction—was the willingness of CIW leaders to encourage applications of their research. Wright found no conflict in studying gravity data at sea as a test for isostasy, compared to its more practical applications in survey work and navigation; for him both were faces of the same coin. Likely this offers an insight into the culture of geophysics during this period: certainly the differences between "pure" and "applied" research were more apparent and troubling to researchers in physics and chemistry in the 1920s [Kevles, 1979; Servos, 1990; Oreskes, this volume]. But it also reveals that CIW leaders were comfortable with utilizing their research for practical ends. John C. Merriam, then CIW president, frequently justified research in oceanography for its potential applications to fishery and—himself a conservationist and former president of the Save-the-Redwoods League—the application of the life sciences to the problems of forestry [Merriam, 1926]. The culture of the CIW thus facilitated expeditionary collaborations with federal agencies with little of the professional discomfort that occasionally intruded into university-government collaborations.

DESIDERATA: TOWARDS A HISTORY OF EXPEDITIONARY SCIENCE

The example of the CIW can be used to illustrate what we need to ask about the role of expeditionary science in a larger historical context. There is much that we do not yet know about how scientific expeditions were organized and funded, or how their hard-won acquisitions—data, specimens, and photographs—were used and evaluated. Lacking too is an integrated approach to the study of expeditionary science, or its influence on laboratory-based experimental traditions and theory.

Expeditions launched by the CIW can serve as case studies for these important questions. To this end, I address below several topics that merit additional attention. It is not my purpose here to review the historiography of modern expeditionary science in detail. Nevertheless, I believe that broad attention to these issues will lead to an improved understanding of the intellectual, institutional, and social practice of science in the U.S. during the late nineteenth and early twentieth centuries.

EXPEDITIONS: INTEGRATING THEORY, PRACTICE, AND INSTITUTIONS

Expeditionary science is not a distinct branch of scientific practice, but rather a style of data collection that differs, sometimes in significant ways, from experiments done at the laboratory bench. Historians must consider not only the evolution of particular research problems and instruments designed to address them, as with laboratory science, but also a host of ancillary issues: access to a vehicle or other means of transport, acquisition of support funds for logistical and related costs, and the recruitment and training of field workers. This is not to say that they cannot be accommodated within the traditional frameworks of disciplinary and institutional history, only that appropriate questions must be asked of expeditionary science and its influence on the practice of science.

For instance, one may question how "transient institutions," such as expeditions, evolve into permanent institutions of science. In the nineteenth century, exploring expeditions catalyzed the development of permanent institutional structures: abundant collections brought back by the U.S. Exploring Expedition, led by Charles Wilkes, aided the founding of a national museum. Similarly, the Wheeler, Powell, Hayden, and King surveys led to the establishment of the U.S. Geological Survey in 1879 [Stanton, 1975; Pyne, 1986, p. 78; Manning, 1975]. Except in a limited number of cases, we have little information about how CIW-sponsored expeditions influenced subsequent research programs, or whether they led to permanent institutional successors. More work on this topic is needed, both within the CIW family and elsewhere on the institutional canvas.

Historians, however, should not unduly emphasize institutional issues. Also important is how individual expeditions, such as those of the *Carnegie*, were influenced by, and themselves influenced, research in geomagnetism and oceanography conducted by relevant scientific communities. Traditional histories of oceanographic expeditions often overemphasize particular voyages of discovery at the expense of intellectual history. To remedy this, Mott Greene [1982 and 1992] has suggested that historians construct a new history of oceanography based on the development of theoretical concepts, perhaps by comparing this branch of geophysics with its better studied neighbor, meteorology. By focusing on what problems motivated scientists on the *Carnegie*, for example, historians could better understand how expeditionary science contributed to theory growth, and to the development of new research problems [Good, 1988].

Increased concentration on the role of scientific expeditions launched by the CIW already suggests ways to adjust the traditional historiographical framework of modern American science. Despite the heightened attention historians have paid to the history of physical chemistry, geology, and geophysics in recent years, our perception of modern American science still reflects the preeminence of models drawn from physics and biology [Kevles 1979, Allen, 1975; Pauly, 1984; Maienschein, 1991]. Most historians unhesitatingly declare World War II the watershed for major federal investment in science, and they have many valid reasons for doing so. Yet too great an emphasis on this saltation disguises the considerable federal support already available to disciplines such as geology and geophysics well before 1940. Moreover, by concentrating on expeditionary science, historians can address little-explored questions about the practice of field science, the training of field workers, and the evaluation of expedition-generated data that were central to members of the geophysics community. The scientific enterprise looks very different from the deck of Amundsen's ship or the S-21 submarine than it does from a laboratory workbench [Goetzmann, 1972; Eliasson and Sörlin, 1990; Jonsson, 1987].

Lastly, as David DeVorkin has demonstrated, many profitable questions arise from focusing upon accommodations scientists made to obtain access to vehicles of exploration, be they the *Beagle*, military submarines, the *Carnegie*, or captured German V-2 rockets [DeVorkin, 1989; DeVorkin, 1992]. Continued efforts to evaluate the CIW's support for expeditionary science can help us better evaluate the significance of federal support for science in the period before World War II, and achieve a clearer overview of the political economy of science during this critical period in the growth of American science [Dupree, 1957; Dupree, 1990; Rosenberg, 1979].

SCIENTISTS AS MANAGERS: CHOOSING EXPEDITIONS TO INTERDISCIPLINARY RESEARCH

Despite the CIW's tremendous endowment, its funds were not limitless: Institution leaders still faced difficult decisions about which proposals to fund. Who made these decisions, and how were these decisions reached? Although influential trustees such as Billings and Walcott steered CIW policies during its earliest years, its presidents soon gained dominance. Between 1904 and 1938 there were just two presidents: Robert S. Woodward (1904-20), and John Campbell Merriam (1920-1938). Among their most significant moves was to steer Carnegie support towards interdisciplinary fields, rather than the disciplines nurtured at universities. Yet with limited exceptions, we know more about the work of the leaders of the CIW's principal branches, the DTM and the Geophysical Lab, than about the individuals who ran the CIW as a whole [Good, this volume; Yoder, 1989; Cornell, this volume]. This may owe to a tendency among historians to apply to the CIW a historiographical framework borrowed from studies of German institutes in the nineteenth and twentieth centuries. Directors of these institutes were seen as having authority in determining research programs, and, as specialists in their research fields, exercised intellectual control over these programs [Cahan, 1988]. What we do know of Woodward and Merriam is largely about their relations with the broader community of science: for instance, their efforts to increase the influence of the National Academy of Sciences and the National Research Council [Reingold, 1991; Kohler 1991].

It is particularly unfortunate that we have no in-depth studies of Merriam's long, eighteen-year reign as CIW president [for an introduction, see Dennis, 1990]. Merriam exercised greater influence over the CIW than is commonly recognized. Influenced by CIW traditions as well as George Ellery Hale's vision of interdisciplinary research, which found concrete expression in the observatory and astrophysical laboratories Hale constructed in California with CIW funds, Merriam became a dedicated advocate of cooperative research. Despite his limited training in the physical sciences—he was a paleontologist who had taught at Berkeley—Merriam delighted in organizing cooperative "attacks" on problems whose solution demanded expertise from several disciplines. Taking advantage of the diverse range of disciplinary talent available in the various CIW departments, Merriam organized, funded, and nurtured interdisciplinary research programs on such subjects as human behavior, seismology, and the origin of lunar craters [Merriam, 1921; Doel, 1993]. He encouraged the ionospheric research program pursued by Lloyd V. Berkner, supported the rocket research of Robert Goddard, immersed himself in the cosmic ray controversy between Robert Millikan and Arthur Compton, and promoted the exploration of cyclic phenomena in nature [DeVorkin, 1989; 1990, pp. 129-130; this volume]. No less importantly, Merriam shaped the expeditionary programs sponsored by CIW scientists. In one instance, Merriam initially rejected an expedition proposed by a CIW scientist to help test Alex du Toit's ideas about continental drift; he approved the plan only when multiple working hypotheses were formally included in its proposal [Oreskes, 1990]. Too much ought not be read into a single example, but it is extremely significant that Merriam imposed methodological criteria on the design of expeditions, and influenced the type of field practice performed by CIW scientists.

Much can be gained therefore from studying Merriam's approach to expeditionary science. His career offers to historians a means of moving beyond the limiting confines of traditional disciplinary histories. What other expeditions did he approve, for instance, and why? Equally interesting, which ones did he turn down? Future investigations of

Merriam might benefit from existing studies of science philanthropy in America, particularly Rockefeller Foundation support for science in the early twentieth century. Woodward and Merriam, to borrow the evocative language of Robert Kohler, were not administrators but managers of science, occupying a powerful organizational role that largely did not survive the massive shift to federal patronage for science after World War II. Recently Kohler, Finn Aaserud, Pnina Abir-Am, and Good among others have produced important studies of the research programs and policies that the best-known managers of American science, Wickliffe Rose and Warren Weaver, promoted in oceanography, geophysics, biophysics, and molecular biology within the Rockefeller Foundation's various operating agencies [Kohler, 1991; Aaserud, 1990; Abir-Am, 1982; Good, 1990]. Similar investigations of CIW leaders as managers of science are needed to evaluate the expeditions it launched, and to contrast the philanthropic cultures of these two major patrons of American science. It is equally important to note that such studies would add to our small but growing body of knowledge about hybrid, interdisciplinary fields, which despite their significant role in modern science have been scrutinized more by sociologists than by historians [Geison, 1981, Abir-Am, 1992].

IMAGES AS ARTIFACTS: THE PHOTOGRAPHY OF EXPEDITIONARY SCIENCE

Finally, a matter often neglected by historians of science is the role played by photography in scientific expeditions. Photographers accompanied CIW scientists on many of the Institution's more significant expeditions, a common practice with many expeditions from the late nineteenth century forward. Not infrequently CIW scientists themselves became photographers and photo-documentarists. Thousands of their images are preserved in the archives of the CIW's DTM. They reveal myriad subjects: the *Carnegie* in distant ports, portable scientific instruments, unusual landforms and landscapes, meteorological phenomena, and villagers in distant lands. These prints also document CIW scientists participating in Amundsen's arctic explorations and the 1928 S-21 submarine voyage. Among the most interesting images are those of CIW scientists and observers engaged in field work, or simply gathered before the camera. They constitute a veritable archival collection in their own right, comparable to the extensive holdings of the American Museum of Natural History. Yet relatively few historians scan such photograph collections systematically, which owes more to methodological practice than to occasionally limited cataloguing of source materials. Compared to the written documents, narratives, and correspondence that form the traditional foundations of narrative history, they are an under-utilized resource.

Why were these photographs made? On one level, the answer seems entirely straightforward: like postcards, they reminded expedition participants of where they had been. They documented the landforms or artifacts encountered and visually supplemented narrative accounts. Yet historians can use these images in several ways: for example, to probe how scientists visualized the subjects they investigated. As historians of photography have recently pointed out, photographs of geological landscapes are less objective and transparent than one might suspect. Rick Dingus has convincingly argued that Timothy O'Sullivan's famous landscape photographs, taken during Clarence King's Geological Survey of the Fortieth Parallel of 1867, were not simply objective illustrations of geologic phenomena, but instead show the influence of King's catastrophist theory of geology in their design [Dingus, 1982, pp. 62-63; Green, 1984]. Evaluating the training of expedition photographers is therefore important for understanding how scientific concepts and cultural values are transmitted and reinforced.

Images can be examined as evidence in other ways, as several historians have recently noted. Robert Levine has employed historical images to examine subjects poorly documented by traditional archival sources (a problem encountered by historians of expeditionary science), and Martin Rudwick has examined visual representations of deep geologic time to better understand the role of visualization in scientific practice [Levine, 1989; Rudwick, 1989, 1992]. Moreover, one can employ expedition photographs to assess better the accuracy and limitations of narrative accounts of laboratory practice [Latour, 1987; Golinsky, 1990, p. 498].

Posing the question of why expedition photographs were made suggests other lines of investigation. Did CIW officials use these photographs to persuade others to cooperate in scientific research, particularly patrons or agencies outside the inner sanctum of science? What ideological agendas did they reflect or inculcate, and what kinds of arguments were framed in accompanying captions? Many of the surviving photographs from CIW-sponsored expeditions fit into familiar patterns: explorers frequently are seen standing alone in hostile terrains, ranging from dense jungle to polar ice. Typically they are accompanied by their equipment, with their ships or transport visible in the distance. Such views were not merely scenic. A distinctly American iconography of exploration had emerged by the late nineteenth century, reflecting prevailing attitudes about the frontier, the notion of Manifest Destiny, and belief in the inevitability of technological process [Fowler, 1989; Hales, 1988; Ostroff, 1976]. These images

Images from scientific expeditions often feature a distinctive American iconography. Elements of familiarity in strange lands helped to reinforce ideas of technological progress for popular audiences. Expedition photographs also reveal unexpected details of explorations themselves, and can raise new questions about scientific and technical practice. The Gobi Desert expeditions of Roy Chapman Andrews for instance were the first to use motored vehicles in this remote region.

Top left: Black Canyon of the Colorado, Looking Down from Camp 8, Above, 1871; Timothy H. O'Sullivan, photographer. George M. Wheeler's Exploration of the Territories West of the 100th Meridian. (Photograph courtesy of Prints and Photographs Division, Library of Congress.)

Bottom left: Magnetic and electrical variation observatory at Bowdoin Harbor, Baffin Island, Canada, Spring 1922. This facility, erected by the MacMillan Baffin Land Expedition together with the CIW's Department of Terrestrial Magnetism, also permitted observations of tides, polar lights, and meteorological conditions. (Photograph courtesy of Carnegie Institution, Department of Terrestrial Magnetism.)

Top right: Expedition encampment, Tsetsenwang, Outer Mongolia, June 1922. From Roy Chapman Andrews archeological-paleontological exploration of the Gobi Desert, 1922-1930. (Negative no. 251088, Photograph courtesy Department of Library Services, American Museum of Natural History.)

of exploration thus had deep resonances with American culture by the twentieth century [Krause, 1982; Novak, 1980; Pyne 1986; Jussim and Lindquist-Cock, 1985; Goldberg, 1991; Hyde, 1990]. In the late nineteenth century, expedition leaders recognized the potential significance of photographs as instruments of persuasion: Wheeler spent a tenth of his budget for his 1871-73 Exploration of the Fortieth Parallel on photography in the hope of securing funds for future expeditions [O'Sullivan, 1983, p. vi]. Dramatic images of scientific exploration also helped to underwrite the cost of stratospheric and geophysical research from manned balloons in the 1930s, which flew under National Geographic Society sponsorship [DeVorkin, 1989]. Although Michael Dennis [1990] has recently demonstrated that CIW interest in photo-documentation continued through the post-World War II era, regrettably few studies of this kind are available. It is important to learn how these images were used to communicate the CIW's mission to a larger public, and to extend such studies to other institutions and to other fields of science.

Much work remains to be done on these topics. Examining the Carnegie Institution of Washington's support of science in the field promises to yield rich insights about the role of expeditions in the building of modern science.

ACKNOWLEDGEMENTS

I wish to thank Thomas D. Cornell, Denis Cosgrove, David H. DeVorkin, Rick Dingus, Don D. Fowler, Peter Hales, Estelle Jussim, John Lankford, and Spencer R. Weart for comments on earlier versions of this manuscript, and Shawn Hardy for assistance in locating one accompanying illustration. This work was made possible in part by NSF grant DIR-9112304.

REFERENCES

Aaserud, F., *Redirecting Science: Niels Bohr, Philanthropy and the Rise of Nuclear Physics*, Cambridge Univ. Press, New York, 1990.

Abir-Am, P., The Discourse of Physical Power and Biological Knowledge in the 1930s: A Reappraisal of the Rockefeller Foundation's "policy" in Molecular Biology, *Soc. Stud. Sci.*, *12*, 341-382, 1982.

Abir-Am, P., The Politics of Macromolecules: Molecular Biologists, Biochemists, and Rhetoric, *Osiris* [second series], *7*, 1664-191, 1992.

Abramson, H. S., *National Geographic: Behind America's Lens on the World*, Crown Publishers, New York, 1987.

Allen, G. E., *Life Sciences in the Twentieth Century*, Wiley, New York, 1975.

Amstutz, G. C., Pumpelly, Raphael, in C. C. Gillispie, ed., *DSB*, *11*, 209-211, 1975.

Anonymous, Proposed Explorations and Investigations on a Large Scale, *CIW Year Book*, *1*, 239-284, 1902.

Anonymous, *Annual Register of Princeton University*, p. 18, 1932.

Ault, J. P., *Scientific Results of Cruise VII of the Carnegie During 1928-1929 under the Command of Captain J.P. Ault*, CIW, Washington, D.C., 1942-1944.

Bruce, R. V., *The Launching of Modern American Science, 1846-1876*, Knopf, New York, 1987.

Bugos, G. E., Managing Cooperative Research and Borderland Science in the National Research Council, 1922-1942, *Hist. Stud. Phys. Biol. Sci.*, *20*, 1, 1-32, 1989.

Bunker, F. F., Cooperative Research, Its Conduct and Interpretation, in *Cooperation in Research*, CIW, Washington, D.C., 713 pp., 1938.

Cahan, D., *An Institute for an Empire: The Physikalisch-Technische Reichsanstalt, 1871-1918*, Cambridge Univ. Press, New York, 1988.

Cannon, S. F., *Science in Culture: The Early Victorian Period*, Dawson and Science History Publications, 1978.

Champlin, P., this volume.

Champlin, P., *Raphael Pumpelly: Gentleman Geologist of the Gilded Age*, Univ. of Alabama Press, Birmingham, 1993.

Cornell, T. D., this volume.

Day, A. L., An Adventure in Scientific Collaboration, in *Cooperation in Research*, CIW, Washington, D.C., 3-35, 1938.

Dennis, M. A., A Change of State: The Political Cultures of Technical Practice at the MIT Instrumentation Laboratory and the Johns Hopkins University Applied Physics Laboratory, 1930-1945, Ph.D. dissertation, Johns Hopkins Univ., Baltimore, 1990.

DeVorkin, D. H., *Race to the Stratosphere: Manned Scientific Ballooning in America*. Springer Verlag, New York, 1989.

DeVorkin, D. H., Defending a Dream: Charles Greeley Abbot's Years at the Smithsonian, *J. Hist. Astron.*, *21*, 121-136, 1990.

DeVorkin, D. H., *Science with a Vengeance: The Military Origins of Space Science*, Springer Verlag, New York, 1992.

DeVorkin, D. H., this volume.

Dingus, R., *The Photographic Artifacts of Timothy O'Sullivan*, Univ. of New Mexico Press, Albuquerque, 1982.

Doel, R. E., *Science on the Periphery: Interdisciplinary Research and Solar System Astronomy, 1918-1960*, Cambridge Univ. Press, New York, 1993.

Dupree, A. H., *Science in the Federal Government: A History of Policies and Activities to 1940*, Harvard Univ. Press, Cambridge, MA, 1957.

Dupree, A. H., Smithsonian Astrophysical Observatory: From Washington to Cambridge, *J. Hist. Astron.*, *21*, 107-110, 1990.

Eliasson, P. and S. Sörlin, Travel: Part of the Empirical Knowledge Program, *Uppsala Newsletter*, *13*, 5-7, 1990.

Fowler, D. D., *The Western Photographs of John K. Hillers*, Smithsonian Institution Press, Washington, DC, 1989.

Furner, M. O., *Advocacy and Objectivity: A Crisis in the Professionalization of American Social Sciences, 1865-1905*, Univ. of Kentucky Press, Lexington, 1975.

Geiger, R., *To Advance Knowledge: The Growth of American Research Universities, 1900-1940*, Oxford Univ. Press, New York, 1986.

Geison, G. L., Scientific Change, Emerging Specialties, and Research Schools, *Hist. Sci.*, *19*, 20-40, 1981.

Goetzmann, W. H., *Exploration and Empire*, Knopf, New York, 1972.

Goldberg, V., *The Power of Photography: How Photographs Changed our Lives*, Abbeville Press, New York, 1991.

Golinsky, J., The Theory of Practice and the Practice of Theory: Sociological Approaches in the History of Science, *Isis*, *81*, 492-505, 1990.

Good, G. A., The Study of Geomagnetism in the Late 19th Century, *Eos Trans. AGU*, *69*, no. 16, 218-228, 1988.

Good, G. A., The Rockefeller Foundation, the Leipzig Geophysical Institute, and National Socialism in the 1930s, *Hist. Stud. Phys. Biol. Sci.*, *21*, part 2, 299-316, 1991.

Green, J., *American Photography: A Critical History 1945 to the Present*, Harry N. Abrams, New York, 1984.

Greene, M., Oceanography's Double Life, *Earth Sci. Hist.*, forthcoming, 1992.

Greene, M., *Geology in the Nineteenth Century*, Cornell Univ. Press, Ithaca, 1982.

Hales, P. B., *William Henry Jackson and the Transformation of the American Landscape*, Temple Univ. Press, Philadelphia, 1988.

Hyde, A. F., *An American Vision: Far Western Landscape and National Culture, 1820-1920*, New York Univ. Press, New York, 1990.

Jonsson, U., Historical and Current Perspectives on Polar Research, *Uppsala Newsletter*, *8*, 1-3, 1987.

Jussim, E., and E. Lindquist-Cock, *Landscape as Photograph*, Yale Univ. Press, New Haven, 1985.

Kargon, R. H., Temple to Science: Cooperative Research and the Birth of the CIT, *Hist. Stud. Phys. Sci.*, *8*, 3-32, 1977.

Kohler, R. E., *Partners in Science: Foundations and Natural Scientists, 1900-1945*, Univ. of Chicago Press, Chicago, 1991.

Kohler, R. E., The Ph.D. Machine: Building on the Collegiate Base, *Isis*, *81*, 638-662, 1990.

Krause, R., Photography's Discursive Spaces: Landscape/View, *Artjournal*, 311-319, 1982.

Latour, B., *Science in Action*, Harvard Univ. Press, Cambridge, MA, 1987.

Levere, T. H., The Continental Shelf, and a New Arctic Continent, *Brit. J. Hist. Sci.*, *21*, 233-247, 1988.

Levine, R. M., *Images of History: Nineteenth and Early Twentieth Century Latin American Photographs as Documents*, Duke Univ. Press, Durham, 1989.

Livingstone, D., *The Geographical Tradition: Episodes in the History of a Contested Enterprise*, Blackwell Publishers, Cambridge, MA, 1993.

Maienschein, J., *Transforming Traditions in American Biology, 1880-1915*, Johns Hopkins Univ. Press, Baltimore, 1991.

Manning, T. G., *Government in Science: The U.S. Geological Survey, 1867-1894*, Univ. of Kentucky Press, Lexington, 1975.

Merriam, J. C., Researches on New Problems Considered, *CIW Year Book*, no. 20, 9-10, 1921.

Merriam, J. C., Earthquake Study as a World Problem for All Time, *CIW Year Book*, no. 25, 7-9, 1926.

Merriam, J. C., The Carnegie Institution of Washington, in *Cooperation in Research*, CIW, Washington, D.C., 2504-2527, 1938.

Miller, S., *Dollars for Research*, Univ. of Washington Press, Seattle, 1970.

Novak, B., *Nature and Culture: American Landscape and Painting, 1825-1875*, Oxford Univ. Press, New York, 1980.

Oreskes, N., Uniformitarianism and unity in American Historical Geology, 1922-1933, [draft presented at History of Science Society meeting, 1990; under review at *Isis*].

Oreskes, N., this volume.

Ostroff, E., *Photographing the Frontier*, Smithsonian Institution Press, Washington, D.C., 1976.

O'Sullivan, T., *Wheeler's Photographic Survey of the American West*, Dover reprint edition, New York, 1983.

Pauly, P., The Appearance of Academic Biology in late Nineteenth-Century America, *J. Hist. Bio.*, *17*, 369-397, 1984.

Perkins, J., *To The Ends of the Earth: Four Expeditions to the Arctic, the Congo, the Gobi, and Siberia*, Pantheon Books, New York, 1981.

Pyenson, L., *Cultural Imperialism and Exact Sciences: German Expansion Overseas, 1900-1930*, Peter Lang, New York, 1985.

Pyne, S. J., *The Ice: A Journey to Antarctica*, Ballantine Books, New York, 1986.

Reingold, N., National Science Policy in a Private Foundation: The Carnegie Institution of Washington, in N. Reingold, *Science: American Style*, pp. 190-223, Rutgers Univ. Press, New Brunswick, 1991.

Rosenberg, C., Toward an Ecology of Knowledge: On Discipline, Context, and History, in A. Oleson and J. Voss, eds., *The Organization of Knowledge in Modern America, 1860-1920*, pp. 440-455, Johns Hopkins Univ. Press, Baltimore, 1979.

Rudwick, M. S., Encounters with Adam, or at least the Hyenas: Nineteenth Century Visual Representations of the Deep Past, in J. R. Moore, ed., *History, Humanity, and Evolution: Essays for John C. Greene*, Cambridge Univ. Press, New York, pp. 231-251, 1989.

Rudwick, M. S., *Scenes from Deep Time: Early Pictorial Representations of the Prehistoric World*, Univ. of Chicago Press, Chicago, 1992.

Servos, J. W., To Explore the Borderland: The Foundation of the Geophysical Laboratory of the Carnegie Institution of Washington, *Hist. Stud. Phys. Sci.*, *14*, 1, 147-185, 1983.

Servos, J. W., *Physical Chemistry from Ostwald to Pauling: The Making of A Science in America*, Princeton Univ. Press, Princeton, New Jersey, 1990.

Silverman, S. and M. E. Smith, this volume.

Spjeldnaes, N., Sverdrup, Harald Ulrik, in C. C. Gillispie, ed., *DSB*, *13*, 166-167, 1976.

Stanton, W., *The Great United States Exploring Expedition*, Univ. of California Press, 1975.

True, W. P., *Explorations and Field-work of the Smithsonian Institution in 1927*, Smithsonian Institution, Washington, D.C., 1928.

Willis, B., *Research in China, Volume II: Systematic Geology*, CIW, Washington, D.C., 1907.

Yoder, H., Geophysical Laboratory, *CIW Year Book*, 144-197, 1989.

Sharing a Mountaintop: The Smithsonian Astrophysical Observatory on Mount Wilson

R. S. BRASHEAR
Huntington Library, 1151 Oxford Rd., San Marino, CA 91108

INTRODUCTION

For the first half-century after its 1904 inception, the Mt. Wilson Observatory of the Carnegie Institution of Washington (CIW) shared its mountain site with a field station of the Smithsonian Astrophysical Observatory (SAO). During this time, the chief SAO astronomer, Charles Greeley Abbot, and his assistants took advantage of the excellent Mt. Wilson observing conditions to conduct their program of solar research. Abbot's goal was to understand the amount of solar radiation and its patterns of variation and to relate these to climatic, meteorological, and agricultural conditions on Earth. Abbot's initial success was due in no small part to the efforts of the Mt. Wilson Observatory director, George Ellery Hale, in getting the SAO to come in the first place and in constantly providing much needed assistance during their early years.

The benefits of studying the establishment of the SAO station on Mt. Wilson are twofold. First, we can learn something of the nature of solar research during the early twentieth century. In particular, we can see the interplay between the use of solar studies in Hale's work on stellar evolution and Abbot's research on solar-terrestrial relations. The second benefit, and the one on which this study concentrates, is that we can learn more about Hale and his program for research at Mt. Wilson Observatory. By illuminating Hale's role as the prime mover in this episode, it becomes clear that the SAO's presence was a result of his constant efforts to fulfill his plans for solar research on Mt. Wilson.

Earlier studies of the SAO have not examined in detail the relationship between the SAO's Mt. Wilson field station and the Mt. Wilson Observatory. These works have either focused on other issues or covered too long a time span to examine closely the establishment of the SAO's Mt. Wilson field station [Jones, 1965, pp. 189-196]. Conversely, historical studies of Hale and the Mt. Wilson Observatory have not discussed why the SAO appeared on the mountain so quickly after the CIW established the Mt. Wilson Observatory [Adams, 1954; Wright 1966, pp. 159-196].

The Earth, the Heavens and the Carnegie Institution of Washington
History of Geophysics Volume 5

By concentrating on the relationship between the SAO and Mt. Wilson it becomes apparent that the origins of their mountaintop cooperation lay in 1902, the first year of the CIW. The presence of Hale and Samuel Pierpont Langley, the Smithsonian Institution secretary and SAO director, on the CIW's Astronomy Advisory Committee forged a conception of a solar observatory that would attack the problems which both found to be significant. When the solar observatory became a reality at the end of 1904, it was only natural that Hale, the chosen director, would turn to the SAO staff to fulfill the obligations of Langley's research program. Hale and Langley's backgrounds in solar physics will place this scenario in proper context.

SOLAR PHYSICS BACKGROUND

Astrophysics, the new astronomy of the late nineteenth century, was dominated by research in solar physics. As Meadows [1984] has pointed out, the Sun was a natural target of astrophysical investigators: it was bright enough for the new and often inefficient instruments, it was close enough to be studied in detail, it might lead to more knowledge about stars in general, and it was of prime concern for solar-terrestrial relations. The 1850s discovery of a correlation between sunspot frequency and changes in the Earth's magnetic field spurred interest in solar-terrestrial relations [Herrmann, 1984, p. 106]. Ensuing solar-terrestrial research proved fertile with a great deal of collaboration among solar physicists and those involved in terrestrial magnetism, an area on which much more historical work needs to be done [Good, 1985; Multhauf and Good, 1987; Good, 1990].

One of the first researchers in the U.S. who took up the study of the Sun and its possible effects on the Earth was Langley. Langley's interest in the Sun came about shortly after he became the director of the Allegheny Observatory near Pittsburgh, Pennsylvania, in 1867. During his first few years at Allegheny, Langley worked in the traditional manner. By 1873, however, he began to concentrate on the Sun. This was probably due to various factors, including poor observing conditions at Pittsburgh, the practicality of daytime observations, and the spectacular effect of solar eclipses which he had seen on various expeditions [Obendorf, 1969, pp. 65-78]. Langley's reputation in solar physics quickly grew, aided by his invention of the bolometer, which measured small amounts of heat, and his discovery of the vast extent of the infrared solar spectrum. By 1900, Langley, who had moved to the Smithsonian, and the Princeton astronomer Charles Augustus Young were the leading American solar physicists of their generation [Abbot, 1906; Walcott, 1912; Barr, 1963; Obendorf, 1969; Moyer, 1973; Hufbauer, 1991, pp. 67-68].

The discoveries in solar physics proved to be exciting to a new generation of astrophysicists as well, chief among them being Hale. Hale's doting father, a wealthy Chicago industrialist, encouraged his natural curiosity. Hale studied physics and astronomy at the Massachusetts Institute of Technology and quickly took to studying the Sun, due largely to the influence of Young [HL, Hale Collection, Hale to H. M. Goodwin, June 6, 1889, Oct. 1, 1890]. By age twenty, Hale, thanks to his father, had a fully-equipped astrophysical observatory at the family home and was becoming the shining light on the astronomical stage [Wright, 1966].

Hale's reputation grew quickly and he soon was professor at the new University of Chicago, helped in part by making his private observatory a part of the University. Hale soon moved within wealthy circles and had convinced the streetcar magnate Charles Yerkes to fund the world's largest telescope for the University. At the same time, Hale also founded the *Astrophysical Journal* with his good friend and fellow astrophysicist, James Edward Keeler. By 1900, Hale, though thirty-four years younger than Langley, had equaled if not surpassed Langley's stature within the astronomical community.

The foundation of the Carnegie Institution of Washington in 1902 was the next step in Hale's ascendancy in solar physics. Andrew Carnegie's bequest to the new Institution produced a major foundation for scientific research. The CIW's Trustees appointed committees in many fields to advise them on the best way to disburse their money. The committees included leading experts who prepared reports for the Trustees [Miller, 1970, pp. 166-181; Reingold, 1979].

The CIW appointed the leading American astronomer Edward Charles Pickering to chair the Advisory Committee on Astronomy. Pickering, director of the Harvard College Observatory, selected four eminent astronomers to complete the committee, a balanced group composed of two classically trained astronomers and two astrophysicists. The former were Lewis Boss, director of the Dudley Observatory in Albany, New York, and Simon Newcomb, former director of the U.S. Naval Observatory, recently retired from the Nautical Almanac Office, and a major figure in the early development of American astronomy. The two astrophysicists were Hale and Langley. Hale was chosen to be the secretary of the committee.

The two groups in the Advisory Committee did not always agree on the most pressing research problems facing astronomers. The key combatants were Boss and Hale. Newcomb, at the end of his career, did not actively debate. Langley had lost interest in active astronomical research and was trying to develop the first successful powered airplane. His only major astronomical concern was to provide for solar research. One important project that the Committee considered was the establishment of an observatory in the southern hemisphere. However, the astronomical project that most interested Hale was an observatory that would house his long-dreamed-of 60-inch (1.5 m) reflecting telescope. Boss, on the other hand, saw little need for another large reflector like those that had failed at Melbourne, Australia, and Lick Observatory. He also doubted the value of a new, large observatory even if it was located below the equator when many existing American observatories needed monetary support. However, Boss eventually supported the southern observatory scheme. The one uncontentious project was the solar observatory [CITA, Hale Papers, folder "Lewis Boss 1902"; HLMtW, Hale Papers, LCB10; LOA, Lewis Boss Correspondence, 1902-1904].

One reason the Advisory Committee favored a solar observatory was that Carnegie himself was partial to such a scheme. When Charles Doolittle Walcott, secretary of the CIW Trustees, visited Hale in August 1902 he told Hale of Carnegie's interest in astronomy. Carnegie had told Walcott that if the astronomers wanted a solar observatory and a large reflecting telescope and if they could achieve results comparable to those from the Lick and Allegheny Observatories, he would be willing to furnish as much as three million dollars. In particular, Carnegie showed an interest in a solar observatory idea proposed by Langley [HLMtW, Hale Papers, LCB10, Hale to Pickering, Aug. 15, 1902]. Shortly after the establishment of the CIW, Walcott contacted Langley at the Smithsonian informing him of the possibility of Langley receiving funding in order to assist him in his research. Langley replied that Carnegie might be interested in funding research on "the determination of the heat the Sun sends the Earth and the causes of its probable variation." Langley pointed out that this radiation affects every organism on Earth and that even great harvests and famine might be predictable from knowledge of its variation [SIA, RU34, 6.11, Langley to Walcott, Feb. 28, 1902]. Langley's letter probably had an impact on Carnegie and the Trustees even if they professed the importance of science in terms of basic research [HLMtW, Hale Papers, LCB10, Hale to Pickering, Aug. 15, 1902]. Langley's proposal was certainly in line with an important CIW guideline: the promotion of "[p]rojects of broad scope that may lead to the discovery and utilization of new forces for the benefit of man . . . " [Walcott, 1903, p. xxxvi]. Langley, however, did not propose that the research on solar variation be done at the Smithsonian Astrophysical Observatory. As secretary of the Smithsonian Institution, Langley was automatically on the CIW's Board of Trustees. He therefore felt that it would be a conflict of interest if he proposed a research project that would directly benefit himself. Instead, Langley proposed that, in accordance with Carnegie's plan, the "exceptional man" should be funded directly by the CIW to do the research [SIA, RU31, Langley to D. C. Gilman, Mar. 26, 1902].

Knowing that Carnegie and some of the CIW's trustees (James S. Billings and Walcott in particular) favored solar research, the members of the Advisory Committee were inclined to support them. Hale was certainly willing to go to bat for a solar observatory, but he preferred to develop a broader scope of solar research than just Langley's plan to study solar radiation in terms of solar-terrestrial relations. Over the years, Hale's interest in the Sun had focused on the Sun as a typical star. By understanding the Sun, Hale reasoned, we can begin to understand all the stars. The study of the changes that stars undergo, stellar evolution, began to dominate Hale's research agenda [HLMtW, Hale Papers, folder "International Solar Union," circular letter [1903]; Hale, 1908]. To Hale, then, the building of the 60-inch telescope (Hale already had its mirror since 1896, thanks to his father) was a more significant step forward for astrophysical research than Langley's proposed solar observatory. Nevertheless, Hale was quite willing to allow for Langley's solar radiation research for at least two reasons: Hale's basic interest in solar physics and his concern in having Langley, Carnegie, Billings, and Walcott as valuable allies in obtaining funding. At that time, however, Hale and the other Committee members felt no need to ask for an elaborate observatory to do pure solar research. Hale felt that both his and Langley's solar work could be done as a special expedition or even as part of a permanent southern hemisphere observatory, but not as a dedicated three-million-dollar solar observatory. As Hale explained to Boss,

> . . . I believe in providing means on a sufficient, though not extravagant, scale, for Langley to investigate the solar constant. Of course he would not go to South America to do this, but he might be able to send his chief assistant Abbot, who is thoroughly qualified to do the work in the very best way. The solar investigations which I shall propose for the southern observatory will not require a very elaborate equipment. I doubt if there would be any special advantage in making the observations in the southern hemisphere and would hope to find a suitable place in this country [HLMtW, Hale Papers, LCB10, Hale to Boss, second letter of Aug. 29, 1902].

In the report of the Advisory Committee to the CIW, therefore, the astronomers acknowledged the importance of a solar observatory, but they stayed with Langley's original plan of an eleven-year program to study solar radiation for a total cost of a half a million dollars [Pickering et al., 1903, p. 96].

The CIW Trustees, based on the astronomers' report, funded a site survey for the proposed solar and southern observatories. At the November 25, 1902, meeting, the Trustees presented five thousand dollars to a committee consisting of Boss, Hale, and William Wallace Campbell, director of the Lick Observatory [Walcott, 1904, p. xix]. William J. Hussey of the Lick Observatory was selected to explore potential observing sites. In April 1903 he set out to California and Arizona to examine locations for the proposed solar observatory. Two months later, Hale went west and together with Campbell, visited Hussey to take a first-hand look at the skies of southern California. When the three men met on June 25, 1903, Hussey was encamped with his telescope on top of Mt. Wilson, which rises 1742 m above sea level over Pasadena and the San Gabriel Valley. Hale was quite impressed with the observing conditions and, despite a few misgivings, he felt that it would be the best site for the solar observatory [Wright, 1966, pp. 166-170].

Hale's concept of a solar observatory crystallized after his trip to Mt. Wilson. Considering the favorable day and night observing conditions, Hale saw the solar observatory as the best location for an all-out attack on solar and stellar astrophysics. Both Hale's and Langley's solar research could be accommodated there and, most importantly for Hale, it was an excellent location for his 60-inch telescope. As a result the committee of Boss, Hale, and Campbell recommended the establishment of a Solar Observatory in order to accomplish three aims:

> (1) To measure the intensity of the solar heat radiation, and to determine whether it varies from perfect constancy during at least one sun spot period of eleven years. . . .
> (2) To bring to bear upon the solution of solar problems various modern methods of research, principally of a spectroscopic nature, which have not hitherto been applied with adequate facilities. . . .
> (3) To provide, through the construction of a large reflecting telescope, for the investigation of various problems of stellar evolution, intimately related to solar work, which existing instruments are inadequate to solve [Boss, Campbell, and Hale, 1904, p. 50].

This passage describes a combination of Hale's and Langley's research plans into one observatory, evolved from Langley's original temporary expedition into a permanent site for astrophysical research. Indeed, the proposal also included a recommendation to establish two auxiliary stations at different altitudes in order to study high altitude atmospheric absorption as a part of Langley's solar constant research [Boss, Campbell, and Hale, 1904, p. 66].

To the committee's disappointment, however, the CIW Trustees did not fund the proposed solar observatory. The funding needed to start the observatory ($150,000 for the first year) was deemed too large in view of the number of projects proposed by all the other committees. All that could be spared was $15,000, given to Hale in April 1904 to bring the Snow telescope from the Yerkes Observatory to Mt. Wilson [Adams, 1954, pp. 288-290; Wright, 1966, pp. 170-172]. Nevertheless, Hale was encouraged by some of the Trustees to continue pressing for the solar observatory. With this encouragement, and his growing desire to move his sickly children to the invigorating climate of California, Hale moved to Pasadena in December 1903 and soon set up shop on Mt. Wilson [HL, Hale Collection, Hale to H. M. Goodwin, Dec. 24, 1903; HLMtW, Hale Papers, folder "International Solar Union," Hale to A. Bélopolsky, Apr. 7, 1904]. Using a temporary telescope at first, he soon brought out the Snow telescope from Yerkes along with some of his staff: Walter Sydney Adams, George Willis Ritchey, Ferdinand Ellerman, and Francis Gladheim Pease. Before long, Hale and his staff were performing their initial solar investigations and building laboratory space and living quarters on the mountain.

Charles Greeley Abbot

Having established a presence on Mt. Wilson, Hale began looking at ways to begin Langley's research program. For this purpose, Hale contacted Langley's assistant, Charles Greeley Abbot, who was the acting head of the Smithsonian Astrophysical Observatory. Not surprisingly, Hale contacted Abbot instead of Langley, since Langley was no longer active in solar research. Equally important, were Hale's close relationship with Abbot and direct experience with Abbot's skill with instrumentation.

Abbot was born in Wilton, New Hampshire, in 1872 and educated in chemistry at MIT, Hale's alma mater. Both men obtained master's degrees there. In 1895, five years after Langley had founded the SAO, Abbot was recommended to him as an assistant. Abbot came to the SAO and quickly proved himself adept at working with Langley's quirky bolometers. Abbot made several improvements to the bolometer and its recording apparatus and was soon made "aide acting in charge" of the Observatory after Langley became secretary of the Smithsonian Institution [Warner, 1976; DeVorkin, 1990].

During the campaign to establish a solar observatory on Mt. Wilson, Hale knew that Abbot was the logical choice to carry out Langley's solar research plans. They had come to know each other well at the 1900 SAO solar eclipse expedition to Wadesboro, North Carolina. Hale noted Abbot's observational skills and was impressed with his cooperation generally. The two men were soon corresponding about coronal research and a lifelong friendship ensued. In August 1903 after Hale told Abbot about Mt. Wilson, Abbot visualized a program to study solar radiation. The main questions would concern the solar constant and a bolometric study of the sun's visible disk. He pictured three bolometric observing stations with two men at each site. He would be stationed at Mt. Wilson. The three sites would be Mt. Wilson, Mt. Whitney, and Lone Pine, all in California. The last two stations would study atmospheric absorption of solar radiation at two different altitudes. Abbot wanted Charles E. Mendenhall to supervise these two stations. An infrared physicist at the University of Wisconsin, Mendenhall had been Abbot's assistant in Washington [HLMtW, Hale Papers, Abbot to Hale, Aug. 10, 1903]. After several months on Mt. Wilson, Hale became concerned about the effect of atmospheric dust on solar constant measurements. Hale believed that solar constant work might have to be scaled down. He proposed that the Mt. Whitney and Lone Pine stations might be worthwhile only for a brief expedition and that daily solar constant measurements might be useless, except when there was little dust. However, Hale noted that bolometric work on radiation distribution over the Sun's disk would be unaffected by the dust. This last program was also more along the line of Hale's own research interests [HLMtW, Hale Papers, Hale to Abbot, Feb. 24, 1904].

In late 1904 the chances for the solar observatory improved. Hale learned that Alexander Agassiz had turned down a major CIW grant earmarked for him. Walcott confirmed this rumor and suggested that Hale renew the earlier request for the solar observatory. Hale personally presented the proposal to the CIW Executive Committee in October and was informally told afterwards by Walcott that they would soon recommend $150,000 for a solar observatory [Adams, 1954, p. 300; Wright, 1966, pp. 192-193].

With full funding, Hale asked Abbot to join him full-time as his second-in-command on the mountain. For expertise on solar radiation and general companionship, there was no other choice for Hale. Abbot, though tempted to join his friend on Mt. Wilson, could not leave Washington permanently:

There being no advantage to be gained as regards salary or health . . . the question is, can I really do enough better work with you to justify me for leaving friends, aged mother, home, city comforts, companionship with my wife; and for grieving Mr. Langley in his old age after he has really made me what I am, and for upsetting this Observatory just as it is growing to its greatest usefullness [sic] [CITA, Hale Papers, Abbot to Hale, Nov. 14, 1904].

Not inconsequentially, Abbot also expected to become director of the SAO upon Langley's death. Abbot had a scientific justification, too, which clearly differentiated between Hale's and the SAO's research agenda:

As regards [the question of location] . . . the advantage of your location depends in part on the view one takes of solar research. You look upon the sun as the nearest star, and the most interesting investigation to be that of the evolution of the universe. I grant the absorbing interest of the theme, but must confess to a growing and more absorbing interest in the earth as an abode of life, so that I rather look upon the sun as a source of energy whose variability deserves attention, but whose nature is an unnecessary consideration to my problem, which is to discuss the temperature and moisture of the earth as a function of solar radiation, terrestrial radiation, reflection, absorption, emission of air, dust, water vapor, CO_2, clouds, emission and absorption of gases as a function of temperature; convection as a function of temperature quality and density of the atmosphere, etc. In all this I see a study that will take at least all my life, and for most of which a cloudy station is as good as a clear one [CITA, Hale Papers, Abbot to Hale, Nov. 14, 1904].

Hale's plan to have Abbot join him on Mt. Wilson failed. However, Hale's program required research on solar radiation at Mt. Wilson, and he could think of no better person than Abbot to do it. So if Abbot would not join him as a staff member, perhaps he could come under the auspices of the SAO. For Hale, this plan had its advantages since he could still "meet the obligation of doing radiation (solar constant) work at little expense, as Langley would probably furnish both the men and the apparatus" [HL, Hale Collection, Hale to H. M. Goodwin, Feb. 22, 1905].

THE SMITHSONIAN ON MT. WILSON

On December 20, 1904, Hale learned that the CIW Trustees had approved his proposal and appropriated $150,000 for the first year. Hale was to direct the new CIW Solar Observatory on Mt. Wilson. Hale could begin bringing the SAO people in to attack the problem of solar radiation.

Hale soon tried to convince Langley to mount an expedition there. Hale wrote to Langley in January 1905 and put his cards on the table. He pointed out that he needed to carry out the solar research he originally proposed to the CIW and that Abbot was the only man qualified to do the job. As a result, Hale wrote,

I therefore wish to inquire whether you would consider the possibility of sending an expedition from the Smithsonian Institution to Mt. Wilson for such a period of time as might seem to you desirable, with the understanding, in case you wish to have this done, that we will arrange to continue the work when it becomes advisable for your men to return to the Smithsonian Institution [HLMtW, Hale Papers, folder "Smithsonian Institution," Hale to Langley, Jan. 6, 1905].

That same day, Hale wrote to Abbot of his scheme. He enclosed a copy of his letter to Langley and argued strongly that Abbot should go to Mt. Wilson, at least to start up the research program. After pointing out the advantages of Mt. Wilson, Hale concluded by saying:

Please let me know as soon as possible what you think of the scheme, and if you favor it, do your best to induce the Secretary to send an expedition. I think we can contribute some funds, if necessary, but I am, of course, anxious to keep the expenditure as low as possible, especially as the plan assumed that the Smithsonian Institution is to be given full credit for the work [HLMtW, Hale Papers, Hale to Abbot, Jan. 6, 1905].

Hale here attacked Langley directly and via Abbot.

Hale, normally very diplomatic, had gone a little too far in his invitation to Langley and the SAO. Abbot told Hale that the latter's suggestion to continue the solar radiation work after the SAO expedition ended was tantamount to asking for free instruction in bolometry and all the glory from the research to be gleaned from their association with the SAO. As a result, Langley considered sending an expedition for only three months in order to determine the true value of the solar constant, and not a longer one to determine the variability of the solar radiation [HLMtW, Hale Papers, folder "Smithsonian Institution," Langley to Hale, Jan. 18, 1905]. Abbot confided to Hale: "I don't understand his mental attitude, unless he thinks it is unsafe to leave you and I together too long" [CITA, Hale Papers, Abbot to Hale, Jan, 17, 1905]. Abbot thought that the only result of such an expedition would be to show that

Langley's value for the solar constant was incorrect. Langley had long maintained that the solar constant was three calories per square centimeter per minute. Abbot felt, and Langley admitted that the method of data reduction was flawed. Abbot believed two calories was closer to the mark, but he did not challenge Langley's value until after the latter's death [HLMtW, Hale Papers, Abbot to Hale, Feb. 12, 1905].

Langley decided slowly whether to send an expedition to Mt. Wilson. Meanwhile, Abbot took the liberty of making plans and preparing instruments without telling Langley. This surreptitious preparation also promised to keep Langley out of Abbot's way because, as Abbot told Hale,

> I would find great difficulty in getting ready, what with 2 or 3 hours conference every day with him and some more time lost in writing him letters and getting up fake rehearsals out on the lot.

Hence, Abbot planned to wait until March to push Langley to decide [HLMtW, Hale Papers, Abbot to Hale, Feb. 12, 1905].

While Langley took his time, Hale moved quickly with his research program. Hale, unlike Abbot, felt confident that Langley would accept his invitation and so his planning assumed this [HL, Hale Collection, Hale to H. M. Goodwin, Feb. 22, 1905]. Knowing that the chances of Abbot mounting a full-scale solar radiation program at Mt. Wilson were slim, Hale approached Mendenhall about joining him. Mendenhall had recently turned down the directorship of the Allegheny Observatory. Hale quickly made him an offer:

> I want to ask whether you would take a position in the Solar Observatory at a salary of $3,000. Abbot cannot come here on account of the demands of the Washington work, though he may come out for a short expedition this summer. I want a man who can take up bolometric and spectroscopic work on the Sun, under the splendid conditions that exist on Mt. Wilson, and I think you are just the man for the place [CITA, Hale Papers, Hale to Mendenhall, Mar. 3, 1905].

If Mendenhall accepted the offer, Hale's program would be in excellent shape. Abbot could bring one of his absolute continuous pyroheliometers with him in the summer and this instrument could be set up on the mountain and left there for a full solar cycle of solar constant data. The pyrheliometer, which measured the heat the Earth receives from the Sun, absorbed the incoming heat and, in its various guises, measured it with a mercury thermometer or with a Wheatstone Bridge circuit. With the solar constant taken care of, the remaining solar radiation work, as Hale noted to Abbot,

> could then be limited to special studies of the Sun's disk. I have asked Mendenhall to take charge of this work, but as he has recently declined the Allegheny position I fear he will not come. So far, however, I have not received his reply. It will be a decided disappointment if he does not accept my offer [HLMtW, Hale Papers, Hale to Abbot, Mar. 14, 1905].

Unfortunately, Mendenhall did refuse. Mendenhall was getting married soon and he did not want to be on the mountain away from his new wife [CITA, Hale Papers, Mendenhall to Hale, Mar. 26, 1905].

As Hale expected, Langley finally decided to send an expedition to Mt. Wilson. Langley had also changed his mind about the nature of the expedition's research. He no longer was interested in redetermining the value of the solar constant and instead wanted to concentrate on its variations,

> so far as this can be done in the limited time in disposal. To determine the constant anew would require a prolonged study of the considerations which brought about the values of the Mt. Whitney expedition [of 1881]. To determine its variations, it is comparatively immaterial, I think, whether the absolute values be correct, within a considerable range [SIA, RU34, 6.13, Langley to Abbot, Mar. 6, 1905].

Langley apparently realized that an absolute measurement would indicate that his value of the solar constant was incorrect. Not wanting to go to great lengths to defend or retract his value, Langley felt that the expedition could at least obtain evidence of solar variability, regardless of the solar constant's true value.

Abbot immediately wrote to Hale telling him that Langley's letter of acceptance was forthcoming. Langley's letter, Abbot explained, would contain details about the expedition and if Hale wanted to make any changes, he should send them directly to Abbot. Langley's letter was the fourth draft and Abbot felt that there was no point in making a fuss over details with Langley at that point [HLMtW, Hale Papers, Abbot to Hale, Mar. 22, 1905]. Hale finally received Langley's letter and saw the details to which Abbot had alluded. The SAO plan called for Abbot and an assistant to arrive sometime after May 1. Their equipment needed two temporary shelters, one for the spectrobolometer (which utilized a bolometer to measure the amount of radiation in various regions in the solar spectrum) and another for the pyrheliometer. If the shelters could be ready when Abbot arrived, the expedition would run more smoothly. The SAO would eventually bear the construction expense [HLMtW, Hale Papers, folder "Smithsonian Institution," Langley to Hale, Mar. 22, 1905]. Hale quickly guaranteed room and board for Abbot and his assistant at the "Monastery," the astronomers' lodging on Mt. Wilson. But heavy rains and other factors ended up delaying the construction [HLMtW, Hale Papers, folder "Smithsonian Institution," Langley to Hale, Mar. 27, 1905].

Despite this delay, the shelters illustrate an important point in the relationship between Mt. Wilson and the SAO. The Mt. Wilson Solar Observatory, already in place on the mountain, was an invaluable resource for the SAO expedition. Abbot was impressed with the workers there, from the research staff down to the carpenters. He noted that Hale "has been very successful in finding young enthusiastic workers in all lines, so that things fairly fly along when they take hold" [SIA, RU31, Abbot to Langley, May 20, 1905]. The SAO, which depended on funding from the U.S. Government, had to plan well in advance to have Congress approve their budget for the upcoming fiscal year. Whatever funding the SAO did not get had to be made up for by private donations. With Hale's assistance, the SAO could have the people at Mt. Wilson prepare its last minute arrangements as well as provide lodging and the SAO could pay them back after its budget was approved. If the SAO attempted a large-scale expedition on its own, last minute changes or additions were not nearly as easy. For the 1905 and future SAO expeditions, the financial issue often rose to prime importance. The SAO expedition was also a financial boon to Mt. Wilson because Hale could count on the former to pay for the expense of men and apparatus to do the research [CITA, Hale Papers, Hale to R. S. Woodward, May 15, 1905].

THE 1905 SMITHSONIAN EXPEDITION

With Abbot coming to Mt. Wilson in the summer, Hale felt some relief. He would not have to start up a solar radiation research program from scratch. Rather, Abbot would start it and Hale tend to other questions. He told a friend,

> I think this a very advantageous scheme for all concerned, as it will enable us to find what advantages may be expected from a continuation of this work on Mt. Wilson, and just how much we ought to do ourselves at the conclusion of the expedition [CITA, Hale Papers, Hale to H. H. Turner, Mar. 29, 1905].

Abbot and his assistant Leonard R. Ingersoll arrived on Mt. Wilson on May 10 and found their equipment waiting for them. Originally, Abbot had asked Mendenhall to accompany him on the expedition [CITA, Hale Papers, Mendenhall to Hale, Mar. 26, 1905]. Mendenhall had recommended Ingersoll, one of his students, to take his place. Even though their shelters were not yet constructed, they started their observations with the pyrheliometer two days after their arrival [SIA, RU31, Abbot to Langley, May 11, 1905]. The spectrobolometric outfit took longer to prepare, but it was operating on June 5 (Figure 1). The pyrheliometer was used up to twice a week for five to eleven hours at a time for solar constant measurements (Figure 2). The energy spectrum of the radiation from the solar surface and sunspots was examined with the spectrobolometer. The weather was exceptionally good and Abbot thought that, as far as lack of clouds, lack of wind, and excellent seeing, Mt. Wilson was an ideal location [Abbot et al., 1908, pp. 7-8].

Abbot found the location desirable in ways other than scientific. With no large reflecting telescopes on Mt. Wilson at that time, there were no night observations. Consequently, the solar astronomers collected in the Monastery in the evening and spent the night in conversation. Hale and Abbot, both extraordinary raconteurs, delighted their colleagues with various stories and parodies. Abbot made a few life-long friendships in 1905, perhaps none closer than that with Walter S. Adams. Adams, who eventually became Hale's number-two man and the Observatory's director on Hale's retirement in 1923, was a New Hampshire Yankee, just like Abbot. They also found that they had both attended Phillips Academy in Andover, Massachusetts. Their friendship was cemented when Adams came to Abbot's defense in an argument on the proper way to eat breakfast cereal (they were of the "sugar before cream" variety of cereal eaters) [Abbot, 1958, p. 90]. All in all, the 1905 expedition was a happy one for Abbot and an experience which he looked forward to repeating.

Near the end of July, Abbot felt that he needed longer than the three months originally specified by Langley and he requested to stay until November. Fortunately for Abbot, Langley was away in Europe and the acting secretary, Richard Rathburn, noting that Langley would probably emphasize completing the observations over an arbitrary time limit, told Abbot to use his own discretion [SIA, RU34, 6.13, Rathburn to Abbot, Aug. 5, 1905]. Abbot accepted this offer and replied that he would stay on Mt. Wilson until November 1, unless he heard otherwise [SIA, RU31, Abbot to Rathburn, Aug. 11, 1905]. When Langley returned in October, he wrote to Abbot confirming the extension until November. Indeed, Langley mentioned that if the solar variation seemed great, Abbot might stay longer though this seemed unlikely [SIA, RU34, 6.13, Langley to Abbot, Oct. 11, 1905]. Nevertheless, the door

Fig. 1. The SAO spectrobolometer shelter on Mt. Wilson, Aug. 8, 1905. All photographs are reprinted with permission of the Huntington Library and the Observatories of the CIW.

Fig. 2. The SAO pyrheliometer shelter on Mt. Wilson, Aug. 8, 1905.

seemed to be open for future Smithsonian work on Mt. Wilson.

The 1905 SAO expedition was not yet over before Hale began planning for Abbot's return. While writing Langley to tell him of the successes enjoyed by Abbot, Hale was also enlisting the CIW President, Robert S. Woodward, to lobby on behalf of a second SAO expedition. Hale told Woodward that

> I feel sure that a letter from you will accomplish the desired purpose in view of the friendly relations so fortunately established between the two Institutions by your conference with Prof. Langley last winter . . . The importance of continuing the work is great, and I trust Prof. Langley will appreciate this [CITA, Hale Papers, Hale to Woodward, July 24, 1905].

Woodward was only too happy to continue the cooperation between the CIW and the SAO. Woodward and Walcott were veterans of the U.S. Geological Survey. As Kevles has pointed out, both Woodward and Walcott brought with them to the CIW the Survey's emphasis on institutional cooperation [Kevles, 1978, p. 110]. That spirit, together with the CIW's desire not to undertake "anything that is being well done . . . [or] better done by other agencies," made the SAO's expeditions to Mt. Wilson very amenable to Woodward [Walcott, 1903, p. xxxvii]. Woodward quickly wrote to Langley after visiting Mt. Wilson in July, and asked if the latter would consider sending Abbot back "during the summer season for a series of years." Woodward also appealed to Langley's pride by mentioning the importance of Abbot's work and that, after all, it was Langley who first thought of establishing a CIW solar observatory [SIA, RU31, Woodward to Langley, Aug. 4, 1905]. After Abbot returned to Washington in November, Hale advised him to see Woodward, whom he called "our best ally," to talk over the matter of returning to Mt. Wilson [HLMtW, Hale Papers, Hale to Abbot, Dec. 19, 1905]. In the meantime, Hale and Abbot decided that some of the Smithsonian's materials and equipment should remain on the mountain in the likelihood of a return visit [Abbot, et al., 1908, p. 9].

RETURNING TO MT. WILSON

Upon Abbot's return to Washington, Langley's health suddenly took a turn for the worse when he suffered a stroke in November. Langley was confined to bed, the right side of his body paralyzed although his mind was unaffected. Abbot was not allowed to discuss the Mt. Wilson expedition with him at all and decided to wait until February when it was believed that Langley might recover to some degree. At that point Abbot could recommend that the SAO return to Mt. Wilson in the summer of 1906 [HLMtW, Hale Papers, Abbot to Hale, Dec. 27, 1905]. As the new year began, however, Langley had not left his home. Nevertheless, it appeared that he was still involved in running the Smithsonian. As a result, Abbot told Hale that it was time to plan the summer expedition and avoid a last-minute rush. Abbot advised Hale or Woodward to write Langley and pointed out the details that should be raised with the secretary. As Abbot orchestrated matters in Washington, however, he started worrying about the Mt. Wilson Observatory taking too much advantage of the SAO. Knowing that Hale was willing to take up the solar radiation work if the SAO did not continue with its expeditions, Abbot cautioned Hale:

> [I t]hink if you propose to continue the work you ought to get *your* man trained this time. Also why should the S.I. put more money into the buildings to be given to you if you intend to use them [HLMtW, Hale Papers, Abbot to Hale, Jan. 18, 1906, emphasis in original]?

Hale acknowledged his friend's admonition and asked him to watch for an assistant who would be willing to remain behind when Abbot left [HLMtW, Hale Papers, Hale to Abbot, Jan. 30, 1906].

Planning for the 1906 SAO expedition became even more complicated when Langley suffered a second stroke and died on February 27. Abbot began to worry about the existence of the SAO and expected to get opposition for its continuation from the U.S. Weather Bureau. Even with this new pressure, however, he still tried to get the 1906 expedition started by lobbying acting secretary Rathburn [HLMtW, Hale Papers, Abbot to Hale, Mar. 21, 1906]. Not getting any word from either Rathburn or Woodward, however, Abbot began to despair about going to California [HLMtW, Hale Papers, Abbot to Hale, Mar. 28, 1906]. What Abbot did not know was that Woodward had convinced Rathburn to renew the SAO presence on Mt. Wilson. At the end of March, Hale received a letter from Rathburn announcing that the SAO expedition would begin around May 1 and continue through July and perhaps longer [HLMtW, Hale Papers, folder "Smithsonian Institution," Rathburn to Hale, Mar. 29, 1906].

Abbot returned to Mt. Wilson in May, along with his assistant of the year before, Ingersoll. They continued their pattern of observations of 1905 along with a study of the reflecting power of clouds. For this latter study, Abbot had a temporary tower built on a promontory just below the Monastery which let him point his instruments down at the cloud banks in the canyons below. The expedition lasted well into the autumn with the last observations being done in October. Once again, Hale, Adams, and Abbot (who had

Fig. 3. The Monastery on Mt. Wilson and the SAO observing station with its fifty-foot tower (at lower right). Note the runway tracks between the two structures. May 19, 1932.

been appointed acting director of the SAO on July 1) enjoyed each other's company and all felt that some good scientific work was being done [Abbot et al., 1908; Abbot, 1908, pp. 152-153].

A Permanent Station

After the two successful expeditions Abbot, who was named director of the SAO on March 1, 1907, prepared for the second volume of the Observatory's *Annals*. Abbot and his assistants assembled their results since the first volume, published seven years earlier. The effort required for this endeavor proved enormous and Abbot felt constrained to remain in Washington throughout 1907. In the meantime, Hale carried out observations with Abbot's instruments. H. K. Palmer, of the Lick Observatory, was assigned to continue studying the radiation pattern over the solar disk. Charles M. Olmsted was assigned to carry out solar constant observations. Some of Palmer's observations were sent to Abbot for use in the preparation of the *Annals*. Olmsted's work did not fare as well, especially when he discovered that the pyrheliometer was leaking mercury and his observations were discontinued [HLMtW, Hale Papers, Hale to Abbot, Mar. 29, 1907; Aug. 28, 1907; Abbot to Hale, Apr. 9, 1907; Sep. 4, 1907]. Palmer and Olmsted were only temporary staff members, however, and Hale apparently had no plans at that time for a complete takeover of the SAO work. Hale expected Abbot to return soon, considering that Abbot was quite pleased with his results at Mt. Wilson. After examining the 130 solar constant observations made in 1905 and 1906, Abbot found that the value of the solar constant ranged from 1.93 to 2.14 calories per square centimeter per minute, to within one percent. Abbot considered this good evidence for the variability of the solar radiation output [Abbot, 1908, pp. 152-153]. But he had to continue these observations to remove any doubt [Abbot et al., 1908, pp. 6-7].

As director of the SAO, Abbot began planning for a long-term presence on Mt. Wilson. The plans for a permanent field station became apparent with the third expedition which Abbot scheduled for 1908. Abbot was again in charge but this time his assistant was Loyal B. Aldrich, another student of Mendenhall's. They arrived on the mountain on May 11, and occupied the same 1905 shelters. Soon, however, the SAO would have a new permanent structure. Hale, whose Observatory was on land leased from the Mt. Wilson Toll Road Company, arranged with them for a lease which would give the SAO 930 m^2 of land for ninety-five years for one dollar, on the promontory where the temporary tower had been erected in 1906 to study reflection from clouds. The lease was signed by all parties in the autumn of 1908. That November the construction of the SAO's permanent structure was completed: an L-shaped building with four tall piers for the eventual placement of a tower for a vertical telescope. The building contained enough room for the spectrobolometer, pyrheliometer, a darkroom, and an office.

Starting in 1908, every year saw an SAO presence on Mt. Wilson during Abbot's tenure as Observatory director. The SAO station continued to grow. The next year a small wooden cottage was completed as observer's quarters. This allowed Mrs. Abbot to come too since no quarters were provided for women at the Monastery. In 1912, the planned fifty-foot (15.2 m) tall metal tower with a coelostat at the top was finally erected (Figure 3). As in Hale's sixty-foot (18.3 m) and 150-foot (45.7 m) tower telescopes, the height of the coelostat above ground reduced the disturbing effect of convection currents in the air caused by ground heating. That same year, in order to move equipment more easily up and down the steep mountainside, Abbot had a steel runway installed between the station and the Monastery. Most of these construction projects involved the staff of the Mt. Wilson Observatory to some degree. Observatory staff helped effect these improvements to the SAO station at cost and Abbot did not need to pay the money up front for the workers.

The research done by Abbot and his staff at the Mt. Wilson station from 1908 to 1910 concentrated primarily on studying solar variability. During the summers and autumns for these three years they measured the solar constant every day they could. Their results confirmed Abbot's belief in solar variability:

> the most probable conclusion is that the sun actually varies from day to day in its output of radiation within limits of from 5 to 10 per cent in quantity and in irregular periods of from 5 to 10 days [Abbot et al., 1913, pp. 14-15].

Abbot tried hard to remove any doubts about errors in his observations. In 1909 and 1910 he went on an expedition to the summit of Mt. Whitney, the highest mountain in the United States at that time, to see if he was correcting properly for atmospheric absorption at Mt. Wilson. He concluded that the Mt. Wilson results were valid [Abbot et al., 1913, pp. 11-12]. Abbot then led expeditions to Algeria in 1911 and 1912 in order to make sure that the local conditions had no effect on the Mt. Wilson results. The Algerian results, Abbot wrote, "prove conclusively, as we think, the variability of the sun" [Abbot et al., 1913, p. 18]. Abbot also attempted to correlate the solar variation with his bolographic studies of the distribution of radiation over the visible solar disk. The fifty-foot tower gave him an excellent image of the Sun for the bolographic work. The results were difficult to interpret and seemed to be contradictory. Eventually, however, Abbot believed he had seen through the data and found a correlation separate from presumed long-period changes in solar radiation (equivalent to the eleven-year sunspot cycle) and the short-period irregular fluctuations. In the case of long-period changes, an increase in contrast of the brightness between the center of the solar disk and its limb was associated by Abbot with high solar constant values, and for short-period changes the opposite was true. Abbot thought that the long-period relation could be explained by a change in solar temperature and the short-period relation by a change in the Sun's opacity [Abbot et al., 1922, pp. 17-18; HLMtW, Hale Papers, Abbot to Hale, Mar. 28, 1914].

After 1910 Abbot began to doubt the conditions on Mt. Wilson. He kept noticing an increase in haze and cirrus clouds during the day there. Abbot found the sky to be particularly cloudy in 1917 and 1918 as "rain fell frequently and a great many clouds came up" [Abbot et al., 1922, p. 34]. The cloudiness continued in 1919 and Abbot became discouraged. He found that "the station is quite unsuitable for 'solar-constant' work in winter and spring months owing to cloudiness." Wanting to compare solar constant observations regularly with simultaneous observations at Washington, Abbot decided to look for a new permanent station [Abbot, et al., 1922, p. 42]. Abbot believed that he had found a good location at Harqua Hala Mountain, Arizona, and in 1920 started moving the solar constant equipment. Abbot, however, did not completely dismantle the Mt. Wilson station, for it was still an excellent location in which to perform bolographic work, stellar energy spectra research, and solar cooker investigations. The Mt. Wilson facilities were so well equipped and provided for by Hale and his staff compared to the primitive surroundings at Harqua Hala that only solar constant observations were done at the latter site [Jones, 1965, pp. 208-209]. Abbot also wanted to keep visiting Hale and Adams. Ironically, Abbot found the weather even worse for solar constant research at Harqua Hala, and in 1925 moved that field station to Table Mountain, California, only fifty kilometers northeast of Mt. Wilson. Eventually, Abbot set up other field stations on Mt. Montezuma in Chile, Mt. Brukkaros in Namibia, and Mt. Saint Katherine in the Sinai, to obtain year-round measurements of the solar constant. The Mt. Wilson station, however, was maintained for Abbot or his assistants to perform other non-solar constant research.

Abbot's research on solar variation and his applications of it to the Earth's weather came under attack from a number of critics, mainly meteorologists and statisticians. As DeVorkin [1990] has pointed out, their chief criticism was his interpretation, not his observations themselves. Nevertheless, Abbot maintained his program throughout his tenure as SAO director and when he became secretary of the Smithsonian Institution in 1928. Abbot's extensive solar-terrestrial research program finally ended in 1954 when the SAO was moved from Washington and merged with the Harvard Observatory [Doel, 1990]. The observing station on Mt. Wilson remained until its demolition around 1960 by the U.S. Forest Service. Abbot's measurement of the solar constant, 1.94 calories per square centimeter per minute, has stood the test of time, but his evidence of solar variability has not. Ironically, the Sun has been found to be variable in its solar radiation output, but the amount of variation, determined by instruments onboard spacecraft, was too small for Abbot's earthbound instruments to detect. His results were undoubtedly not real but due to instrument noise and the fact that corrections for atmospheric extinction cannot be made with sufficient precision to detect the actual variations [Newkirk, 1983; Hufbauer, 1991, pp. 259-305].

CONCLUSION

The Mt. Wilson observing station of the SAO was an important part of Abbot's program to study the variation in the Sun's solar radiation. Although it was at first only to be a site for an expedition, the excellent observing conditions and Hale's assistance led to its being made a permanent station. Nevertheless, it was Hale who was responsible for the SAO coming in the first place. Although Langley knew of the value of a mountain station, he was convinced that the most urgent need was a very high station and a lower station near it to take into account the effects of atmospheric absorption. And even when Abbot showed that a single observing station sufficed, he was not convinced that Mt. Wilson would be so much better than Washington for solar constant studies. It was Hale's persistence along with his growing friendship with Abbot that finally brought about the SAO expedition to Mt. Wilson. Hale had two major objectives in bringing Abbot to Mt. Wilson. First was his obligation to the CIW to fulfill his proposal for a solar observatory. Second, and perhaps more important, was Hale's belief that Abbot's bolometric work would prove valuable to his research on the Sun and stellar evolution. While solar constant studies were not particularly important to Hale, the study of the radiation over the Sun's disk was, and he knew that Abbot could not perform this work under the poor sky conditions at Washington, D.C. The SAO might have eventually established permanent field stations in various locations, but it was Hale who enabled them to begin as early as they did and show Abbot the importance of mountain observing.

ARCHIVAL SOURCES

For this article I used the George Ellery Hale Papers which are split between the California Institute of Technology Archives (CITA) and the Huntington Library's

Mt. Wilson Collection (HLMtW). These papers are, for the most part, available together on a microfilm edition. The items on microfilm reels 1-80 are at CITA and those on reels 81-100 are at HLMtW. The Hale Papers are organized by correspondent; a particular letter to or from Hale and a correspondent will therefore be located in the folder with that correspondent's name. In the case of exceptions to this rule, I have noted in the citation the folder in which the letter can be found (e.g., letters to and from Langley in the "Smithsonian Institution" folder). Due to the original organization of Hale's and Adams's Papers, it is not unusual to find correspondence between Adams and others in Hale's Papers and vice-versa. The ten letterpress copy books in the Hale Papers have been abbreviated as "LCB" together with the appropriate number. For the letterpress copy books in RU34 of the Smithsonian Institution Archives, I have included the series number and volume number (e.g., series 6, vol. 11 is rendered as 6.11). The Hale Collection at the Huntington Library is separate from the Hale Papers and the two should not be confused. A microfilm copy of the Hale Collection is on deposit at the Niels Bohr Library of the American Institute of Physics.

REFERENCES

Abbot, C. G., Samuel Pierpont Langley, *Astrophys. J.*, *23*, 271-283, 1906.

Abbot, C. G., Researches of the Smithsonian Expedition on Mt. Wilson, *CIW Year Book [for] 1907*, No. 6, pp. 152-153, 1908.

Abbot, C. G., *Adventures in the World of Science*, 150 pp., Public Affairs Press, Washington, D. C., 1958.

Abbot, C. G. et al., *Annals of the Astrophysical Observatory of the Smithsonian Institution*, 2, 245 pp. Government Printing Office, Washington, D. C., 1908.

Abbot, C. G. et al., *Annals of the Astrophysical Observatory of the Smithsonian Institution*, *3*, 241 pp., Government Printing Office, Washington, D. C., 1913.

Abbot, C. G. et al., *Annals of the Astrophysical Observatory of the Smithsonian Institution*, *4*, 390 pp., Government Printing Office, Washington, D. C., 1922.

Adams, W. S., The Founding of the Mount Wilson Observatory, *Publ. Astron. Soc. Pacific*, *66*, 267-303, 1954.

Barr, E. S., The Infrared Pioneers—III. Samuel Pierpont Langley, *Infrared Phys.*, *3*, 195-206, 1963.

Boss, L., W. W. Campbell, and G. E. Hale, Report of Committee on Southern and Solar Observatories, *CIW Year Book [for] 1903*, No. 2, pp. 5-70, 1904.

DeVorkin, D. H., Defending a Dream: Charles Greeley Abbot's Years at the Smithsonian, *J. Hist. Astron.*, *21*, 121-136, 1990.

Doel, R., Redefining a Mission: The Smithsonian Astrophysical Observatory on the Move, *J. Hist. Astron.*, *21*, 137-153, 1990.

Good, G. A., Geomagnetics and Scientific Institutions in 19th-Century America, *Eos Trans. AGU*, *66* (27), 521, 524-526, 1985.

Good, G. A., Scientific Sovereignty: Canada, the Carnegie Institution and the Earth's Magnetism in the North, *Sci. Canadensis*, *14*, 3-37, 1990.

Hale, G. E., *The Study of Stellar Evolution*, 252 pp., Univ. of Chicago Press, 1908.

Herrmann, D. B., *The History of Astronomy from Herschel to Hertzsprung*, 220 pp., Cambridge Univ. Press, 1984.

Hufbauer, K., *Exploring the Sun, Solar Science since Galileo*, 370 pp., Johns Hopkins Univ. Press, Baltimore, 1991.

Jones, B. Z., *Lighthouse of the Skies*, 339 pp., Smithsonian Institution, Washington, D. C., 1965.

Kevles, D. J., *The Physicists*, 489 pp., Alfred A. Knopf, New York, 1978.

Meadows, A. J., The New Astronomy, in *Astrophysics and Twentieth-Century Astronomy to 1950: Part A*, edited by O. Gingerich, pp. 59-72, Cambridge Univ. Press, 1984.

Miller, H., *Dollars for Research: Science and Its Patrons in Nineteenth-Century America*, 258 pp., Univ. of Washington Press, Seattle, 1970.

Moyer, D. F., Samuel Pierpont Langley, in *DSB*, *8*, edited by C. C. Gillispie, pp. 19-21, Charles Scribners' Sons, New York, 1973.

Multhauf, R. P., and G. A. Good, *A Brief History of Geomagnetism and a Catalog of the Collections of the National Museum of American History*, 87 pp., Smithsonian Institution Press, Washington, 1987.

Newkirk, G., Jr., Variations in Solar Luminosity, *Ann. Rev. Astron. Astrophys.*, *21*, 429-467, 1983.

Obendorf, D. L., Samuel P. Langley: Solar Scientist, 1867-1891, Ph.D. dissertation, Univ. of California, Berkeley, 1969.

Pickering, E. C., et al., General Plan for Furthering Special Researches in Astronomy, Appendix A to Report of Advisory Committee on Astronomy, *CIW Year Book [for] 1902*, No. 1, pp. 94-104, 1903.

Reingold, N., National Science Policy in a Private Foundation: The Carnegie Institution of Washington, in *The Organization of Knowledge in Modern America, 1860-1920*, edited by A. Oleson and J. Voss, pp. 313-341, Johns Hopkins Univ. Press, Baltimore, 1979.

Walcott, C. D., Proceedings of Executive Committee, *CIW Year Book [for] 1902*, No. 1, pp. xxx-xl, 1903.

Walcott, C. D., Report of Executive Committee on the Work of the Year, *CIW Year Book [for] 1903*, No. 2, pp. xv-lix, 1904.

Walcott, C. D., Samuel Pierpont Langley, 1834-1906, *Biog. Mem. Nat. Acad. Sci.*, *7*, 245-268, 1912.

Warner, D. J., Charles Greeley Abbot (1872-1973), in *Am. Phil. Soc. Year Book, 1975*, pp. 111-116, American Philosophical Society, Philadelphia, 1976.

Wright, H., *Explorer of the Universe*, 480 pp., E. P. Dutton & Co., New York, 1966.

A Fox Raiding the Hedgehogs: How Henry Norris Russell Got To Mt. Wilson

David H. DeVorkin
National Air and Space Museum, Smithsonian Institution
Washington, D.C. 20560

A Discipline in Transition

American astrophysics around 1900 was rapidly becoming the world's leader in the collection of data about the stars. It was a cartographic, data-driven effort best described as astronomical spectroscopy. Countless solar and stellar spectra were classified into groups, studied for line identifications to determine the presence of the chemical elements, and measured for line shifts to reveal motions in the line of sight [Menzel, 1972; Meadows, 1972; DeVorkin and Kenat, 1983; Hufbauer, 1991]. Akin to the naturalist, the typical American professional astronomer was a collector and classifier. Instead of museum shelves and cases, astronomers stored their systematic observations in plate vaults and letterpress log books, and displayed them in catalogues sponsored by universities and observatories.

Astronomers of that day had assimilated Kirchhoff's laws, and so were able to equate celestial spectra to qualitative form and substance; they could tell that iron existed in the atmospheres of the sun and stars, for instance, but they could not say how much was there, relative to other elements. There was no robust model for how stars themselves were related, nor were there commonly accepted laws of how radiation and matter interacted that could act as guide.

The evident success of the Darwinian theory made it a provocative model for other disciplines, and astronomy was no exception [Hofstadter, 1955; Russett, 1976; Numbers, 1977, p. 100; Kohn, 1985]. For one thing, it caused the problem of the development of stars and stellar systems, present since Newton, to become a central issue in astronomy. The astronomer George Ellery Hale, who more than anyone else was the architect of the transformation of American astronomy in the 20th Century, defined his astronomical agenda in 1908 as the solution to the problem of how stars developed. Along with his contemporaries, Hale appropriated the term evolution to point out that "the idea of evolution and development is by no means new to the astronomer . . . it may nevertheless be maintained that it has occupied a more important position since Darwin published his great work." [Hale, 1908, p. 2]

Within the American astronomical community, however, even by 1920, one would be terribly hard put to find someone pursuing problems like the development of stars using state of the art mathematical physics. Although it was paid lip service as the rationale behind collecting astronomical data, there was no equivalent to the Cambridge School, embodied in Arthur Stanley Eddington, that was rapidly showing how stars behaved as rational physical systems. And there were only a very few places following the English/German tradition of comparing celestial spectra to laboratory spectra to derive the physical characteristics of the stars by experimental means [Meadows, 1972; Herrmann, 1984; Becker, forthcoming].

By the 1930s, however, American astrophysics was definitely changing into a highly rational, theory-driven enterprise, embodying elements of both the English and German schools, where celestial spectra were compared to laboratory phenomena and were interpreted in terms of how light interacts with matter, using the most modern physical theory available. Although a robust theory of stellar evolution still eluded astronomers, there were now many new useful paths to take, guided by discoveries in quantum and nuclear physics. Astronomers were by then confident that they were on the right track. [DeVorkin and Kenat, 1983, Hufbauer, 1981]

How this transition took place is the central question that has been directing my research into the scientific life of the Princeton theoretical astrophysicist Henry Norris Russell. Russell, notorious as a nervous bundle of boundless energy, was both an agent and beneficiary in this transition. Already upon obtaining his PhD in 1900, Russell's career was distinguished by his "fox" mentality, seeking out hot topics ripe for plunder. He enjoyed fruitful professional associations first with E. C. Pickering of Harvard College Observatory, and later with George Ellery Hale, the builder of the Yerkes and the Mt. Wilson Observatories. As a result, Russell had access to the world's greatest repositories of astrophysical data, and became possibly the most effective consumer of those data painstaking collected by their single-minded staff hedgehogs. My discussion

The Earth, the Heavens and the Carnegie Institution of Washington
History of Geophysics Volume 5

centers on Russell and Hale, how their shared vision for the future of astronomy differed from Pickering's, and how this shared vision brought Russell to Mt. Wilson.

TAPPING THE WEALTH

Russell's view of what was "hot" in American astronomy comes from his only job application. In April 1905, 27-year old Russell, recuperating in Capri after an extended postdoctoral stay at Cambridge University, knew that a faculty position in astronomy at Princeton would soon become open. He wrote to his old Princeton instructor and now university president Woodrow Wilson that what Princeton really needed was a good astrophysicist:

> It is in this branch of Astronomy that the greatest advances of recent years have been made . . . and the most important unsolved problems of the science are at present astrophysical; - for example, the constitution of the Sun, the explanation of the different types of stellar spectra, . . . All these problems lead up to the greater one of stellar evolution. . . . [WW, Russell to Wilson, April 6, 1905]

Russell argued that his experiences at Cambridge, where he had been deeply influenced by George Darwin and where he had learned the "modern methods of research," placed him in the best possible position to answer these ultimate questions; he was just the man to harvest the vast store of astronomical data that lay fallow in America:

> Thousands of photographs of the stars are being taken in America every year . . . But only a very few of the plates are being measured. In this respect European astronomers are far ahead of us . . . Plenty of people in America can take good photographs, but there are not more than half-a-dozen men in the country who are familiar with the modern ways of measuring them, and of reducing the result to a useful form . . . [WW, Russell to Wilson, April 6, 1905]

Russell wished to emulate George Darwin, who argued that observational studies were best designed with the solution of particular problems in mind [HNR/P, Box 70.1, Darwin, Dynamical Astronomy lecture notes, 22 January 1903], but also Jacobus Kapteyn, director of Groningen, who was rapidly organizing a network of observatories into a cooperative program to conduct samplings of selected areas in the sky to explore problems in the structure and dynamics of the sidereal universe. Russell particularly admired Kapteyn's ability to get others to gather up the data he wanted, something Russell, from his proposal to Wilson, wished to do in his own career.

Princeton hired Russell, and indeed, he set out to do just what he said he'd do. Invited by E. C. Pickering to exploit Harvard's vast spectroscopic and photometric resources, Russell combined these data with his stellar parallaxes measured at Cambridge to test a new theory of stellar evolution, later to be embodied in his famous Diagram identifying the existence of giant and dwarf stars [DeVorkin, 1984]. The important point for us here is that Russell acquired data to verify a theory - his Giant to Dwarf theory of stellar evolution which he had adopted from the speculations of another fox, the Briton Norman Lockyer. This was how Russell worked: his research was directed to the solution of particular problems. But it was not Pickering's mode of practice, nor was it of the average astronomer of that day.

HALE AND PICKERING: CONTRASTING AGENDAS FOR ASTRONOMY

Pickering was a revolutionary in the 1880s and 1890s, organizing the first large-scale systematic spectroscopic and photometric programs in the United States as Harvard College Observatory's fourth director [Jones and Boyd, 1971]. Trained as a physicist, Pickering was appalled at the paucity of astrophysical data after he took the directorship in 1877, and so dedicated his energies to filling that void. By the turn of the century his staff had amassed more spectra and magnitudes than the rest of the world's observatories combined, yet there was no agreement on which of the two dozen spectroscopic classification systems, or which of the several systems of photometric magnitudes, would survive [DeVorkin, 1981].

Possibly because of his need to protect his investment, or possibly because he knew that Harvard could not do the job alone, Pickering worked to establish a national philanthropic fund for the support of astronomical research in America [Miller, 1970; Jones and Boyd, 1971, p. 418ff; Plotkin, 1978]. Although Pickering's financial efforts were moderately successful [Plotkin, 1978, 1990; Reingold, 1977; Kohler, 1991] he also labored to supply American observatories with standardized photometric equipment, and thus by the turn of the century had garnered enormous equity among American astronomers working under moderate means. In 1895, Pickering's rationale for supporting astronomy generally was identical to his agenda for Harvard. The strength of American astronomy, Pickering argued, centered upon:

> . . . the undertaking of large pieces of routine work, and the employment of numbers of inexpensive assistants whose work is in a great measure mechanical, such as copying and routine computing [Pickering, quoted in Plotkin, 1990, p. 47]

Pickering's ideal "factory observatory" was not unlike those found elsewhere at the time, for data accumulation was the rule of the day [Lankford (private communication); Osterbrock, et al., 1988, ch. 8; Smith, 1991; Boltzmann, 1992]. In 1903, Pickering was elected President of the Astronomical and Astrophysical Society of America, succeeding Simon Newcomb. He also became chairman of the Advisory Committee on Astronomy to the new Carnegie Institution of Washington. In his new national role, Pickering redoubled his efforts to establish a means to support routine operations of observatories led by established senior astronomers by raising a multi-million dollar fund which would be administered by the Harvard Corporation for the good of astronomical research. Predictably, directors of other major observatories, like Hale, and Lick's W. W. Campbell, objected to the plan, as did the venerable Newcomb, and Pickering lost out on Carnegie's largess. He fared no better in his proposals to the Rockefeller Foundation in 1906, which led him to modify his Harvard-centrism by adopting the plan created by Science editor and publisher James McKeen Cattell whereby a "Committee of One Hundred" under the auspices of the American Association for the Advancement of Science would both raise and dispense funds [Plotkin, 1978; 1990; Reingold, 1977]. Although this latest gambit fared little better, by 1910, at age 64, still Astronomical Society president, still chairman of the core committees constituting the Committee of One Hundred, and now a rising force in the American Philosophical Society as well as gatekeeper for numerous funds for astronomical research administered by Harvard, the American Academy and the National Academy of Sciences, Pickering was clearly the most influential astronomer in America.

Unlike Pickering, George Ellery Hale believed circa 1900 that the way to astronomical progress lay through linking the laboratory to the stars. At Yerkes he invited physicists to adapt their tools to astronomy, and at Mt. Wilson, made possible by his successful tapping of the vast resources of the Carnegie Institution of Washington, Hale hired physicists to build spectroscopic laboratories. Hale also differed from traditional practice in American astronomical institutions by arguing that observational programs should be directly related to the solution of significant problems, like the evolution of the sun and stars. Although observatory directors argued that their programs would inform such ultimate questions, few like Hale actually designed observational programs with such considerations in mind.

In 1906, Hale told Simon Newcomb that astronomers were too specialized and had to be encouraged to move away from "the details and the merely technical elements of investigation." [LC, Newcomb Papers, Box 25, Hale to Newcomb, March 21, 1906] Hale's new world in southern California was one within which Carnegie's "exceptional man" was sought out and supported. Carnegie officers created a multi-tiered program to establish large-scale observatories and laboratories fostering cooperative research, led by these visionaries, and, to stimulate their growth, added "a system of research associates comprising a considerable number of picked men who might spend all or nearly all of their time in investigations under the auspices of the Institution" [CIW, Moulton Papers, Woodward to Moulton, September 19, 1906; Kevles 1978, p. 83; Reingold, 1979; Dennis, 1991, pp. 136-137].

Like Pickering, Hale thought about priorities for American astronomy at the turn of the century. In time, this interest expanded until the organization of science became his overriding passion. Hale's drive to revive the National Academy of Sciences is well known [Kevles, 1968; Reingold, 1977]. In his push to establish the National Research Council as primary agent mobilizing the nation's scientists, wherein he had to absorb Cattell's and Pickering's Committee of One Hundred, and circumvented Edison's Naval Consulting Board at the threshold of World War I, Hale managed to create not only a new mechanism to organize science for wartime needs, but he managed as well to find a new mechanism whereby American science itself might be organized in peacetime.

To illustrate how the latter was accomplished, we focus here on how Hale worked to change the future of astronomy in America. Hale's ascendancy in astronomy, made possible largely by the vast resources of the Carnegie Institution of Washington, coupled with his political acumen, not only tipped the balance of power in American astronomy, but changed the nature of astronomy itself. To appreciate how astronomy changed, we need to know a bit about Hale's agenda for his Mt. Wilson Solar Observatory.

HALE'S MT. WILSON

Continuing a plan started at Yerkes, Hale attracted specialists to his observatory not only to exploit their knowledge of useful techniques and broader insights, but to allow them the chance to influence the research agenda of the observatory itself. [Wright, 1966, p. 139; Brashear, Sharing a Mountaintop, this volume; Brashear, 1993]. Appropriate to his Carnegie patronage, Hale sought out competent staff for his observatory to engage in large-scale cooperative programs which combined the talents of laboratory spectroscopists with instrument builders and telescope observers adept in spectroscopy, photometry, radiometry, and photography.

Beyond fostering the application of new instrumentation and technique, Hale wanted to strengthen his observatory's connection to the new physics. There were many lines of research opening up at Mt. Wilson, and Hale knew that more contact with modern physics was required to make physical sense of what his staff were finding from purely empirical observations. He knew these efforts had to be connected somehow into a coherent picture, and chose stellar evolution as his cal to arms. His original charge for Mt. Wilson was to attack the problem of stellar evolution "along converging lines, leading from solar, stellar, and laboratory investigations" [Hale, 1908, p. x; Wright, et al. 1972, p. 117].

But prior to World War I there was little convergence at Hale's observatory. His bureaus worked more or less in parallel but with little cohesion. Try as he might, during this time Hale was unable to attract prominent European physicists versed in the new physics to be staff members at Mt. Wilson [Wright, 1966, p. 139]. The agent Hale sought out, who would systematically enhance and fundamentally refashion the research agenda of the Mt. Wilson spectroscopic staff, had to speak the language of astronomy, but had to also know modern physics. The man he had in mind for this job was Henry Norris Russell. To Hale, Russell became pivotal to this closer alliance with physics and chemistry.

Just how and why Hale chose Russell as his torchbearer is explored in the remainder of this paper. Real contact between the two began in 1917 during Hale's campaign to establish the National Research Council not only as the wartime agent for mobilizing science, but as a post-war forum for supporting and directing American science. Here he found in Russell a kindred spirit and ally, especially in his protestations with the old guard in astronomy, still led by Harvard's E. C. Pickering.

"SOME PROBLEMS IN SIDEREAL ASTRONOMY . . . "

In 1916, Pickering was still President of the Astronomical and Astrophysical Society of America, was still chairman of the Committee of One Hundred, chairman of its Committee on Research, and chairman of its Sub-Committee on Astronomy, which had just canvassed American astronomers asking them what they would do with $1000 per year for five years [ECP/HUA, HUG 1690.6.5 "Corresp, Chiefly with Committees and Societies 1911-1922," Box 1, folder 325A]. True to Hale's plan to absorb these committees into the NRC structure, Pickering became chair of an NRC astronomy committee that would assess preparedness among astronomers, and report on the needs of astronomy after the war. Pickering, who shared Cattell's lack of enthusiasm for involvement in the war, decided to carry over his Committee of One Hundred activity by conducting another survey of observatory directors first. His 1917 NRC survey revealed to him just what an earlier survey had, that:

> In almost every case, the demand was for more assistants to aid in extensive routine observations. A relatively very large increase of output could thus be secured [Pickering, 1917].

In short, Pickering's 1917 draft report for the NRC committee was unchanged from his plan for Harvard astronomy in 1895. Although this view was shared by other members of his committee, not all agreed with him. His two most passionate detractors were Hale and Russell. Both disagreed with Pickering's conclusions after seeing preliminary drafts, and in the process, found that they thought alike. Russell and Hale argued that Pickering emphasized routine work far too much, and Russell, responding to Pickering's earlier survey for the Committee of One Hundred, argued further that the type of manpower required to do work in astronomy was changing. In December 1916 Russell told Pickering that "One Harlow Shapley"—a "good and experienced man"—was better than "two or three ordinary girls." [HNR/P, Russell to Pickering, December 14, 1916] And in February 1917, responding to Pickering NRC questionnaire, Russell's opinion of his former graduate student increased; now Shapley was worth 10 computers [HNR/P, Russell to Pickering, February 27, 1917].

Hale told Pickering that the report had to put "at least as much emphasis . . . upon the formulation and solution of the chief problems of astronomy, as upon the collection of data relating mainly to the bright stars . . . " [CIT/GEH, Hale to Pickering, March 16, 1917]. Responsible observatory directors, Hale warned, had to be ready to take advantage of new opportunities, especially "in those closely related fields from which important suggestions may be derived." Many new lines of work in physics were just opening up, and "should not be ignored in preparing schemes of astrophysical research." What was needed, Hale went on, was a "radical change" in outlook . . .

> these and many similar advances warn us that the empirical procedure of the recent past must continue to yield place to studies based upon a physical foundation . . . " [CIT/GEH, Hale to Pickering, March 16, 1917].

Pickering didn't budge in the face of this criticism, and published his report as he saw fit. But Pickering's NRC circular was an embarrassment to Hale, who took quick steps to remedy the situation. First, Hale asked Russell to follow up with an essay that was ultimately titled "Some Problems of Sidereal Astronomy" and then planned to follow this with reports by specialists who were generally not from Harvard. Pickering was deeply angered by this move, but apparently did not fight it directly, although he tried at least once to influence the Carnegie Institution against Hale [Charles W. Eliot Papers, UAI.5.150 Box 376, Pickering to Eliot, August 8, 1917].

In another year, Pickering would be dead, and Russell would write his obituary several times. Russell hailed Pickering as a pioneer in the new astronomy, building up a mass of data that he shared willingly with everyone, Russell most of all. But Russell's NRC essay "Some Problems in Sidereal Astronomy," prepared for Hale in 1917 and published in 1919 after wartime pressures abated, left little doubt about its author's view of astronomical research:

> The main object of astronomy, as of all science, is not the collection of facts, but the development, on the basis of collected facts, of satisfactory theories regarding the nature, mutual relations, and probable history and evolution of the objects of study [Russell, 1919, p. 212].

After outlining in detail the many areas of astronomy in need of theoretical elucidation, Russell concluded that:

> . . . it must not be forgotten how important a place *theoretical investigations* will occupy in the solution of the larger problems of sidereal astronomy. The increasing observational data are already furnishing just those guides which point the skilled mathematician in the right direction, and these indications have been very successfully followed, especially by certain members of that "Cambridge School" which combines keen mathematical analysis with a thorough knowledge of modern physics. Results of remarkable generality have already been obtained." [Russell, 1919, p. 273]

By late 1917, knowing that they were of one mind on many things, largely through their wartime experiences working at the NRC, Hale and Russell corresponded frequently and personally. Usually, Hale wrote to Russell for advice about Mt. Wilson's research programs, and Russell on two occasions wrote asking for Hale's advice about taking the directorships first of the Yale University Observatory and then of the Harvard College Observatory.

RUSSELL AT MT. WILSON

Hale consistently advised Russell not to accept the offers from Yale and Harvard. He knew well that Russell's health was barely better than his own; both suffered frequent breakdowns and required extended periods away from all pressure [HNR/P, Hale to Russell, October 20, 1920]. In response, Russell, who was never too excited about moving but thought the offers would provide him with better access to data he desired, started to ask Hale if he might obtain the data he wanted from Mt. Wilson. In June 1919, Hale said yes, and repeated the invitation the following year when Russell was faced with the Harvard offer.

Hale proposed Russell for a Carnegie Research Associateship in late 1919, but his budget was already overtaxed by Michelson's projects, and R. S. Woodward turned the request down. [GEH/CIT, Hale to Russell, January 19, 1920] Hale knew, however, that Woodward was about to retire, and that the incoming president John C. Merriam would be more supportive of any move to increase attention to the analysis of data using mathematical and physical theory [HNR/P, Hale to Russell October 20, 1920; Dennis, 1991, 149-153; Good, Vision of a Global Physics, this volume]. Indeed, when Merriam succeeded Woodward in January 1921, the climate changed. Hale, always the manipulator, also found a new source of funds: Shapley's salary would be available because, largely through Russell's and Hale's efforts, Shapley secured the very job at Harvard that Russell had turned down [HNR/P, Hale to Russell, January 19, 1921; Gingerich, 1988].

The question remained: what would Russell actually do during his visits? Throughout their discussions, Russell and Hale talked about cooperative research only in generalities, although it was clear that Hale wanted Russell to acquaint his staff with modern physical theory. Beyond that, Russell would have the run of the place [HNR/P, Russell to Hale, March 23, 1921]. The rhetoric both had used to define the future of astronomy—solving problems like the life-cycles of the stars—remained largely that, just rhetoric. Great strides had been taken by Russell and Eddington, who, stimulated by Russell, had "proposed a problem-solving strategy in which astrophysical theory would mediate between physical hypotheses and astronomical evidence" [Hufbauer, 1981, p. 303]. But there was little astronomers at Mt. Wilson could do immediately after World War I to attack the problem of stellar evolution.

On the other hand, a series of papers just appearing from a wholly different quarter demonstrated how Mt. Wilson's store of astrophysical and laboratory data could at last be combined. By December 1920, Russell's discussions with Walter Sydney Adams, Hale's second in command at Mt. Wilson, focussed on a series of papers by an unknown Calcuttan physicist named Megh Nad Saha. After Russell had read Saha's October 1920 *Philosophical Magazine* paper on ionization in the solar atmosphere, he became extremely excited and predicted to Adams that "I believe that within a few years we may utilize knowledge of ionizing potentials . . . to obtain numerical determinations of stellar temperatures from spectroscopic data—or at least of relations between temperature and pressure" [HNR/P, Russell to Adams, December 18, 1920; DeVorkin and

Kenat, 1983, pp. 116-117]. Saha had created a means to equate the total pressure and temperature of a gas to the degree of ionization of a particular element within that gas and the ionization potential of that element. What Russell immediately seized upon, after reading Saha, was that two main lines of effort at Mt. Wilson could now converge: Adams' technique of spectroscopic parallaxes could be linked to A. S. King's furnace temperature classifications. Both could at last be put on a rational footing.

The coincidental appearance of Saha's theory when Russell planned his first Mt. Wilson visit gave him just the focus he and Hale had hoped for. Russell knew that high quality, high dispersion solar and stellar spectroscopic data were required to both exploit and test the theory. Access to the spectroscopic data and expertise at Mt. Wilson was essential, and Russell had the inside track. After he arrived at Mt. Wilson in the summer of 1921 Russell wrote to Saha that "When I was appointed as a Research Associate of this Observatory I planned at once to do what I could to carry on work on the lines you have suggested" [HNR/P, Russell to Saha, August 3, 1921]. The first thing he did was search for rubidium in sunspot spectra in Mt. Wilson's plate vaults, where Saha predicted it could be found [DeVorkin and Kenat, 1983, p. 117]. Russell quickly found the predicted lines, and used them as a springboard to refine and extend Saha's theory.

Two days after Russell wrote to Saha, Russell told the Pacific Division of the AAAS that with the appearance of Saha's theory, astronomy had at last a means to apply Bohr theory to the stars. Russell echoed Hale's plan for a large coordinated attack on spectroscopic problems, explaining that:

> This is but a single illustration of the immense possibilities of the new field of investigation which opens up before us. A vast deal of work must be done before it is even prospected—much less worked out, and the astronomer, the physicist, and the chemist must combine in the attack, bringing all their resources to bear on this great problem, which is of equal importance to us all [Russell, 1921, p. 280].

Following rhetoric Hale used to combine his observatory staff with the new expertise gathering at Caltech, Russell added that:

> It is not too bold to hope that, within a few years, science may find itself in possession of a rational theory of stellar spectra, and, at the same time, of much additional knowledge concerning the constitution of atoms [Ibid].

This was the agenda Russell set for himself and a good fraction of Hale's spectroscopic staff. By 1924, Russell had published over a dozen papers extending Saha's work starting with simple confirmations of predictions and then moving on to refinements in theory, which led to discussions of inconsistencies between observation and theory, and subsequent modification and generalization of the theory [DeVorkin and Kenat, 1983, pp. 180- 222; Kenat and DeVorkin, 1990]. When Miguel Catalan's theory of multiplet structure appeared in the early 1920s, Russell immediately incorporated it into his Mt. Wilson spectroscopic program, helping Hale's staff to do the same.

Russell's first visit to Mt. Wilson so impressed Hale that he lobbied strongly on Russell's behalf, convincing Princeton's President Hibben that Russell had to be granted maximum possible freedom to return to Mt. Wilson. Hale argued that among the many illustrious Research Associates at his observatory:

> . . . no one has ever shown so intelligent and active an interest in every branch of our work, or done so much to stimulate the members of the staff to wider thought and greater effort. Russell's encyclopedic knowledge, and his extraordinary ability to grasp a new idea and apply it effectively, render him invaluable to us. In fact, the alliance we have thus established with Princeton seems to me a perfect one because it affords just such contact with the most advanced observational work as is necessary for the full development of Russell's rare talents [HNR/P, Hale to Hibben, September 30, 1921, clipped to November 4, 1921].

Russell at Mt. Wilson went beyond being a mere beneficiary to an agent of change. He redirected a good portion of the Mt. Wilson laboratory and solar spectroscopic staff to an agenda informed by the latest advances in modern physics. Laboratory physicist Arthur S. King revealed how Russell did this when he reported Russell's arrival to Hale in the summer of 1923:

> Henry Norris Russell arrived, 'sailed in high,' and (to continue the disrespectful comparison) with plenty of oil in his crankcase. The talking became a solo and continued unabated during his stay. He gave us three or four talks a week on spectral series applied to atomic structure . . . Most of his time for a while before he left was devoted to working out the titanium series, according to the selections made in the furnace classifications. The complexity proved greater than he expected, and he is still at the job, but the fundamental sorting out was made and it became clear what the character of the multiplets is. Between the furnace and

> the sun-spot lists, there is now getting to be a pretty satisfactory amount of material for the selection of series groups, and it looks as if the way to go at the many-lines spectra has been found. It is a large subject in which only the barest beginning has been made [CIT/GEH, King to Hale, July 22, 1923].

King was clearly delighted that Russell became sufficiently involved in his work to direct not only its astronomical application, but to show as well how astronomical data could be utilized to better understand atomic structure. In Russell's hands, Saha's and Catalan's methods were used synergistically to inform not only the nature of stellar atmospheres, but the structure of atoms. By early 1926, he was sure that his analyses of celestial and laboratory spectra were both informed by, and in return confirmed, Hund's theory of the structure of atoms producing complex spectra [Russell and Meggers, 1926, abstract; HNR/P, Russell Papers, Russell to Webb, January 26, 1926].

MT. WILSON STAFF REACTS TO RUSSELL

Both Russell and Hale were driven, insatiable workers. Both shared the same hyperactive enthusiasm for astronomy. This hyperactivity, however, could leave members of Hale's staff sometimes a bit bewildered, as one might infer from King's report to Hale. Without doubt, Russell fulfilled Hale's desires as an agent of change, and as well helped many on Hale's staff like King inform their empirical research. But Russell's tenure was also challenging for Mt. Wilson staff; Hale's second in command, his chief observational hedgehog Walter Sydney Adams, revealed to a confidant how the staff reacted to the presence of a scientific fox:

> We have had a strenuous summer with Russell who has just gone. He is a tremendously able fellow but I sometimes think that the man who secures none of the results upon which theories have to be based has rather an unfair advantage. It is possible to think of many schemes in the time it takes to secure and study a few stellar or solar spectra. However, he was full of suggestions and we enjoyed his visit while he himself had a great time [HL, Adams Papers, Adams to Abbot, August 9, 1921].

Unlike King, who was delighted that Russell helped him to see so much in his hard-won laboratory data, Adams harbored a strong sense of proprietary ownership of those data. Among senior Mt. Wilson staff members, Charles St. John was probably the most likely person to benefit from exploiting Saha's theory, followed by H. D. Babcock and King. St. John was madly taking solar spectra and examining facular spectra when Russell first arrived at Mt. Wilson. In addition to his observing routine, St. John also had numerous programmatic tasks to attend to, such as revising the Rowland solar tables, which had occupied much of his time and energy since 1919. He was therefore not as free as Russell to take direct advantage of Saha's discoveries, even though he indicated to Hale that "The ionization theory of Saha's is certainly going to give us a new tool" [HL, Adams Papers, St. John file 59.1031, St. John to Hale, August 27, 1921]. But St. John's contributions remained within the bounds set by his ongoing Mt. Wilson responsibility for the Rowland table project.

Russell, however, brought atomic theory to the Rowland tables project by suggesting that it be enlarged to take advantage of the new understanding of spectra brought about by Saha's and Catalan's work. Specifically, in addition to improving solar line intensities and wavelengths, putting the latter on the international standard, by the mid-1920s the revision also included a temperature classification, based upon King's almost two decades of effort, and was interpreted in terms of Saha's exploration of marginal appearances and rate of growth of spectral lines with increased temperature. Russell also campaigned to include excitation potentials, partly because complete multiplet designations for each line were too bulky to publish, and it was relatively easy to derive them using an appendix of term designations for each energy state in elements found in the sun [St. John, et al., 1928].

The ultimate revision of the Rowland Atlas bore Russell's imprint. In turn, Russell became a powerful advocate for the project. When a manuscript by the Dudley Observatory's Sebastian Albrecht reached Russell's desk for comment in 1926, he found that it used the old Rowland designations for wavelengths. Russell lodged a "very emphatic protest" against the use of the old system "in any new work of any sort whatever" mainly because, as he told Albrecht, St. John was revising the system to make it conform to the modern international system of wavelengths. "One of our main purposes in this work" Russell told Albrecht, "is to bury [?] the old Rowland scale in oblivion" [DO, Albrecht Papers, Russell to Albrecht, September 14, 1926]. Buried it was.

As much as they all respected Russell, and followed his lead in designing new research programs in spectroscopy at Mt. Wilson, Hale's staff nevertheless did not like being around Russell too much. St. John again said it best in 1928 when he learned that they would be on the same ocean liner travelling to Europe: "Russell is going on the same boat. I hope we have not been by chance assigned the same cabin!" [HL, Adams Papers, St. John file 59.1031, St. John to Adams, June 12, 1928].

RUSSELL'S VIEW

Russell, for his own part, was immensely satisfied with the arrangement, modelling the future of his own observatory after the relationship he had established at Mt. Wilson. When asked by the Princeton University Executive Secretary to assist efforts to raise an endowment for a research fund at Princeton in 1926, he replied that the department of astronomy would profit more from faculty and staff exchanges with other observatories than from additions to the Princeton staff or to its bricks and mortar [PUA, Department of Astronomy Records, Box II, section 1, "Astronomy at Princeton" folder 1925-1927, Russell to Smith, April 20, 1926]. Cross-fertilizing exchanges did take place from time to time between Russell's staff and those at various western observatories, and continued among Russell's successors well into the 1950s [DeVorkin and Kenat, forthcoming]. As Harlow Shapley planned for the revitalization of the Harvard College Observatory in the 1920s, Russell's influence was felt there as well, mainly in the continued use of Harvard's collection of spectra by Russell's graduate students. But it was Russell's long connection with Mt. Wilson that remained the embodiment of Hale's vision for the pursuit of astronomy not only at his astronomical institutions, but by the international scientific community Hale devoted his life to building.

Russell spent an average of three months per year for 25 years as Carnegie Research Associate at Mt. Wilson. Gradually, he placed his own people there, such as research assistant Charlotte E. Moore, who helped to make Russell's hyperactive presence more tolerable to Mt. Wilson staff, and former students like Theodore Dunham, who remained receptive to Russell's ideas and suggestions. Over the first ten years of Russell's tenure, the centroid of publications of the observatory staff migrated slowly from purely empirical studies in the lab and at the telescope, to observational studies informed by physical theory, and even to theoretical studies [Mt. Wilson Contributions, 1917-1930]. Many of these efforts were collaborations designed and led by Russell. Hale's goal for Russell's visits, as well as visits of many young physicists like Albrecht Unsold, indeed did much to fill the perceived gap between the laboratory bench and the observatory dome. Closing that gap was Russell's reason for existence at Mt. Wilson, and provides insight not only into his professional life, but into a science in transition.

ACKNOWLEDGEMENTS

I owe a great debt to the many archivists and archives which have made my research into Russell's life and times possible, and which are identified elsewhere in this book. I also acknowledge the support of a sabbatical at the Institute for Advanced Study at Princeton, and grants from NSF, the Dudley Observatory, the American Institute of Physics, and the Smithsonian's Scholarly Studies Program, and Women's Committee. During the conference, conversations with Frank Edmondson, Russell's son-in-law, about Isaiah Berlin's essay "The Hedgehog and the Fox," were especially delightful.

ARCHIVAL SOURCES

Princeton: the manuscripts division of its Firestone Library [HNR/P], and the Seeley G. Mudd Manuscript Library of Princeton University Archives [PUA]; The Mt. Wilson archives at the Huntington Library [HL]; Carnegie Institution of Washington [CIW]; Library of Congress Manuscript Division [LC]; California Institute of Technology [CIT], Dudley Observatory [DO]. Harvard University Archives [HUA] E. C. Pickering Papers, and the Charles W. Eliot Papers. [WW]: Link, A. S., et al, eds., The Papers of Woodrow Wilson, Princeton, Princeton, N.J., 1971.

REFERENCES

Becker, B., William Huggins, Ph.D. diss., Johns Hopkins, Baltimore, MD, forthcoming.

Boltzmann, L., A German Professor's Trip to El Dorado, *Physics Today*, *45*, 44-51, 1992.

Brashear, R. S., A Brewing Storm: Hale, Gale, and Researcher Selection at the Mount Wilson Observatory (abstract), American Astronomical Society, Phoenix, January 1993.

Dennis, M. A.. A Change of State: The Political Cultures of Technical Practice at the MIT Instrumentation Laboratory and the Johns Hopkins University Applied Physics Laboratory, 1930-1945, Ph.D. diss., 451 pp., Johns Hopkins Univ., 1991.

DeVorkin, D. H., Community and Spectral Classification in Astrophysics: The Acceptance of E.C. Pickering's System in 1910, *Isis*, *72*, 29-49, 1981.

DeVorkin, D. H., Stellar Evolution and the Origins of the Hertzsprung-Russell Diagram in Early Astrophysics, in *Astrophysics and Twentieth-Century Astronomy to 1950*, edited by M. A. Hoskin and O. Gingerich, pp. 90-108, Cambridge Univ. Press, Cambridge, MA, 1984.

DeVorkin, D. H., Henry Norris Russell, *Sci. Amer.*, *260*, 126-133, May 1989.

DeVorkin, D. H., and R. Kenat, Quantum Physics and the Stars I: The Establishment of a Stellar Temperature Scale, *J. Hist. Astron.*, *14*, 102-132, 1983.

DeVorkin, D. H., and R. Kenat, Quantum Physics and the Stars II: The Abundances of the Elements in the Atmospheres of the Sun and Stars, *J. Hist. Astron.*, *14*, 180-222, 1983.

DeVorkin, D. H., and R. Kenat, Stellar Evolution, 1938-1955, in *General History of Astronomy 4b*, edited by M. A. Hoskin and O. Gingerich, Cambridge Univ. Press, Cambridge, MA, forthcoming.

Gingerich, O., How Shapley Came to Harvard or, Snatching the Prize from the Jaws of Debate, *J. Hist. Astron.*, *19*, 201-207, 1988.

Hale, G. E., Mount Wilson Observatory, in *CIW Yearbook*, No. 20, pp. 215-295, CIW, Washington, D.C., 1922.

Hale, G. E., *The Study of Stellar Evolution*, 252 pp., Chicago, 1908.

Herrmann, D. B., *The History of Astronomy from Herschel to Hertzsprung*, translated by K. Krisciunas, 220 pp., Cambridge, 1984.

Hofstadter, R., *Social Darwinism in American Thought*, Second Ed., 248 pp., Beacon Press, Boston, 1955.

Hufbauer, K., Astronomers take up the stellar-energy problem, 1917-1920, *Hist. Stud. Phys. Sci.*, *11*, 277-303, 1981.

Hufbauer, K., *Exploring the Sun: Solar Science Since Galileo*, 370 pp.,

Johns Hopkins, Baltimore, 1991.
Jones, B. Z., and L. G. Boyd, *The Harvard College Observatory 1839-1919*, 495 pp., Belknap Press, Cambridge, MA, 1971.
Kenat, R., and D. H. DeVorkin, Quantum Physics and the Stars III: Towards a Rational Theory of Stellar Spectra, *J. Hist. Astron.*, *21*, 157-186, 1990.
Kevles, D. J., George Ellery Hale, the First World War, and the Advancement of Science in America, *Isis*, *59*, 427-437, 1968.
Kevles, D. J., *The Physicists: The History of a Scientific Community in Modern America*, 489 pp., Vintage, New York, 1978.
Kohler, R. E., *Partners in Science*, 415 pp., Univ. of Chicago, Chicago, 1991.
Kohn, D., ed., *The Darwinian Heritage*, Princeton Univ. Press, Princeton, 1985.
Meadows, A. J., *Science and Controversy: A Biography of Sir Norman Lockyer*, 331 pp., MIT Press, Cambridge, MA, 1972.
Menzel, D. H., The History of Astronomical Spectroscopy, in R. Berendzen, ed., *History of and Education in Modern Astronomy*, *Annals of the New York Academy of Sciences*, *198*, 225-244, 1972.
Miller, H. S., *Dollars for Research: Science and its Patrons in Nineteenth-Century America*, 258 pp., Univ. of Washington, Seattle, 1970.
Numbers, R. L., *Creation by Natural Law: Laplace's Nebular Hypothesis in American Thought*, 184 pp., Univ. of Washington, Seattle, 1977.
Osterbrock, D. E., J. R. Gustafson, and W. J. S. Unruh, *Eye on the Sky, Lick Observatory's First Century*, 295 pp., Univ. of California Press, Berkeley, 1988.
Pickering, E. C., Report of the Committee on Astronomy, *Proc. Nat. Acad. Sci.*, circular, National Academy Press, Washington, D.C., June 1917.
Plotkin, H., E. C. Pickering and the Endowment of Scientific Research in America, 1877-1918, *Isis*, *69*, 44-57, 1978.
Plotkin, H., Edward Charles Pickering, *J. Hist. Astron.*, *21*, 47-58, 1990.
Reingold, N., The Case of the Disappearing Laboratory, *Amer. Quart.*, Spring, 1977, reprinted in N. Reingold, *Science, American Style*, pp. 225-246, Rutgers Univ. Press, New Brunswick, 1991.
Reingold, N., National Science Policy in a Private Foundation: The Carnegie Institution of Washington, in *The Organization of Knowledge in America 1860-1920*, edited by A. Olesson and J. Voss, pp. 313-341, Johns Hopkins, Baltimore, 1979.
Russell, H. N. and W. Meggers, Analysis of Arc and Spark Spectra of Scandium, abstract, 1926.
Russell, H. N., Some Problems in Sidereal Astronomy, *Popular Astronomy*, *28*, 212, 1919.
Russell, H. N., The properties of matter as illustrated by the stars, *Publ. of the Astr. Soc. of the Pacific*, *33*, 275-90, 1921.
Russett, C. E., *Darwin in America: The Intellectual Response, 1865-1912*, 228 pp., W. H. Freeman, San Francisco, 1976.
Smith, R. W., A National Observatory Transformed: Greenwich in the Nineteenth Century, *J. Hist. Astron.*, *22*, 5-20, 1991.
St. John, C. E., C. E. Moore, L. M. Ware, E. F. Adams, and H. D. Babcock, *Revision of Rowland's Preliminary Table of Solar Spectrum Wave-lengths*, CIW, Washington, DC, 1928.
Wright, H., *Explorer of the Universe*, 480 pp., Dutton, New York, 1966.
Wright, H., J. Warnow, and C. Weiner, (Eds.), *The Legacy of George Ellery Hale*, 293 pp., MIT, Cambridge, MA, 1972.

Converting an Hypothesis into a Research Program: T. C. Chamberlin, his Planetesimal Hypothesis, and its Effect on Research at the Mt. Wilson Observatory

N. S. Hetherington
Office for the History of Science and Technology
University of California, Berkeley, CA 94720
and Institute for the History of Astronomy

Much has been written about astronomical research at the Mt. Wilson Observatory, but little has been written about the geologist Thomas Chrowder Chamberlin's influence on some of that research. A tantalizing hint of a story long has existed. In his annual report for 1915 to the Carnegie Institution of Washington (CIW), Chamberlin wrote that

> July and August were spent at Mount Wilson Observatory at the invitation of its Director in the preparation of a system of tentative working interpretations of nebulæ based essentially on dynamic considerations, with a view to serviceability in systematic researches on the nature and history of nebulæ and their evolutionary place in the stellar system [Chamberlin, 1916a].

With the recent opening of the archives of the Mt. Wilson Observatory, it now is possible to fill out the story [Brashear, 1990; this volume].

After Chamberlin developed his planetesimal hypothesis of the origin of the solar system, he pointed to spiral nebulæ as possible analogies. A few years later, he and George Ellery Hale, the director of the Mt. Wilson Observatory, attempted to draw from the theory ideas to help focus a research program on spiral nebulæ at Mt. Wilson.

The development of the planetesimal hypothesis and Chamberlin's initial attempts to obtain assistance from astronomers have been thoroughly described in an excellent study by Stephen G. Brush [1978; Hetherington, 1975a, 1990a]. This part of the story need be reviewed only briefly here.

The primary concern of this paper is the later attempt at cooperation between Chamberlin and Hale. Consciously, they could contemplate only partially potentialities and difficulties inherent in any attempt at cooperation between scientists from different fields of science. The failure of their exemplary effort emphasizes just how much ahead of their time was their vision of cooperation in research. Their joint effort now reveals to historians much about the nature of science, particularly disciplinary formation.

The Planetesimal Hypothesis

In the course of his work on the changing climates of the geological past, Chamberlin, chairman of the geology department at the University of Chicago, had begun to question the plausibility of Laplace's nebular hypothesis, in which the solar system was thought to have been formed by separation and condensation of parts of a whirling cloud of hot gas. A cold and arid climate in the past, however, as indicated by glaciation and salt deposits, was incompatible with the exceedingly extensive, dense, warm, and moist atmosphere postulated for earlier times by Laplace's theory. Examining the nebular hypothesis critically, Chamberlin noted that photographs of the Andromeda Nebula, originally hailed as supporting Laplace's theory, offered only a vague and general analogy. Nor did Saturn's rings (by 1897 known to be composed not of gas but of solid particles) any longer furnish a helpful analogy for Laplace's ideas [Chamberlin, 1897].

Furthermore, if a molten planet had condensed from the gaseous mass, the temperature of the Earth's atmosphere probably would have been so great that water vapor would have evaporated and been lost into space. This likelihood was shown in tables prepared by Forest R. Moulton, an assistant in astronomy at the University of Chicago and a member of Chamberlin's interdisciplinary research team investigating related problems in geophysics and astronomy [Chamberlin, 1903]. Beginning in 1903, half of Moulton's salary was covered by a CIW grant [Gasteyer, 1970; Brush, 1978, p. 14]. Alternatively, if water vapor had not been evaporated and lost into space, it would have dissociated into hydrogen and oxygen, and the hydrogen would have escaped quickly into space. Thus Laplace's hypothesis of a molten Earth seemed untenable to

The Earth, the Heavens and the Carnegie Institution of Washington
History of Geophysics Volume 5

Chamberlin.

Chamberlin's initial doubts concerning Laplace's nebular hypothesis rested on geological indications of a formerly cold and arid climate and on physical questions concerning retention of an atmosphere. Astronomical tests further weakened faith in the theory and forestalled acceptance of a modified form of the Laplacian hypothesis in which meteoroids slowly aggregated. The theory could not explain why much of the solar system's momentum resides in the outer regions but the mass is heavily concentrated near the center. This unsymmetrical distribution of matter and momentum suggested to Chamberlin that the solar system had been formed by a collision between a small nebula with large momentum and the periphery of a large nebula with very little momentum. Spiral nebulæ might represent such collisions, though little was known of their dynamics, and speculations thus were premature [Chamberlin, 1900]. Moulton similarly concluded that the original solar nebula probably had been similar to a spiral nebula, like the ones for which James Keeler at the Lick Observatory was obtaining and exhibiting such remarkable photographs [Moulton, 1900; Keeler, 1900, 1908; Osterbrock, 1984]. Ideas based on the form of spiral nebulæ were, however, only conjecture. As Chamberlin and Moulton continued to admit, nothing was yet known of the dynamics of spiral nebulæ [Chamberlin and Moulton, 1900].

Chamberlin made an early effort to enlist astronomers in the aid of geology. Early in 1900, he wrote to Keeler, telling him that he was considering a theory that the Earth had originated from a spiral nebula and asking for copies of photographs of such nebulæ [January 30, 1900; quoted in Brush, 1978, p. 15]. A month later, Chamberlin expressed his hope that Keeler

> might be able to come to the relief of geologists in the dilemma in which we seem to be finding ourselves as the result of the apparent breaking down of the accepted form of the nebular hypothesis [February 28, 1900; quoted in Brush, 1978, p. 15].

Keeler's photographs were promising, and Chamberlin subsequently stated that spiral nebulæ offered "the greatest inherent presumption of being the ancestral form" of the solar system, possibly formed by "the collision of a small nebula upon the outer portion of a large one" [1900]. A few months later, Chamberlin and Moulton argued that consideration of spiral nebulæ should have precedence in attempts to find analogies for the origin of the solar system [1900]. Keeler also found attractive the idea that the solar system had evolved from a spiral nebula [1900].

Other types of disruptive encounters between astronomical objects also attracted Chamberlin's interest. Solar eruptions, he began to think, might be understood in the context of disruption by an approaching body [Brush, 1978]. And novae might be attributed to the collision of a star with another body. Indeed, shortly after Chamberlin wrote down this thought, Nova Persei was observed, on February 22, 1901, and Chamberlin mentioned this new observation in a footnote added to his paper before it was published [1901; Brush, 1978, p. 18].

Keeler and Chamberlin shared an interest in the planetesimal hypothesis, based on Keeler's photographs of spiral nebulæ taken at the Lick Observatory between 1898 and 1900. In 1907, with the prospect of similar photographs soon forthcoming from the 60-inch reflecting telescope then under construction at the Mt. Wilson Observatory, Hale initiated a discussion with Chamberlin about the planetesimal hypothesis and spiral nebulæ.

Chamberlin and Hale had known each other at the University of Chicago. Hale had moved to the Mt. Wilson Observatory from Chicago's Yerkes Observatory, in whose site selection Chamberlin had participated in 1893 [Wright, 1966, p. 111]. And in 1897, while still at Chicago, Hale had shared with Chamberlin an advance copy of a paper on the retention of water vapor in the Earth's atmosphere that was to appear in Hale's *Astrophysical Journal* [Chamberlin to G. J. Stoney, December 10, 1897; quoted in Brush, 1978, p. 33, footnote 47].

Furthermore, striking similarities between the two men may have provided common ground for mutual understanding and may have facilitated their attempt at cooperative research. They shared common professional duties: Chamberlin was the founder-editor of the *Journal of Geology*, begun in 1893, as was Hale of the *Astrophysical Journal*, begun two years later. Both scientists recognized potential advantages from combining different disciplinary approaches: Chamberlin's explicit recognition of links between geology and astronomy was fully matched by Hale's thorough and farsighted appreciation of potential benefits in interdisciplinary scientific investigations, most manifest in his combining astronomy and physics into the new science of astrophysics, and also apparent in his cooperative effort with Chamberlin [Chamberlin, 1900, p. 58; Brush, 1978, p. 22; Wright et al., 1972, p. 3].

Not least among their similarities was the fact that both Chamberlin and Hale had been singled out by the CIW, among whose aims was "to discover the exceptional man in every department of study whenever and wherever found, inside or outside schools, and enable him to make the work for which he seemed especially designed his life work" [Carnegie, 1903, p. xiii]. Between 1902 and 1910, Chamberlin received more CIW money than any other individual grantee and Hale was enabled to found the Mt. Wilson Observatory [Brush, 1978, p. 23; Wright, et al.,

1972, pp. 42-43; Hale, 1906, p. 56; Chamberlin, 1934; Wright, 1966; Reingold and Reingold, 1981]. Although the historian of American science Nathan Reingold has noted that Chamberlin and Hale were very far apart on matters of science policy [1981, p. 57], on many other matters their beliefs were similar.

In 1907, Hale wrote to Chamberlin to request help in planning a research program for the Mt. Wilson Observatory:

> If all goes well, the 60-inch reflector will be erected on Mt. Wilson next autumn. . . . I have been at work on the observational programme for this instrument, and I am anxious to obtain your view on the subject, in order that we may endeavor to obtain some of the material required in the further tests and development of the Planetesimal Hypothesis. . . . I am now desirous of obtaining the benefit of your ideas and suggestions, in the hope that we can cooperate with you to mutual advantage [March 20, 1907; reproduced in Kevles, 1968, reel 10, frames 31-34].

Chamberlin replied that he was "deeply gratified that you [Hale] should be willing to make a place in your scheme for studies of nebulæ with reference to their bearings on the planetesimal hypothesis and that you should think my suggestions worthy of consideration." Of possible tests of the planetesimal conception of spiral nebulæ, Chamberlin suggested

> A series of observations to determine whether the matter of the arms, the knots, for example, are moving outward or inward along the arms, on the one hand, or somewhat transverse to them, on the other [hand], is therefore of first importance.
>
> Two leading classes of observational tests to determine this are naturally suggested, (1) spectroscopic determinations of motion in the line of sight, and (2) comparisons of photographs taken at intervals [April 20, 1907; reproduced in Kevles, 1968, reel 10, frames 35-48].

Chamberlin, however, had doubts that spectroscopic studies would succeed, because of the faintness of nebular light and the low velocities. Furthermore, the familiar spiral nebulæ probably were several orders of magnitude greater than our solar system, and only very small, nearby nebulæ would be suitable for a test of the planetesimal theory [Brush, 1978, p. 81].

This letter of 1907 is the first evidence we have that Chamberlin suggested photographic comparisons as well as spectroscopic measurements. In published papers, he called only for spectroscopic tests, leading to the subsequent conclusion by one historian that Chamberlin "did not go on to suggest comparison of photographic plates" [Hetherington, 1975]. This conclusion must now be corrected. Nonetheless, attention focused more sharply on spectroscopic measurements than on photographic comparisons.

Spectroscopy was already well-established in astronomy by 1900, while the photographic reflecting telescope was not yet a basic instrument of astronomical research. Astronomical photography, common among amateur astronomers, was not yet accepted by professionals. It soon made the transition, primarily due to the efforts of George Willis Ritchey at Mt. Wilson [Mills, 1966; Hetherington, 1978].

Apparently neither Chamberlin nor Hale pursued the problem of spiral nebulæ much further at this time. Hale did write, in a book whose preface was dated November, 1907, and which was published a year after Chamberlin's April, 1907, letter to Hale, that with regard to the nebular hypothesis, "in testing it, recourse may be had both to theoretical investigations and to observations of various kinds, particularly of nebulæ, which may throw light on the earlier stage of the process of condensation" [1908, p. 187]. But Hale did not yet begin testing the planetesimal hypothesis with observations of spiral nebulæ.

Chamberlin had been pessimistic about spectroscopic observations, and photographic comparisons lay yet in the future, after plates had been obtained over periods of several years. The instruments on Mt. Wilson in 1907 were suitable primarily for solar research; indeed, the observatory's original name was officially the "Mount Wilson Solar Observatory." As Walter S. Adams, Hale's successor as director, later noted, the observatory was known as the "Solar Observatory . . . during the years preceding the completion and operation of the 60-inch telescope" [1947, p. 219]. The telescopes that made spiral nebulæ accessible to investigation were the 60-inch (1.52-meter) reflector, under construction in 1907, and the 100-inch (2.54-meter) reflector, then being planned. Not until well after the 60-inch telescope was producing results and the 100-inch telescope was nearing completion, in 1917, did the *Annual Report of the Director of the Mount Wilson Solar Observatory* become *The Annual Report of the Director of the Mount Wilson Observatory*. A further impediment to Hale's comparison of the planetesimal hypothesis with photographs of spiral nebulæ was his nervous breakdown in 1907, which limited his productive research, especially between the years 1909 and 1913 [Wright, 1966, pp. 255-282].

Still, if Chamberlin's observational suggestions had been promising enough in 1907, and had he been more optimistic about the possible outcome of observations, he might have interested Hale then. Hale, after all, had approached Chamberlin. And just two years later, Hale postponed some of his initial plans for the 60-inch to enable the Dutch astronomer J. C. Kapteyn to test his "star-streaming" hypothesis, of two groups of stars rotating in different directions, both parallel to the plane of the Milky Way [Wright, 1966, pp. 235-236; Hale, 1915, pp. 50-51; Hale, 1908, pp. 201-203]. With Kapteyn, Hale felt that "by the preparation of a suitable scheme . . . observations can be made to serve, not merely for the tabulation of accurate data, but for the solution of the greatest problems of astronomy" [Hale, 1915, pp. 4-5].

Ever the geologist, Chamberlin remained most concerned about the Sun, about a possible disruptive encounter between the Sun and another star, and about the formation of planets from fragments torn from the Sun. Some further investigation by an interdisciplinary research team followed, led by Chamberlin and financed by the CIW [Brush, 1978, pp. 22-23]. But then Chamberlin and Moulton dropped their promising investigation into the origin of spiral nebulæ. They found it "futile" to hypothesize collisions between nebulous bodies as a mode of origin of spiral nebulæ because the resulting orbits of perturbed matter would be too eccentric to fit the case of the solar system [1907; Brush, 1978, p. 17].

Moulton, the astronomer, had promised in a 1905 paper future numerical treatments of possible cases of the origin of spiral nebulæ from encounters between two stars [1905; Brush, 1978, p. 17]. But he never published detailed calculations, probably because the work proved too difficult to complete [Brush, 1978, p. 40, footnote 180]. Moulton's inability to model mathematically collisions between stars is not surprising, since only recently have scientists begun to create simulated galaxies upon which simulated experiments, including collisions, can be performed using supercomputers and new computational techniques [Schweizer, 1986; Miller, 1992].

As Brush has noted, several astronomers, while praising Chamberlin and Moulton for their quantitative criticism of the nebular hypothesis, at the same time decried the lack of quantitative statements in the planetesimal theory. F. J. M. Stratton wrote:

> Of necessity these applications [of the theory] are somewhat vague, and further evidence in favour of the hypothesis seems badly wanted. Spectroscopic evidence from the study of actual spiral nebulæ, of such motions as are indicated by the hypothesis, would give valuable quantitative support to a theory which rests mainly on a basis of qualitative constructive reasoning and quantitative destructive reasoning [in criticizing Laplace's theory] [1910; Brush, 1978, p. 80].

The lack of quantitative predictions, the dubious potential of spectroscopic measurements, and the nascent state of astronomical photography all argued against initiating a nebular research program in 1907 based on the planetesimal hypothesis. Also, a shared sense of priorities was against it. Chamberlin's interest centered on the origin of our planetary system, and he had hoped to use studies of spiral nebulæ to support the planetesimal theory. Hale, too, in 1907 accepted this priority. When Hale later reversed the priority and sought to draw from the planetesimal theory suggestions that might help investigations of the spiral nebulæ, such investigations—now pursued for their own sake—naturally were more attractive. However doubtful it might be that the planetesimal hypothesis would emerge from the investigation with additional strength, some sort of information of value about spiral nebulæ very likely was to be uncovered.

CHAMBERLIN AND HALE: 1915

Chamberlin and Hale's cooperative effort on the problem of the spiral nebulæ, long in gestation, finally saw birth in 1915. On April 5, Chamberlin wrote to Hale that he had noticed, while examining slides of nebulæ, "one case in particular that there were knot-like concentrations in the outer part and that there were strings of nebulosity in a somewhat inclined position suggesting that they were the remnants of spiral arms. . . . " He continued:

> It fits in beautifully with the conception I entertain relative to spiral nebulæ, which is to this effect: The very great spiral nebulæ could not well deploy symmetrically and retain their spiraloid form if their arms were projected thru the spheres of control of a large number of massive stars aggregated as they are in this portion of our stellar galaxy. I therefore assume that the great spiral nebulæ lie outside our cluster, at the sides of the galaxy, where alone they are seen, the presumption being that such as may lie beyond the Milky Way are too distant to be seen distinctly. Such spiraloid dispersions, therefore, as took place within our galaxy would only remain spirally symmetrical near the nucleus and if that were large this portion would soon be wrapped up into discoidal form and the spiral be obscured. I long since noticed that near the nucleus of the large spirals when the nucleus seemed to be massive, the spiraloid structure disappeared in a discoidal structure. In the case of the very mild types of evolution like that assigned as the parent of the solar system, where the nucleus is very massive and the deployed portion reaches only moderate distances, the

spiral, however declared when first formed, would very soon be wrapped up and disguised in a disc tho the knots and other concentrations would persist. If you have other photographs that show this knotted structure I should be delighted to learn of it and if convenient to see them [HLMtW, XI Director's Files, box 153].

Hale invited Chamberlin to come to Mt. Wilson for the summer to work on the problem. There was a clear precedent for the invitation. Kapteyn was at the Observatory to test his star-streaming hypothesis with the help of the 60-inch telescope for five summers, from 1909 until the outbreak of World War I [Wright, 1966, p. 235]. Chamberlin replied to Hale on May 24 that he was "about persuaded to accept your kind invitation to spend some time on Mount Wilson this summer." Hale replied that he expected to see Chamberlin at Mt. Wilson in the latter part of June, and Chamberlin answered on June 19 that he was clearing away odds and ends and had fully decided to come. On July 4, Chamberlin notified Hale that he expected to arrive on Saturday [HLMtW, XI Director's Files, box 153].

Once the two scientists came together, there is, naturally, little correspondence from which we may determine the course of their cooperative effort. But Hale did leave the mountain before Chamberlin had quite finished his stay, and on September 3 Chamberlin wrote to Hale:

I am sending you a paper on the nebular problem which I took a notion to propose after you left + only completed yesterday It occurs to me this morning that possibly it may be better to recast this into the form of a review on the nebular problem—for it is in some sense that—+ make it less individual—

Enclosed with this note was a page dated "The 'Monastery' July 27 1915 T.C.C." and labeled a "*Hypothetical diagram* of the two Kapteyn drifts on the assumption that the stellar galaxy took its present form from passage eccentrically thru another galaxy (= The Magellanic Clouds?)." There were two drawings, one showing a "Typical dispersion due to dynamical encounter based on Moulton's mathematical work" and a second showing "Dominant motions of galaxy after encounter" [HLMtW, Director's Files, box 153].

A month later Hale sent Chamberlin a preliminary report on "the nebular campaign." Chamberlin's planetesimal hypothesis suggested studying motions of concentrations of material (knots) in spiral nebulæ. This could be done either spectroscopically or photographically. Good photographs of spiral nebulæ, however, only were obtained at Mt. Wilson beginning in 1908. Mt. Wilson astronomers were more used to spectroscopic work, from their studies of the Sun, and it was from a spectroscopic investigation that Hale first had new information to report to Chamberlin:

As a first result in the nebular campaign I am glad to report that Mr. Pease has obtained a spectrum of the large bright patch in M. 33 Trianguli about 33 mm from the center of the reproduction on p. 58 of my "Ten Years' Work" (to the right, below center). 34½ hour's exposure (four nights) give a faint spectrum, showing only the chief nebular line and *H*d and *H*n of hydrogen. No continuous spectrum appears, and if the second nebular line is present it is too faint to be seen. The velocity has not yet been measured, but it is of the order of 200 km per second towards the earth.

The center of the nebula will be photographed on Mr. Pease's next trip to the mountain. Meanwhile a multiple slit is being made.

Several of us have read your manuscript with great interest, and I will send you some suggestions regarding it before long [October 11, 1915, HLMtW, XI Director's Files, box 153].

Chamberlin replied that he was "delighted to get the news of the first products of the planetary campaign . . . " [October 21, 1915, HLMtW, XI Director's Files, box 153].

Soon, however, there fell to Chamberlin the task of dampening, at least temporarily, Hale's enthusiasm, because Chamberlin had found a mistake in his calculations. Hale also had temporary bad news to report. Because of cloudy skies, Pease's next attempt to photograph spectra of M 33 had failed, and now the moon was in the way. Still, Hale was optimistic. And just before Hale mailed this letter, he had more to report to Chamberlin; Pease had succeeded in making another observation [November 4 and December 15, 1915, HLMtW, XI Director's Files, box 153].

Another prong of the nebular campaign, studying motions of knots in spiral nebulæ by photographic comparisons, was taking form, and Hale reported some early results to Chamberlin:

In spite of the unfinished state of the work, and the consequent doubt involved as to the validity of the conclusions I must tell you what Dr. van Maanen has found in some differential measures of a number of small condensations in the spiral nebula M 101 Ursae Majoris. Using two photographs made by Mr. Ritchey in 1910 and 1914 [not 1914, but 1915, as later reported; see van Maanen, 1916], and employing a differential method of measurement in the stereocomparator (with monocular attachments), he has found that fully half the points show proper motions ranging from 0.″01 to 0.″1

> *outward* along the spiral arms, or nearly in this direction. Of the other points, three indicate small motions in the opposite direction (inward), while three seem to be moving nearly at right angles to the arms. On the whole, however, the small amount of evidence thus far available is decidedly in favor of outward motion. Of course no final conclusions can be drawn until the mean motions of a large number of points have been obtained, and nothing will be said about the preliminary results before then. But you are certainly entitled to hear how matters stand in all of our work. As for M 33, Mr. Pease is now making another exposure on a point intermediate between the center and the outlying region first photographed [December 28, 1915, HLMtW, XI Director's Files, box 153].

Hale also expressed his enthusiasm for Chamberlin's planetesimal hypothesis, on which Chamberlin had just given a series of lectures before the National Academy of Sciences. When it came to publishing the lectures, Hale urged Chamberlin not to "abbreviate in any degree your discussion of the planetesimal hypothesis, which is far too valuable to be cut down."

Adriaan van Maanen, who made the differential measures of small condensations in the spiral nebula M101, had come to Mt. Wilson in 1912 after working under Kapteyn in the Netherlands measuring motions of stars from comparisons of photographic plates. Though hired primarily to measure proper motions and parallaxes of stars on the photographic plates rapidly accumulating from the 60-inch telescope, van Maanen first helped Hale measure positions of solar markings, both sunspots and flocculi (now called plages). The flocculi, some forty or fifty per plate, underwent rapid changes of form, making their centers and consequently their positions difficult to define precisely. The images from two photographic plates were brought together in a heliomicrometer and the positions of the flocculi read from reference points. Techniques developed during solar investigations were well suited to the measurement in 1915 of positions of knots in spiral nebulæ [Hetherington, 1990a, pp. 28-29].

Ritchey had handed van Maanen two plates of M101 sometime in December 1915 and asked him to see if any motion could be detected. Presumably Chamberlin's planetesimal hypothesis was behind Ritchey's request and also was known to van Maanen. In van Maanen's 1916 paper discussing his work on M101, there is neither mention of Chamberlin nor discussion of the significance of the motions. This omission exists, however, only because three paragraphs in a handwritten draft were cut from the published version [van Maanen, 1916]. In the draft, van Maanen quoted from a personal letter from Chamberlin regarding the motions, stating that Chamberlin did not see "anything inconsistent with the doctrine of dispersion by dynamic encounter if we suppose that this nebula was the product of very close approach of very massive stars, which indeed would be inferred by the enormous deployment" [HLMtW, van Maanen, draft titled "Preliminary Evidence of Internal Motion in the Spiral Nebula Messier 101 . . . "; quoted in Hetherington, 1974].

Van Maanen also commented in the draft, but not in the published version, that "the detection of their [spiral nebulæ's] possible internal motions might indicate that these bodies are not as distant as is usually supposed to be the case." This issue of distance, rather than any testing of the planetesimal hypothesis, soon became a focus of attention. If M101 were even one-fifth as large as the estimated size of our own galaxy and were rotating with a period of 85,000 years, as determined by van Maanen, its outer edge would have to be traveling faster than the speed of light [Shapley, 1919; Shapley and Curtis, 1921; Struve, 1960; Hetherington, 1970; Berendzen et al., 1976, pp. 114-117; Hoskin, 1976; Smith, 1983]. By a *reductio ad absurdum* argument, van Maanen's purported rotation contradicted belief in spiral nebulæ as extragalactic systems (Figure 1).

We now know that van Maanen's measurements of rotation great enough to be detectable from a comparison of photographs taken only a few years apart were spurious. There is a considerable body of historical literature on the purported measurements of rotations of spiral nebulæ and their significance for the debate over the identification of spiral nebulæ as island universes, and it need not be repeated here [Hoskin, 1971; Fernie, 1970; Hetherington, 1972; Berendzen and Hart, 1973; Berendzen et al., 1976; Hetherington, 1974a; Hetherington, 1974b; Hetherington, 1974c; Hetherington, 1975a; Hetherington, 1976; Smith, 1982; Hetherington, 1983; Hetherington, 1988; Hetherington and Brashear, 1991; Hetherington and Brashear, 1992]. The correspondence between Hale and Chamberlin, however, has not been discussed before.

Chamberlin replied to Hale's preliminary report regarding van Maanen's measurements on the last day of January, 1916. His doctor had prescribed rest, and this was his first day in the office since Christmas.

> I was deeply interested in the preliminary results of Dr. Von [sic] Maanen. It is very gratifying to feel that results are attainable. While the recent revival of the notion that spiral nebulæ are more distant constellations has not seemed to me to have any substantial basis, it is a satisfaction to feel that definite evidence is about to give it a quietus.
>
> Dr. Von [sic] Maanen's results are also gratifying in

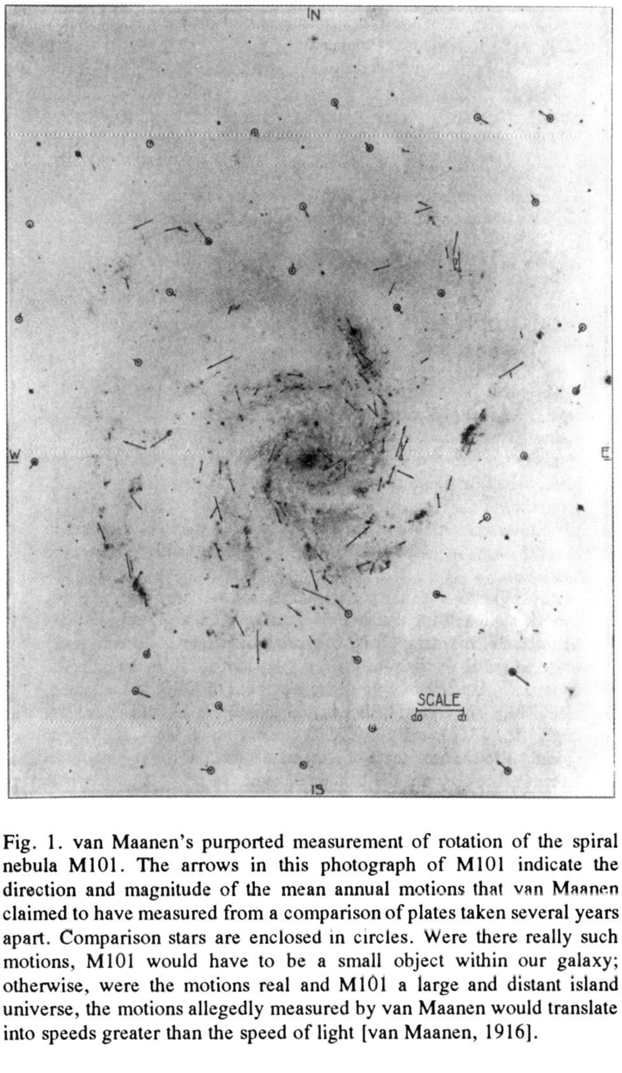

Fig. 1. van Maanen's purported measurement of rotation of the spiral nebula M101. The arrows in this photograph of M101 indicate the direction and magnitude of the mean annual motions that van Maanen claimed to have measured from a comparison of plates taken several years apart. Comparison stars are enclosed in circles. Were there really such motions, M101 would have to be a small object within our galaxy; otherwise, were the motions real and M101 a large and distant island universe, the motions allegedly measured by van Maanen would translate into speeds greater than the speed of light [van Maanen, 1916].

> so far as they strengthen the conviction that spiral nebulæ are essentially the rpoducts [products] of centrifugal action and so forecast the confinement of the problem of their genesis to dispersive, as distinguished from concentrative agencies.

Further on in his letter to Hale, Chamberlin again commented on the question of the dynamics that give birth to spiral nebulæ, remarking that van Maanen's results:

> seem to imply a divided system of motions, suggesting that a portion of the nebula is under its own central control and that a circulatory system is forthcoming from this portion, while the other portion may be destined to diverge indefinitely and be lost to control. This is quite consistent, as I see it, with an origin from dynamic encounter from which a very high dispersive effect arose [January 31, 1916, HLMtW, XI Director's Files, box 153].

Hale was quick to answer Chamberlin, and he had more results to report:

> Van Maanen has repeated his measures of M 101, including a larger number of points, and [is] making rigorous reductions. The results support his preliminary conclusion, but of course they cannot be trusted as they stand because we never feel safe in basing such important conclusions on the measurement of only two plates. It is extremely unfortunate that in taking the photographs Ritchey did not foresee such an application of them. If he had made three negatives at each epoch all doubts might have been cleared away. As it is, we expect to borrow five negatives of M 101, made at various times in a period of some 15 years at the Lick Observatory. Dr. Curtis tells me that some of them are of inferior quality, and of course their scale is smaller than ours. But the longer time interval covered may counterbalance these disadvantages, and permit some definite conclusion to be drawn. The 100 inch telescope [to come into use in 1919] should of course be of the greatest service in the further prosecution of this work, which will be extended to other spirals.
>
> Until we get some confirmation of the Preliminary results, please do not mention them, as it would be unfortunate to have reports abroad which may not be substantiated. In the case of the general magnetic field of the Sun, although some of the earliest plates showed the small displacements due to the Zeeman effect, I did not think it safe to publish the results until over a year later, after hundreds of additional photographs had been taken and measured. In the present problem I hope we can attain a sound conclusion more quickly, but we must make no mistake in a matter of such fundamental importance. I will send you van Maanen's measures, however, as soon as his reductions have been completed. Your discussion of the subject is most interesting, and will be valuable in connection with future work [February 7, 1916, HLMtW, XI Director's Files, box 153].

We now know that Hale's results on the general magnetic field of the Sun also were spurious—apparently another case of the transformation of expectation into purported observation, though perhaps not all historians and astronomers would agree [Stenflo, 1970; Hetherington, 1975b; Hetherington, 1983; Hetherington, 1988]. The issue here, however, is the attempt to draw from Chamberlin's planetesimal theory ideas that could provide focus for a research program on spiral nebulæ at the Mt. Wilson Observatory.

ASTRONOMERS RESPOND TO CHAMBERLIN'S THEORY

Even before van Maanen had begun comparing photographs, Hale was circulating at Mt. Wilson Chamberlin's paper prepared during the summer of 1915 on the nebular problem. Harlow Shapley had some frank and adverse comments, which were not intended "to go to Chamberlin in this form." Shapley found Chamberlin's treatise "an interesting speculative treatment of an almost necessarily speculative problem." Most damning was Shapley's summation: "There appears to be extreme vagueness in most places where the data of observation are referred to" [November 1, 1915, HLMtW, XI Director's Files, box 168]. Hale seems not to have forwarded these comments to Chamberlin immediately. They would not have been easy to communicate to a friend. Apparently before learning of the adverse comments, and on his own initiative, Chamberlin wrote to Hale, on July 1, 1916:

> It is clear that I should discuss more fully the various kinds of consequences that may be supposed to follow from dynamic encounters of different types. I am persuaded that the process is one of quite varied potency, both in kind and degree. This I indicated somewhat more fully than heretofore in the sixth chapter of my little book [Chamberlin, 1916b; chapter six, "Dynamic Encounter by Close Approach," discusses the possible origin of spiral nebulæ in such encounters between stars], but it needs further deployment to forestall misapprehension and to stimulate the proper use of dynamic encounter in the interpretation of the phenomena to which it is applicable. I do not mean by this that I can hope to point out the variety of concrete results that may spring from such encounter, for the subject is one of immense difficulties, and quite beyond my limitations; but I think something more can be suggested than has yet been done, and this may be helpful [HLMtW, XI Director's Files, box 153].

Chamberlin's admission that his discussion of the nebular problem could be more fully developed provided Hale with as gracious a way as possible of advancing much the same opinion, and he now communicated Shapley's frank remarks to Chamberlin only slightly paraphrased. Hale also advised him on the form of presentation for astronomers:

> Most men will throw aside a very long paper for possible examination at some future time, and if you want observational astronomers to attack your questions the best way to secure their aid is through a short paper containing perfectly specific observational suggestions. A list of criteria and an outline of the kinds of observations which would be most useful from your point of view, are the main things the astronomer, crowded for time and overwhelmed with the demands of other observations, is most likely to prize.
>
> Please don't mistake my meaning. I appreciate more than I can tell you the assistance and stimulation you have given us and the great value to all of the Observatory staff of your visit to Mount Wilson. I also appreciate highly the value of this manuscript, and if your important little book had not appeared I should have avoided such comments. As it is, however, I am sure you will get the best results by referring whenever you can to your book and by making the appeal to the observer as concrete as possible [July 27, 1916, HLMtW, XI Director's Files, box 153].

Hale's implicit appraisal of astronomers, in general, as narrowly focused on particular observational problems and with little interest in more abstract theory, was all too true, although Hale, himself, was an outstanding exception. A most egregious example of astronomers' reluctance to link observations and theory was the decision by the editors of the *Astronomischer Jahresbericht* in 1923 to exclude henceforth from their yearly survey of astronomical papers articles on relativity theory judged not to be of astronomical interest [Hetherington, 1973]. Speculations had to be translated into concrete observational consequences before the attention of many astronomers could be awakened. Indeed, well into the 1920s, philosophical and observational strands of relativistic cosmology, now inextricably intertwined, largely developed independently [Kerszberg, 1989].

Hale's and Chamberlin's intellectual breadth and openness were advanced for their time. Later, the CIW explicitly advanced cooperative research, in a collection of essays titled *Cooperation in Research*. The book honored John C. Merriam, retiring CIW president, who stressed cooperative attack on large problems and who initiated projects involving teams of scientists and even research agencies as early as 1920 [Day, 1938]. Edwin Hubble, a Mt. Wilson astronomer, wrote in the volume that:

> Cooperation is an important and distinctive feature of nebular research at Mount Wilson. . . . Among these individuals, there is a marked tendency toward combining resources in particular investigations, and interpreting the results in the light of constructive criticism of the group as a whole [Hubble, 1938, p. 99, footnote 1].

Hubble's vision of cooperative research as bringing together different approaches and ideas marked a considerable advance since the founding of the Institution and the first report of its advisory committee on astronomy in 1902. That committee considered cooperation as sharing work already defined, rather than as bringing in new ideas to broaden the definition. Cooperation was foreseen in "large fields of astronomical work involving masses of observations and computation which cannot be successfully dealt with by any single existing observatory" [Anonymous, 1903a, p. 90]. In the 1920s and 1930s, Hubble's demonstration of the general recession of galaxies depended on cooperation with Milton Humason, a great observer, and Hubble's interpretation of the observations as an indication of an expanding universe depended on cooperation with Richard Tolman, a theoretical physicist [Hetherington, 1990b]. Mrs. Hubble later remembered that during the 1930s:

> About every two weeks some of the men from Mount Wilson and Cal Tech came to the house in the evening . . . astronomers, physicists, mathematicians. They brought a blackboard from Cal Tech and put it up on the living-room wall. In the dining-room were sandwiches, beer, whiskey and sodawater; they strolled in and helped themselves. Sitting around the fire, smoking pipes, they talked over various approaches to problems, questioned, compared and contrasted their points of view—someone would write equations on the blackboard and talk for a bit, and a discussion would follow [HLMtW, Hubble Collection, quoted in Hetherington, 1982, p. 48].

Mrs. Hubble's reminiscences often have not proven reliable in a factual sense [Osterbrock et al., 1991, p. 1]. They can be used, however, as manifestations of intellectual values, including the virtue of cooperation in scientific research.

CONCLUSION

The ideal of vigorous interdisciplinary cooperation was not held by all astronomers at Mt. Wilson. This is evident in the unfolding of the attempt by Chamberlin and Hale to cooperate on the nebular problem. Responding to Hale's gentle indication of a lack of interest by other Mt. Wilson astronomers in the planetesimal hypothesis, Chamberlin replied that he now proposed to prepare a paper:

> into the form of a round-up report on my cosmological studies—undertaken with no other serious purpose than to find a good foundation for geological doctrine—to be offered to the Carnegie Institution, on my own grounds, as such final report [Chamberlin, 1918], and then prepare for you, if it shall still seem desirable, a much briefer paper which shall consist merely of Working suggestions as to observations, experiments, and interpretation.
>
> With the completion of this paper, I should have done all that I have any serious purpose to do in cosmologic lines, and wish now to give my next serious attention to secular climates [July 31, 1916, HLMtW, XI Director's Files, box 153].

Hale expressed hearty approval for Chamberlin's plan:

> It seems to me that a full explanation of your views is essential, and that this should be accompanied by a short paper, say of not more than ten printed pages, giving a series of concrete suggestions suitable for the use of the observers [August 3, 1916, HLMtW, XI Director's Files, box 153].

It is unlikely, however, that any further number of papers from Chamberlin could have had much impact on observational programs at Mt. Wilson. The nebular campaign, begun with hope and enthusiasm, had run up against formidable obstacles.

One obstacle was specific and technical. In retrospect, the task of drawing detailed quantitative observational predictions out of the planetesimal theory was beyond the mathematical and computational capabilities of the age. Admittedly, van Maanen's reported measurements of rotation of spiral nebulæ did follow from Chamberlin's visit to Mt. Wilson, and were this result valid, we might well be telling a very different story here. Nonetheless, quantitative statements from Chamberlin's planetesimal theory were not forthcoming. And without specific suggestions, there was little for observational astronomers to sink their teeth into.

Another more general obstacle confronted the nebular campaign, related to the lack of quantitative predictions. Brush has concluded that "it is surprising not that Chamberlin failed in his difficult and delicate undertaking, but that he came so close to success" [Brush, 1978, p. 97]. Chamberlin was trying to develop an "interfield" theory, a theory generated within geology but requiring for its testing observations from astronomy. As early as the beginning of the Carnegie Institution, its advisory committee on geophysics (of which Chamberlin was a member) had defined the field as "founded on pure physics and chemistry; its data are supplied chiefly by geology; and the ramifications of its superstructure extend into astronomy and astrophysics" [Anonymous, 1903b, p. 28].

Interfield theories can focus attention on areas previously neglected and suggest new kinds of observations. Chamberlin's planetesimal hypothesis did this, in suggesting observations—spectroscopic or photographic—of motions in spiral nebulæ. But in ranging over many diverse fields, interfield theories inevitably perform less well in each individual field than do special theories limited to the single field. And since scientists are more often specialists than generalists, they more often judge a theory primarily by its usefulness in a very narrow context [Brush, 1978, pp. 95-96].

At the University of Chicago, Chamberlin had initiated, with financial backing from the Carnegie Institution, "collaboration on different phases of complex subjects of investigation" [Chamberlin, 1904, p. 261]. Had his interdisciplinary research team continued beyond his lifetime, it might have created a new field of planetary science encompassing geology, astronomy, physics, chemistry, and mathematics decades before this actually happened [Brush, 1978, p. 23]. Now, with planetary science firmly established, we can better appreciate the magnitude of the task that Chamberlin and Hale began, only partially aware of both the potential and the inherent difficulties.

REFERENCES

Adams, W. S., Early days at Mount Wilson, *Pub. Astr. Soc. Pacific*, *59*, 213-231 and 285-304, 1947.

Anonymous, Report of advisory committee on astronomy, *CIW Year Book [for] 1902*, No. 1, pp. 87-93, CIW, Washington, D.C., 1903a.

Anonymous, Report of advisory committee on geophysics, *CIW Year Book [for] 1902*, No. 1, pp. 26-43, CIW, Washington, D.C., 1903b.

Berendzen, R., and R. Hart, Adriaan van Maanen's influence on the island universe theory, *J. Hist. Astron.*, *4*, 46-56 and 73-98, 1973.

Berendzen, R., R. Hart, and D. Seeley, *Man Discovers the Galaxies*, 228 pp., Science History Publications, New York, 1976.

Brashear, R. S., History of modern astronomy sources at the Henry E. Huntington Library, *Am. Inst. Phys. Hist. Newslett.*, *23*, 3-5, November 1990.

Brush, S. G., A geologist among astronomers: the rise and fall of the Chamberlin-Moulton cosmogony, *J. Hist. Astron.*, *9*, 1-41 and 77-104, 1978.

Carnegie, A., Trust deed by Andrew Carnegie creating a trust for the benefit of the Carnegie Institution of Washington, D.C., *CIW Year Book [for] 1902*, No. 1, pp. xi-xiv, CIW, Washington, D.C., 1903.

Chamberlin, R. T., Biographical memoir of Thomas Chrowder Chamberlin 1843-1928, *Nat. Acad. Sci. Bio. Mem.*, *15*, 307-407, 1934.

Chamberlin, T. C., A group of hypotheses bearing on climatic changes, *J. Geo.*, *5*, 653-683, 1897.

Chamberlin, T. C., An attempt to test the nebular hypothesis by the relations of masses and momenta, *J. Geo.*, *8*, 58-73, 1900.

Chamberlin, T. C., On the possible function of disruptive approach in the formation of meteorites, comets, and nebulæ, *Astrophys. J.*, *14*, 17-40, 1901.

Chamberlin, T. C., Study of fundamental problems of geology, *CIW Year Book [for] 1903*, No. 2, pp. 261-270, CIW, Washington, D.C., 1904.

Chamberlin, T. C., Study of fundamental problems of geology, *CIW Year Book [for] 1915*, No. 14, p. 368, CIW, Washington, D.C., 1916a.

Chamberlin, T. C., *The Origin of the Earth*, 271 pp., Univ. of Chicago Press, Chicago, 1916b.

Chamberlin, T. C., Study of fundamental problems of geology, *CIW Year Book [for] 1917*, No. 16, pp. 307-319, CIW, Washington, D.C., 1918.

Chamberlin, T. C., and F. R. Moulton, Certain recent attempts to test the nebular hypothesis, *Science*, *12*, 201-208, 1900.

Chamberlin, T. C., and F. R. Moulton, The development of the planetesimal hypothesis, *Science*, *30*, 642-645, 1909.

Day, A. L., An adventure in scientific collaboration, in *Cooperation in Research*, pp. 3-35, CIW, Washington, D.C., 1938.

Fernie, J. D., The historical quest for the nature of the spiral nebulæ, *Pub. Astr. Soc. Pacific*, *83*, 1189-1230, 1970.

Gasteyer, C. E., Forest Ray Moulton, April 29, 1872 - December 7, 1952, *Nat. Acad. Sci. Bio. Mem.*, *41*, 341-355, 1970.

Hale, G. E., Report of Director of the Solar Observatory, Mount Wilson, California, *CIW Year Book [for] 1905*, No. 4, pp. 56-77, CIW, Washington, D.C., 1906.

Hale, G. E., *The Study of Stellar Evolution. An Account of Some Recent Methods of Astrophysical Research*, 252 pp., Univ. of Chicago Press, Chicago, 1908.

Hale, G. E., *Ten Years' Work of a Mountain Observatory. A Brief Account of the Mount Wilson Solar Observatory of the Carnegie Institution of Washington*, 98 pp., Pub. No. 235, CIW, Washington, D.C., 1915.

Hetherington, N. S., The Shapley-Curtis debate, *Astr. Soc. Pacific Leaf.*, *490*, 1-8, 1970.

Hetherington, N. S., Adriaan van Maanen and internal motions in spiral nebulæ: a historical review, *Q. J. Roy. Astr. Soc.*, *13*, 25-39, 1972.

Hetherington, N. S., The delayed response to suggestions of an expanding universe, *J. Br. Astr. Assoc.*, *84*, 22-28, 1973.

Hetherington, N. S., Edwin Hubble on Adriaan van Maanen's internal motions in spiral nebulæ, *Isis*, *65*, 390-393, 1974a.

Hetherington, N. S., Edwin Hubble's examination of internal motions in spiral nebulæ, *Q. J. Roy. Astr. Soc.*, *15*, 392-418, 1974b.

Hetherington, N. S., Adriaan van Maanen on the significance of internal motions in spiral nebulæ, *J. Hist. Astron.*, *5*, 52-53, 1974c.

Hetherington, N. S., The simultaneous "discovery" of internal motions in spiral nebulæ, *J. Hist. Astron.*, *6*, 115-125, 1975a.

Hetherington, N. S., Adriaan van Maanen's measurements of solar spectra for a general magnetic field, *Q. J. Roy. Astr. Soc.*, *16*, 235-244, 1975b.

Hetherington, N. S., Additional Shapley-van Maanen correspondence, *J. Hist. Astron.*, *7*, 73-74, 1976.

Hetherington, N. S., Observational astronomy in the twentieth century, in *Human Implications of Scientific Advance. Proceedings of the XVth International Congress of the History of Science Edinburgh 10-19 August 1977*, edited by E. G. Forbes, pp. 567-575, Edinburgh Univ. Press, Edinburgh, 1978.

Hetherington, N. S., Philosophical values and observation in Edwin Hubble's choice of a model of the universe, *Hist. Stud. Phys. Sci.*, *13*, 41-67, 1982.

Hetherington, N. S., Just how objective is science?, *Nature*, *306*, 727-730, 1983.

Hetherington, N. S., *Science and Objectivity: Episodes in the History of Astronomy*, 168 pp., Iowa State Univ. Press, Ames, 1988.

Hetherington, N. S., *The Edwin Hubble Papers. Previously Unpublished Manuscripts on the Extragalactic Nature of Spiral Nebulæ*, 250 pp., Pachart Publishing House, Tucson, 1990a.

Hetherington, N. S., Edwin Hubble's cosmology, *Am. Sci.*, *78*, 142-151, 1990b.

Hetherington, N. S., and R. S. Brashear, The Hubble-van Maanen conflict over internal motions in spiral nebulæ: yet more new information on an already old topic, *Vist. Astron.*, *34*, 415-423, 1991.

Hetherington, N. S., and R. S. Brashear, Walter S. Adams and the imposed settlement between Edwin Hubble and Adriaan van Maanen, *J. Hist. Astron.*, *23*, 53-56, 1992.

Hoskin, M. A., Edwin Hubble and the existence of external galaxies, in *Actes V, XIIe Congres International d'Histoire des Sciences*, pp. 49-53, Blanchard, Paris, 1971.

Hoskin, M. A., The 'great debate': what really happened, *J. Hist. Astron.*, *7*, 169-182, 1976.

Hubble, E., Explorations in the realm of the nebulæ, in *Cooperation in Research*, pp. 91-102, CIW, Washington, D.C., 1938.

Keeler, J. E., The Crossley reflector of the Lick Observatory, *Astrophys. J.*, *11*, 325-349, 1900.

Keeler, J. E., Photographs of nebulæ and clusters, made with the Crossley reflector, by James Edward Keeler, Director of the Lick Observatory, 1898-1900, *Pub. Lick Obs.*, *8*, 1-46, 1908.

Kerszberg, P., *The Invented Universe: The Einstein-De Sitter Controversy (1916-17) and the Rise of Relativistic Cosmology*, 403 pp., Oxford Univ. Press, Oxford, 1989.

Kevles, D. J. (Ed.), *Microfilm Edition of THE GEORGE ELLERY HALE PAPERS 1882-1937 at the Mount Wilson and Palomar Observatories Library, Pasadena, California*, 100 reels and 1,000 frames per reel, CIW, Washington, D.C., and California Institute of Technology, Pasadena, 1968.

Miller, R. H., Experimenting with galaxies, *Am. Sci.*, *80*, 152-163, 1992.

Mills, D. J., George Willis Ritchey and the development of celestial photography, *Am. Sci.*, *34*, 64-93, 1966.

Moulton, F. R., An attempt to test the nebular hypothesis by an appeal to the laws of motion, *Astrophys. J.*, *11*, 103-130, 1900.

Moulton, F. R., On the evolution of the solar system, *Astrophys. J.*, *22*, 165-181, 1905.

Osterbrock, D. E., *James E. Keeler, Pioneer American Astrophysicist and the Early Development of American Astrophysics*, 411 pp., Cambridge Univ. Press, Cambridge, 1984.

Osterbrock, D. E., R. S. Brashear, and J. A. Gwinn, Self-made cosmologist: the education of Edwin Hubble, in *Evolution of the Universe of Galaxies. Edwin Hubble Centennial Symposium*, edited by R. G. Kron, pp. 1-18, Astronomical Society of the Pacific, San Francisco, 1991.

Reingold, N., and I. H. Reingold, *Science in America. A Documentary History*, 490 pp., Univ. of Chicago Press, Chicago and London, 1981.

Schweizer, F., Colliding and merging galaxies, *Science*, *231*, 227-234, 1986.

Shapley, H., On the existence of external galaxies, *Pub. Astr. Soc. Pacific*, *31*, 261-268, 1919.

Shapley, H., and H. D. Curtis, The scale of the universe, *Bull. Nat. Res. Coun.*, *2*, 171-217, 1921.

Smith, R. W., *The Expanding Universe: Astronomy's Great Debate*, 220 pp., Cambridge Univ. Press, Cambridge, 1982.

Smith, R. W., The great debate revisited, *Sky & Tel.*, *65*, 28-29, 1983.

Stenflo, J. O., Hale's attempt to determine the sun's general magnetic field, *Solar Phys.*, *14*, 263-273, 1970.

Stratton, F. J. M., Cosmogony, *Mon. Not. Roy. Astr. Soc.*, *70*, 366-368, 1910.

Struve, O., A historic debate about the universe, *Sky & Tel.*, *19*, 398-401, 1960.

van Maanen, A., Preliminary evidence of internal motion in the spiral nebula Messier 101, *Astrophys. J.*, *44*, 210-228, 1916.

Wright, H., *Explorer of the Universe: A Biography of George Ellery Hale*, 480 pp., E. P. Dutton, New York, 1966.

Wright, H., J. N. Warnow, and C. Weiner (Eds.), *The Legacy of George Ellery Hale. Evolution of Astronomy and Scientific Institutions, in Pictures and Documents*, 293 pp., Massachusetts Institute of Technology Press, Cambridge, MA, 1972.

Women and Women's Work at Mt. Wilson Observatory before World War II

JOHN LANKFORD
History Department, West Virginia University
Morgantown, WV 26506-6303

The history of women in science has become an important field of scholarly inquiry [Rossiter, 1982; Abir-Am and Outram, 1989]; both history of science and woman's studies programs offer undergraduate and graduate courses in the area. Women in astronomy is a topic that has attracted a great deal of attention [Kidwell, 1990; Lankford and Slavings, 1990]. Between 1859 and 1940, approximately a third of the membership of the American astronomical community were women, and the rise of American astronomy to world class status was made possible, in part, through the work of women at the great factory observatories.

It should be understood that when historians focus on the role and experiences of women in modern science they are moving beyond an exclusive interest in elite (mostly male) scientists. Any deep understanding of the social history of modern science must begin with an inclusive view of the population of scientific communities [Lankford, forthcoming].

Demand for female workers in the American astronomical community can be traced to the emergence of factory observatories in the late nineteenth century. Table 1 reports data on factory observatories that employed women.

TABLE 1: Employment of Women, 1859-1940. Location of First Job at Nine Leading Factory Observatories

Harvard College Observatory	85
Dudley Observatory●	83
Mt. Wilson Observatory●	53
Lick Observatory/University of California at Berkeley○	22
U.S. Naval Observatory/ Nautical Almanac Office	14
Yerkes Observatory/ University of Chicago○	12
Allegheny Observatory Smithsonian Astrophysical Obsy. Yale University Observatory N=	10 each 30

N=299. Nine institutions account for the first jobs for 70% of Women (N=426) working in American Astronomy.

● Funded by the Carnegie Institution of Washington
○ Includes some women whose first job was as a graduate assistant.

The industrialization of the national economy provided the model for American astronomy. These great observatories came to resemble factories in which knowledge was mass produced in a fashion similar to the mass production of consumer goods. New research technologies, notably photography [Lankford, 1984], dramatically increased data acquisition rates. This was especially true in stellar spectroscopy, photometry, and solar physics. This increase in data demanded an increase in the work force devoted to its reduction and analysis. Expansion of the scientific work force was supported by private patrons and, beginning around 1900, philanthropic foundations like the Carnegie Institution of Washington (CIW). As competition between factory observatories increased, an efficient and economical labor force became key to success. Institutions organized to produce knowledge in a cost effective manner were best able to attract support.

The new factory observatories differed from their nineteenth-century counterparts. Directors became chief-executive officers with duties that ranged from fund raising to the selection of research programs and the organization and supervision of a scientific work force composed of astronomers, computers, and technicians, as well as night assistants, secretaries, bookkeepers, and laborers. Directors of factory observatories followed the example of the business community as they sought to rationalize work. Central to the organization of a factory observatory was the division of labor. In a real sense, the division of labor was gender specific. Many activities were defined as either men's or women's work. For example, men observed at the telescope while women were day workers, measuring and reducing the data collected by men. Generally men wrote-up the results for publication.

In large institutions like CIW's Mt. Wilson, the scientific

The Earth, the Heavens and the Carnegie Institution of Washington
History of Geophysics Volume 5

work force was hierarchically ordered. At the bottom were unskilled workers, generally those with no more than a high school education. The conditions of employment for this group were literacy, numeracy, and good handwriting. Most of the computers were unskilled women, working at routine tasks that required only minimal on-the-job training, thus limiting employer investments in worker education. Women with undergraduate or master's degrees in astronomy and/or mathematics made up the second level: semi-skilled workers. They usually had experience with various forms of computing as well as the use of astronomical instruments. Women with Ph.D.'s represented skilled labor with a detailed knowledge of the craft. Very few female Ph.D.'s were employed by factory observatories. Most became professors in the women's colleges like Mount Holyoke or Vassar.

Factory observatories tended to employ more unskilled than semi-skilled or skilled woman workers. Both Mt. Wilson and the Dudley Observatory, also funded by the CIW, [B. Boss, 1968] illustrated this practice. Benjamin Boss, second director of the Dudley Observatory, wrote, "All of the girls we employ are merely high school graduates with no experience in astronomical work and no scientific training. These girls start with a salary of $750 a year." Advancement depended on "efficiency, longevity of service and vacancies." Boss insisted that "under his system the girls of promise soon acquire skill and training and are encouraged as far as possible through promotion. . . . Only two out of ten girls are paid a salary over $100 per month" [SALO, B. Boss to W. W. Campbell, April 18, 1922]. The previous Dudley director described the work force that prepared the *San Luis* catalogue of southern stars. "I have a well organized corps of young women for the routine work. Everything is always prepared for them to do in large blocks from end to end. It is an extremely economical system" [DOA, L. Boss to D. Gill, March 16, 1911]. The work was done on large printed forms and carried out in duplicate so that errors could be detected. At no point were these women expected to envision, let alone understand, the final product. Dudley directors apparently recruited unskilled female workers through local Albany high school. In no case were these female employees viewed as permanent. The statuses of unskilled and semi-skilled women workers in the great factory observatories were often blurred. The pay differentials between the two were significant.

There were also gender-based differences in hiring and rewarding staff members at the factory observatories. Males appointed as computers or assistants had a wider range of duties and responsibilities. For example, they were hired with the understanding that they could observe. Sometimes men were employed as computers with the clear understanding that this was an entry-level position which would provide mobility up the career ladder. As far as I know, such promises were never made to women.

Table 2 presents data on length of career, controlling for highest level of educational attainment for women who worked at Mt. Wilson before World War II. For the unskilled, more than four-fifths were gone in five years. Among semi-skilled college graduates, three-fifths remained no longer than five years. The number of women with M.A. and Ph.D. degrees is too small to reliably indicate trends.

TABLE 2: Career Length by Level of Educational Attainment for Women at Mt. Wilson Observatory, 1906-1940

High School (N=38)	
1 year or less	18 (47.4%)
2-5 years	14 (37%)
6-10 years	4 (10.2%)
More than 10 years	2 (5.3%)
Baccalaureate Degree (N=15)	
1 year or less	4 (27%)
2-5 years	5 (33.3%)
6-10 years	2 (13.3%)
More than 10 years	3 (20%)
Master's Degree (N=5)	
1 year or less 0	
2-5 years	1 (20%)
6-10 years	2 (40%)
More than 10 years	1 (20%)
Missing data =	1 case
Doctor's Degree (N=3)	
1 year or less	1
2-5 years	1
6-10 years	0
More than 10 years	1•

• Staff Wife. Non-regular appointment. H. Pettit.

High turnover rates among workers are suggestive, indicating a dual labor market [Cain 1975]. A dual labor market is characterized by a two-tier system. At the bottom level are low wages, limited access to on-the-job training, and few promotion ladders leading to greater responsibility, pay or status. This lower tier of a dual labor market is often comprised of ethnics, young people, and/or women. These markets are, in effect, career dead ends.

The schedules of female computers were similar. Generally, the work day began at eight, with a lunch break over the noon hour and ended at five. The factory observatories required women to work from eight to noon on Saturday. In the early days, Mt. Wilson defined Saturday work as extending to three p.m., but soon let the women go at noon.

Women computers all did similar tasks. They measured the positions and magnitudes of stars on photographic plates, determined the position and intensity of spectrum lines, counted sun spots, and determined their intensity and movement across the solar disk. Most of these women performed myriad routine calculations ranging from those connected with astrometric work at the Dudley Observatory to astrophysical investigations at Mt. Wilson. Indeed, the origin of the word computer, as used in this context, relates to the arithmetic operations carried out by assistants.

Ruth Emily Smith (Vassar, 1906) worked at Mt. Wilson for a decade. She assisted George Ellery Hale and Walter S. Adams with solar physics research. She must have been very capable, for she was placed in charge of the photographic darkroom dedicated to solar research. Later she coauthored a paper with the visiting Italian astronomer, George Abetti. Few women were given as much responsibility. Phoebe Waterman (Vassar, 1904) found work as a computer difficult, tedious, and boring. When she sought to negotiate some special projects to be pursued independently, Waterman was rebuffed by the male leadership and soon left for Berkeley to work on a Ph.D. in astronomy. While several women remained at Mount Wilson for many years, the experience of most was closer to Phoebe Waterman than Ruth Emily Smith. Few of these women, however, moved from a computer's job to a Ph.D. program in astronomy. Most women probably traded the post of computer at a great factory observatory for marriage and family.

WOMEN IN THE GEOSCIENCES

If scholars know a great deal about the experiences of women in astronomy and their role in its development, much less is known about women in the geosciences. When did women first become active in the geosciences? How have their numbers grown in the twentieth century? What positions have they occupied? What was their training and were they able to achieve professional mobility, including access to the reward system of modern science as well as to power within a discipline?

These questions also can be asked of the various laboratories supported by the CIW. From Cold Spring Harbor to the Desert Laboratory at Tucson, we know that women were active in various capacities, but have little more than impressionistic evidence with which to work. A good place to begin would be with an historical census of women scientists who worked for the CIW. To be sure, such a census must deal with all women scientists connected with the CIW, not just the distinguished few, whose names are already known to history. The Barbara McClintock's are important, but they are just the beginning of a much larger story.

ARCHIVAL SOURCES

The Dudley Observatory Archives, located in Poughkeepsie, New York are abbreviated DOA. The Mary Lee Shane Archives of the Lick Observatory, located at the McHenry Library, University of California at Santa Cruz, are abbreviated SALO in the text.

REFERENCES

Abir-Am, P. G. and D. Outram (eds.), *Uneasy Careers and Intimate Lives: Women in Science, 1789-1979*, Rutgers Univ. Press, Rutgers, 1989

Boss, B., *History of the Dudley Observatory*, the Dudley Observatory, Albany, 1968.

Cain, G. G., The Challenge of the Dual and Radical Theories of the Labor Market to Orthodox Theory, *Am. Econ. Re.*, *65*, 16-22. 1975.

Kidwell, P. A., Three Women of Astronomy, *Amer. Sci.*, *78*, 244-51, 1990.

Lankford, J., The Impact of Astronomical Photography to 1920, *Astrophysics and Twentieth-Century Astronomy to 1950*, O. Gingerich, ed., pp. 16-39, Cambridge Univ. Press, Cambridge, Eng., 1984.

Lankford, J., *Community, Careers and Power: American Astronomy, 1859-1940.* Chicago, Univ. of Chicago Press. (Forthcoming).

Lankford, J., and R. L. Slavings, Gender and Science: The Experience of Women in American Astronomy, 1859-1940, *Physics Today*, *43*, 58-65, 1990.

Commentary on the Mt. Wilson Papers

OWEN GINGERICH
Harvard-Smithsonian Center for Astrophysics
60 Garden St., Cambridge, MA 02138

Rarely, if ever, have I heard a series of four such high-quality papers back-to-back. If I go on to suggest that we need something more, it is not to criticize the papers as given, but to point to a broader agenda for our historical studies. First, however, let me give some immediate and specific responses to the four papers on the early years of Mt. Wilson Observatory.

A young friend has enthusiastically described for me his activities in the Society for Creative Anachronism, an organization that restages medieval fairs and tournaments, and whose members practice calligraphy and hammer out their own chain mail. The society is devoted to the Middle Ages "not as they were, but as they should have been," namely, the use of English and equal rights for women. I sometimes wonder if we historians are from time to time rewriting history along these lines. For example, when Kenneth Lang and I published the *Source Book in Astronomy and Astrophysics, 1900-1975*, we included the paper on the period-luminosity relation of Cepheid variables with Henrietta Leavitt as author. In reality, although the actual authorship was never in doubt, the original paper was published under the name of Edward Pickering, the Director of Harvard Observatory.

John Lankford's paper, on the role of the female underclass at Mt. Wilson Observatory, is a revelation: not the fact that the women there had no opportunity to aspire to the conspicuous and familiar roles of the male astronomers, but the fact that they were there at all and in such abundance, in numbers second only to the cadre at Harvard. But at Harvard the names of Henrietta Leavitt, Williamina Fleming, Antonia Maury, and Annie J. Cannon come easily to mind, whereas at Mt. Wilson the sea of anonymity extends to the 1950s when Henrietta Swope arrived. As yet the difference remains to be explained.

It would be anachronistic—creative perhaps, but nonetheless anachronistic—to blame Hale for the lack of opportunities for serious astronomical careers for women at Mt. Wilson. Rather, there are other questions to be asked. What were the alternatives? Women with clerical skills could become part of the vast secretarial pool at Sears, Roebuck, shown dramatically on an old stereo card I've collected, but was that more challenging or rewarding? What were the possibilities and rates of pay in biological sciences? In school teaching? John Lankford has given a start by tabulating a few comparative salaries, but a broader base of comparison would help us evaluate the place of these women in the larger turn-of-the century societal structures.

There is another side to the coin of the "factory observatories," as Lankford has called them. Several years ago Howard Plotkin carefully read through the entire, extensive Edward Pickering correspondence in the Harvard Archives, and then came to me in some distress, saying that after around 1900 Pickering was no longer doing any science. He had become a scientific administrator, the CEO of his astronomical factory. My response was to suggest that Pickering had organized enough resources to command a much larger effort than he could accomplish single-handedly. It was not that Pickering was no longer doing science, but he wasn't doing personal science. One has to look far back in history, perhaps to Tycho Brahe, to find a good parallel. Tycho was a scientific entrepreneur of the first rank. Today he gets the credit for building the instruments and carrying out the observations. But he did not personally build the instruments, though he designed them, and his numerous apprentices carried out many of the observations he is credited with. This is not, I believe, unjust, for it was his vision and planning of the observational attack that gave the enterprise direction and meaning. He also recruited theoreticians, and when we now find that he appropriated some of their results, we react with a sense of scandal, but in reality there is a genuine ambiguity.

Like Tycho and Pickering, Hale was a first-class astronomical entrepreneur. It was Hale's remarkable vision that included a physical laboratory with the observatory and that provided a staff position for A. S. King therein. Russell's papers on spectral multiplets did not carry Hale's name as a collaborator, but in a powerful sense Hale made the work possible. Today, the sensitive matter of credits is exacerbated by the immense number of individuals who might be involved in a complex project. Recently at the meeting of the American Astronomical Society, an invited speaker remarked that he should acknowledge his

The Earth, the Heavens and the Carnegie Institution of Washington
History of Geophysics Volume 5

collaborators, and then flashed on the screen an overhead projection crammed full of names to the point of illegibility. Especially with space astronomy, the number of collaborators of a project is so immense that a group synonym like Bourbaki or *la Pleiade* might make good sense. Yet often the vision of a strong organizer, such as Riccardo Giacconi, makes the science a reality, even though he may be far from the details of the actual data. The matter of credits is in a continual state of evolving tension, and the building of Mt. Wilson provides a particularly interesting case study.

Though Pickering and Hale were the two major astronomical entrepreneurs of their age, the contrasts between them are perhaps even more striking than the similarities. When Pickering took the Harvard directorship in 1876, his was a radical appointment; by the 1900s his was the conservative approach, arguing for support for the understructure as opposed to Hale's search for "the exceptional man." It is, in fact, Hale's exceptional men that are the focus of the three earlier papers of this session.

If we look at Hale's recruits, Harlow Shapley and Edwin Hubble head the list; both have been the subject of repeated studies. Among the visitors, Jacobus Kapteyn and Henry Norris Russell are the stars. What about some of the others? How about Walter S. Adams, who seems a figure in the shadows, yet as we have heard, hit it off particularly well with fellow New Englander Abbot (and, as we know, not so well with midwesterners Shapley and Hubble). There is an interesting letter, from Russell to Shapley, in which the Princeton astronomer asks his erstwhile student, who was by then at Harvard, to clarify a point of priority with Ejnar Hertzsprung, who was then visiting the Harvard Observatory, because, writes Russell, "I don't wish to deny to Hertzsprung any credit which he deserves, nor on the other, do I want another Adams-and-Kohlschutter situation." [Harvard University Archives, Russell to Shapley, March 10, 1927] Adams together with Arnold Kohlschutter, who was a young German visitor to Mt. Wilson, is credited with the method of spectroscopic parallax. Did Russell feel that these credits were misattributed? The Mt. Wilson mysteries have not all been settled.

The three papers here, on Abbot, Chamberlin, and Russell, help clarify Hale's goal of identifying "the exceptional man." At age 100 Abbot was still regularly visiting his Smithsonian office. He had become virtually a living fossil, a man who had served that institution as its leader, but who was a man who had stuck so single-mindedly to finding solar variation, a will-o-the-wisp that in the end eluded him, that he had fallen out of the mainstream of science. Ron Doel, in a perceptive piece of archival research, had demonstrated the way in which the Smithsonian Observatory used the onset of the space age to redefine its goals and to unyoke the millstone about its neck, that is, the continued search for solar variation and its terrestrial linkages. It is good to have the earlier Abbot placed in focus, to be reminded that his improved value for the solar constant has indeed passed the test of time, and to see him in a role where he does stand out as "the exceptional man."

In his 1957 obituary in *Publications of the Astronomical Society of the Pacific*, Otto Struve described Henry Norris Russell as the man who follow "the Royal Road to Success" (thereby choosing an expression that Russell himself had used in the first Russell Lecture of the American Astronomical Society) [Struve, 1957]. The vignette of Russell as the fox among the hedgehogs fits into that pattern: Russell, the non-stop talker, sweeping into Mt. Wilson along one of those roads, picking up the lithium in the solar spectrum one week and going after King's spectral lines in another. One sympathizes with Adams' discomfort seen in the report mentioned by David DeVorkin.

While I have been greatly impressed by the excellence of these papers by Lankford, Hetherington, Brashear, and DeVorkin, I would like to suggest that ultimately we need to place these studies against a richer background. We need to hear, systematically, about the exceptional men and the substratum of women supporting their efforts, but we also need to keep in mind the grander canvas against which these events are playing out. Let me illustrate this at two different levels, first by drawing your attention to a wider context for Chamberlin.

One consequence of the Moulton-Chamberlin collision hypothesis for the origin of our planetary system was that such events would necessarily be very rare, and our solar system then virtually unique in the Milky Way. Looking for habitats for alien intelligences was thus highly restricted, practically limited to Mars. It is no wonder that Mars was so central to Percival Lowell's agenda. At the same time, the spirals were objects of curiosity as potential solar-systems-in-the-making, and not just at the Mt. Wilson Observatory. In his *Astronomy and Cosmogony*, James Jeans showed on his Plate XVI two galaxies (similar to the more famous Whirlpool Nebula) in precisely that context. And at Lowell Observatory, V. M. Slipher was assigned the task of studying the spectra of spirals in the anticipation that such a study would offer insights into the evolution of planetary systems. Insights it offered, indeed, but not where they were expected! Instead, Slipher discovered the high radial velocities that, in Hubble's hands at Mt. Wilson, were to provide the jumping-off place for the theory of the expanding universe.

But secondly, there is for Mt. Wilson a worldwide context that has been entirely missing from this discussion.

Sometime between 1890 and 1920 American astronomy came into ascendancy, outstripping the European efforts in its significance. If we look at the chronology of memorable events in astronomy listed by Agnes Clerke in her *Popular History of Astronomy during the Nineteenth Century*, only 3% of those in the decade of the 1860s were linked with America. By the end of the 1890s, the proportion had soared to nearly 40%. When Pickering began his photometric surveys, Oxford provided strong competition. His early spectrographic surveys were rivalled by the work of Vogel at Potsdam. Yet the centrum was noticeably shifting to the New World. In 1870 only two of the six largest refractors were in America. By 1917, the U.S. had two of the three biggest refractors and three of the three biggest reflectors. Surely Mt. Wilson Observatory and the exceptional men there were key to this sea change. Lick Observatory and Harvard Observatory were part of the scene, but the real pivot was in southern California. This ascendancy came with respect to technologically-driven observational astronomy and observational cosmology. In theoretical work America still lagged. Russell mentioned with admiration "the Cambridge school," a name that has not stuck perhaps in part because there was mostly just Eddington and not really a school. Almost singlehandedly, Eddington kept the hub of theoretical astronomy in Europe, though Russell ran a close second.

A seminal account relating to another field and a different era, Simon Schaffer and Steven Shapin's *The Leviathan and the Air Pump*, contains a fascinating discussion of the fact that only Robert Boyle and a very few others had vacuum pumps good enough to confirm or disprove the discoveries and claims that were being made [Schaffer and Shapin, 1985]. Much the same monopolistic situation prevailed when Hale's unrivalled telescopes on Mt. Wilson could observe nebulae beyond the reach of other astronomers. Was there jealousy outside that exclusive circle? Were there doubters? I believe that this potentially fascinating question has gone entirely unexplored.

Am I right to claim that American astronomy came into ascendancy early in this century, and largely through Hale's program at Mt. Wilson? And am I just in declaring that the larger canvas has been neglected in these papers? I did not want these points to be overlooked in an anniversary of this sort, but primarily I set these claims as a provocation and challenge to this active generation of historians. It is my hope that they will not neglect the bigger picture in which Carnegie's philanthropy played such a seminal role.

REFERENCES

Doel, R., Redefining a Mission: The Smithsonian Astrophysical Observatory on the Move, *J. Hist. Astron.*, *21*, 137-53, 1990.
Schaffer, S. and S. Shapin, *The Leviathan and the Air-Pump*, Princeton, 1985.
Struve, O., *Pub. Astron. Soc. Pacific*, *69*, 223-26, 1957.

The Big Story:
Tuve, Breit, and Ionospheric Sounding, 1923-1928

C. STEWART GILLMOR
History Department, Wesleyan University
Middletown, CT 06459-0002

The four decades before World War II saw immense change in numerous aspects of American and World history. The tremendous growth in chemical and electrical industry in the late nineteenth century had brought the need for regulation and standardization and, in this country and abroad, national standards and testing laboratories were founded in the first decade of the new century. Engineering education had been formed also as part of the explosion of late-Victorian industrial and commercial development. It, along with medicine, became "scientificized," required a university degree for most practice, and increasingly entailed the study of university-level physics and mathematics.

Radio, the automobile, and the airplane joined the world circus at the turn of the twentieth century and soon became subject for speculation, exploitation, humor, philosophy, and progress. Visionaries and others realized some of the global implications of these new technological devices and, in the years just before World War I, attempted to establish world scientific and technical organizations. Such attempts largely were stillborn with the conflagration of 1914. New efforts at world peace, science, and technology would begin after the War.

The 1920s was the period of the flowering of modern mass society in America. America was faced with a new technological society and had become more urban than rural. Though the 1920s were the years of Harding and Coolidge, they were more than anything else the Hoover years. Stanford University mining engineer from its first graduating class, Herbert Hoover as Director of War Relief, Secretary of Commerce, and then President was perhaps the single most important man in American government in the decade [St. Germain,1990]. This was a time of readjustment from war, of "Red Scares," the "return to normalcy," "Teapot Dome" scandals, "The Business of America is Business," the Movies, Prohibition, and the "Flapper". As Al Dubin and Joe Burke wrote [1929] in the song for *Gold Diggers of Broadway*:

Now a days, electricity is a dire necessity,
You can buy your violet rays, And sun yourself on rainy days, When clouds are in the sky;
There's the air'plane and radio,
And the talking picture show,
Ev'ry-thing's electrified, Mechanic'ly we fly and ride,
We'll soon make love that way.

Two essays in this volume, by Gillmor, and Hevly, concern the Carnegie Institution of Washington (CIW) and organizational efforts at ionospheric radio and research in the 1920s and 1930s. The present essay, "The Big Story," centers on Gregory Breit and Merle Tuve and ionospheric research at the Carnegie's Department of Terrestrial Magnetism (DTM) in the mid-twenties and their relationships with friends and rivals at the Bureau of Standards (BS), the Naval Research Laboratory (NRL), corporate radio industries, and in England. But first, a brief survey of radio/ionosphere events before 1920.

In the late 1890s the possibilities for long-distance radio communication brought forth speculations of the maximum range obtainable and the physical mechanisms behind such communication. The achieving of distances of about 100 miles by radio in 1899 and then Marconi's trans-Atlantic feats in 1901-02 and ideas about these experiments marked the beginnings of the modern study of the Earth's upper atmosphere. Two now-celebrated hypotheses concerning long-distance radio communications were advanced by Arthur Edwin Kennelly in the U.S. [1902], and by Oliver Heaviside in England [1902], although experimental proof of what was called the upper conducting layer, Heaviside layer, Kennelly-Heaviside layer, or KHL, and would many years later be termed the ionosphere [Gillmor,1976] was not generally accepted by many authorities for more than two decades. Early theorists treated of the long-distance radio phenomenon as with diffraction of light, but diffraction couldn't explain how a radio wave could travel almost a quarter of the distance around the Earth. Some confused the difference between the induction field and the radiation field in electromagnetic waves. Marconi thought the "day-night" effect (the longer transmission distances achieved during nighttime) might be because sunlight

The Earth, the Heavens and the Carnegie Institution of Washington
History of Geophysics Volume 5

striking the radio antenna formed a sort of wave-shield. Kennelly and Heaviside each supposed there was some sort of conducting layer about a hundred kilometers or so above the Earth which could form a reflecting surface for radio waves to travel along as if on a transmission line.

From about World War I until the satellite era began in the late 1950s two different traditions joined together: 1) Geomagnetism and geophysics and 2) Electrical engineering and radio science. Each in effect assisted the other. Radio people and geophysicists used common tools. The geophysical tradition exploited the new radio techniques to study the Earth's upper atmosphere and magnetic field and even the Sun while radio science used new geophysical ideas to aid commercial and military radio to bring the world "broadcasting," and trans-Oceanic message handling, long-distance radar, and "remote sensing." The correlation between long-distance radio propagation conditions and geomagnetic conditions was first discovered and published by Reginald Fessenden [1908]. But Arthur Kennelly *again* wisely concluded as he did in his first speculations in 1902 that "The whole subject of wave contour at great distances must remain in abeyance until sufficient measurements have been made . . . " [Kennelly, 1909, p.168]. Again and again in radio science as in science in general, progress depends alternately upon theory, phenomenology, and technique. And the technique of quantitative measuring of radio waves and radio wave propagation was almost non-existent in the early twentieth century.

The classical radio measurement standards established by W. Duddell and J. E. Taylor [1905] at this time led on to better wavemeters, precise measurement of high frequency inductance and capacitance, and finally to precise frequency generation and measurement using vacuum tube circuits and then quartz crystals. Long-distance radio was off and well-running: the Marconi Company became fairly dominant in the British Isles but fought for business in North America with the newcomers Fessenden and de Forest. In Germany, Count Arco, Dr. Braun, Dr. Slaby, and others combined their firms into the amalgamated *Gesellschaft für Drahtlose Telegraphie*. In Russia, Alexander Popov struggled for years to get the Russian Admiralty to adopt his radio system and died of a stroke in late 1906 brought on by Russian radio's failure against the Japanese in the battle of Tsushima Straits. But even the Marconi companies remained in considerable financial difficulties for some years. Senior engineers were not being paid, and from 1908-1910, Marconi personally took over as Managing Director in an attempt to restore public confidence in his companies [Geddes, 1974, pp. 22-32].

G. W. Pierce at Harvard proposed in a textbook [1910, p.139] that radio waves propagating along the ground might interact with waves from the same transmitter reflected off the supposed upper conducting layer. If the operating wavelength were altered by as much as a half period, he supposed, the interference might be maximized constructively to the advantage of the radio company. Lee de Forest, in 1911, joined C. F. Elwell's Federal Telegraph Company, which had obtained the rights to the Poulsen arc undamped wave (or continuous wave) transmission system of the Dane Valdemar Poulsen. Federal Telegraph operated the first commercially successful point-to-point land communication in competition with the cable and telegraph systems. First, Federal operated low power stations along the West Coast and then a high-power pair for communication between the West Coast and Hawaii [Villard, 1976].

In 1912 de Forest mentioned in print Pierce's suggestion about interference between the ground wave of a transmitter and its space components reflecting off a supposed upper layer. He claimed that recent operating evidence supported Pierce's suggestion. The Federal arc stations operated using frequency-shift-keying (FSK). For technical reasons, it was easiest to send dots and dashes in arc telegraphy by switching an additional inductor into the circuit rather than turning the arc off and on. The "off" was then simply a frequency shifted about 5% from that of "on". De Forest had noted frequency selective fading, which he attributed to differential wave interference effects. In subsequent correspondence with Pierce, September 13, 1912, de Forest wrote that whereas Pierce had suggested (in a letter) a height of 62 miles, his own evidence supported an upper layer at about 27 to 37 miles height, but that "attempts at exact explanation are silly" [Villard, 1976, pp. 851-2]. In a lecture of November 6, 1912 (published 1913), de Forest stated that if the reflecting layer were half-way between the transmitting and receiving stations, his data supported a layer height of 62 miles [de Forest, 1913]. The journal editor commented upon de Forest's paper, stating that the phenomenon of interference of direct and reflected waves was very probably correct and should permit favorable choices of locations of stations and wavelengths to be employed [Goldsmith, 1913, pp. 51-53].

Soon after de Forest's observations, Dr. Leonard F. Fuller, specialist in developing 1000 kW(!) arc generators and Chief Engineer of Federal Telegraph, published propagation curves for the path San Francisco-Hawaii which showed the night-time wave-interference effects clearly, which Fuller considered as due to interference effects between the "upper-air wave" and the "ground wave" [Villard, 1976, p.851]. Others subsequently reported similar findings using different methods: T. L. Eckersley and colleagues at the Marconi Company in England made like suggestions from careful study of long distance transmissions on medium-frequency circuits used just after

World War I. Eckersley suggested also that the plane of polarization of the received waves was rotated due to the presence of electrons in the Earth's magnetic field [Eckersley, 1921]. These studies did not convince the scientific and engineering world. Nor had a theory of the upper layer in long-distance radio communications put forward by W. H. Eccles in 1912 [Eccles, 1912].

But the times changed after vacuum tube prices dropped in 1922. Radio became *the* new technical marvel of society. Women's perfume bottles, for example, were designed to resemble vacuum tubes. Radio now was more than a navigational aid and competitor to the telegraph. Broadcasting became big business. Government panels, such as the British Radio Research Board, funded research into the problems of "Skipping, Swinging, and Strays." Skipping and Swinging, respectively, referred to the "dead" zones within which radio signals were not receivable, and the fading and distortion which afflicted radio reception. Strays, also known as X's or Atmospherics were meteorologically-linked propagation phenomena such as crashing and hissing noises. These phenomena brought new faces into radio science and gave old faces new chances for research support. For example, both Robert Watson Watt and Edward Appleton came into radio science after World War I when the British Radio Research Board funded studies of X's and thunderstorms.

Amateur radio "hams" had astounded professionals in sending and receiving signals over distances of thousands of miles on the supposedly valueless "short waves" with wavelengths less than 200 meters (frequencies of 1.5 MHz and higher) and using only a few watts of transmitting power. Radio professionals began re-thinking the idea of long-distance radio communication, short waves, and the upper conducting layer.

The idea of a direct (ground) radio wave interfering with a sky wave reflected off of an upper conducting layer had been suggested some years earlier as witness Pierce's comments in 1910, and de Forest's and Fuller's empirical observations in 1912-14. Such interferometry in optics was known as the Lloyd's Mirror. But the data from 1912-14 did not catch sufficient support, nor did Eckersley's propagation studies in 1921. Part of the problem was that the most well-known propagation studies, established by the Americans L. W. Austin and L. Cohen from 1911 to 1914 using the U.S. Naval Atlantic Fleet, produced an empirical formula which suggested that the *lower* the frequency, the farther a given radio signal could propagate. Of course, the properties of the ionosphere at frequencies higher than two or three Megahertz had not been searched by Austin since nearly all his measurements had been taken at frequencies from 30 to 300 *Kilohertz* (wavelengths from 10,000 to 1,000 meters!), due to equipment limitations and common operating practices before World War I.

But *somebody* was re-thinking the possibility of direct measurement of the height of the conducting layer: It is little known that the instigation of experiments in direct, pulse-echo measurements of the existence and height of the upper conducting layer may have been due to the English physicist William Francis Gray Swann. Swann was chief of the Physics Division of the DTM (1913-18), invited to work on cosmic rays. He then became a physics professor at the Universities of Minnesota (1918-1923), Chicago (1923-34), and Yale (1924-27), and finally went with the Bartol Research Foundation as advisor and then director until his retirement in 1959. (Here is where the ionospheric networking begins, but more about that below.)

At the annual meeting of the American Geophysical Union on April 18, 1921 in Washington, D.C., Swann mentioned "an experiment in progress at the University of Minnesota designed with the object of measuring the distance of the supposed conducting layer by measuring the time taken by wireless waves to reach that layer and return" [Swann, 1921, p. 73]. Swann and his Minnesota colleagues evidently found the project a bit complicated. Reporting two years later to the same body, Swann stressed the importance of knowledge of the study of geomagnetism, geoelectricity, and the upper atmosphere and said:

> Another avenue of attack on the conductivity of the upper atmosphere is to be found through the medium of wireless waves. If the transition from practical non-conductivity to high conductivity is sufficiently sharp, marked reflection of electro-magnetic waves should result; and, by suitable arrangements, the details of which I must not enter into, it should be possible to measure the time taken for wireless waves to travel to the layer and return, and so deduce the altitude of the layer [Swann, 1923, p. 65].

That same year, 1922-23, University of Minnesota physics student Merle Tuve heard of the pulse-echo experiments and was much impressed with Swann [LC, Tuve, 2], whom he had met soon after his arrival at Minnesota in August 1922, as did Gregory Breit as National Research Fellow and assistant professor of physics at the University of Minnesota, 1923-24, studying the Zeeman effect and VHF oscillations in vacuum tubes. Working with J. G. Frayne, Swann's experiments failed due to receiver blanking problems. The radio receiver input circuits became overloaded by the transmitted pulse and did not recover in sufficient time to receive any echoes [Tuve interview, 1972].

Gregory Breit and Merle Tuve first met in the summer of 1923 and soon became good friends, with Breit helping

Tuve on his M.A. thesis on UHF oscillator tubes. Tuve's M.A. thesis topic had been suggested by Swann and Frayne, to involve short EM waves, vacuum tubes, and piezo-electric crystals at radio frequencies. John T. Tate, editor of the *Physical Review*, then the prime U.S. academic physics journal, was Tuve's M.A. thesis advisor. In his initial research conference on August 11, 1922, Tuve had suggested his first interest in research was in radio short waves and related phenomena including the Heaviside Layer.

Both Breit and Tuve had radio experience. Although Gregory Breit was a brilliant theorist, his M.A. thesis (1920) at Johns Hopkins (under the guidance of E. O. Hulburt, who received *his* Ph. D. from Hopkins in 1915 and who taught radio engineering at Hopkins in 1920!) had been on the transmission of EM waves in wireless telegraphy, and his Ph.D. thesis at Hopkins (1921) had been done on the distributive capacity of inductive coils while he was a worker in the Radio Section of the Bureau of Standards (BS). Though the degrees were from the physics department at Hopkins, electrical engineering influenced Breit's approach to physics throughout his career. He had studied mathematics at Hopkins with L. S. Hulburt, his undergraduate advisor, and had worked on vacuum tubes with LS's son, E. O. Hulburt. (More about the social network, below). Breit authored or co-authored 13 papers on radio and electronics in the period 1920-1929.

Tuve had been interested in radio since his Boy Scout days in South Dakota [LC, Tuve, 384] and had been a noted ham operator, along with his pal E. O. Lawrence. Tuve was also in the ham radio club (station 9NB) at the University of Minnesota. In fact, a few weeks after Tuve received his B.S. in electrical engineering from Minnesota, Lawrence wrote to him from the University of South Dakota with a scheme for the two of them to make a wad of money selling and installing radios [LC, Tuve, 2]. Tuve graduated in electrical engineering from Minnesota in 1922 (as Lloyd Berkner did five years later), studying with C. M. Jansky (who had contacts with the BS), then stayed on to do an M.A. in physics, where he sat in on Swann's course, "The mathematical theory of electricity and magnetism." When Tuve applied in the winter of 1923 to Karl T. Compton for a graduate General Electric fellowship at Princeton, he mentioned that he was studying the mathematical theory of electricity and magnetism with Swann [LC, Tuve, 4]. In the spring of 1924, in fact, Swann offered Tuve a job at the Bartol Research Foundation [LC, Tuve, 2]. When Tuve lost his scholarship at Princeton to Henry D. Smyth, who was returning to take a permanent instructorship, Tuve wrote around trying to find a position. Breit recommended him to Johns Hopkins. On May 1, 1924, Tuve received his offer of an instructorship from Johns Hopkins [LC, Tuve, 2].

Tuve succeeded to a position as instructor in physics (and graduate student) at Johns Hopkins in 1924, receiving tuition and $1800 per year, after a year at Princeton. On July 6, 1924 Gregory Breit became Mathematical Physicist at the DTM and immediately began to plan a program to measure the height of the conducting layer, *and* to involve Tuve, and Johns Hopkins. Tuve had written to Swann (then at Chicago) for advice on May 3, 1924 and also to Breit, concerning doing his Ph.D. thesis research under Robert Williams Wood at Johns Hopkins. Breit, on May 21, 1924 replied that he was not keen on Tuve working with Wood [LC, Tuve, 2]. Breit would see to it that Tuve's Ph.D. thesis would be on the ionospheric experiments.

Tuve spent the summer of 1924 as a student engineer at Western Electric in New York City, working with Clinton Davisson. As Breit wrote to Tuve July 25, 1924:

> I have been calculating lately the effect of the earth on the inductance of longlines of cables connected in series so as to encircle the earth and also the effect of the Heaviside layer. The purpose is to see whether by measurements of inductance and resistance one can learn something as to the conductivity and permeability of the earth's interior. . . . It may be that we shall try to reflect Hertzian waves from the Heaviside layer sending in Washington and receiving in Baltimore or in the opposite direction. I am also thinking of getting the W. E. & G. E. to cooperate [LC, Tuve, 401].

Breit wrote to many radio people for suggestions, men such as inventor Stuart Ballantine, Ernest Merritt of Cornell, experimenter Greenleaf Whittier Pickard, and people at N.Y.U. and GE. Breit made several experiments in the early fall of 1924 and attempted to secure funds for construction of a very large parabolic antenna. DTM could have been the Sugar Grove, West Virginia of 1925. Whatever, Breit soon proceeded on to detail plans for radio echoes from the ionosphere. L. A. Bauer wrote to Kennelly on September 12, 1924 how very interested Breit was in the "supposed" conducting layer. From October 6 to 10, 1924, Breit conducted radio tests in cooperation with Johns Hopkins, having secured their permission to involve Tuve and a younger graduate student. From notes in Breit's hand dating from about September 1924 [LC, Tuve, 11] it seems he assumed the ionospheric layer height to be 62 miles (the same figure mentioned by de Forest in his 1913 publication noted above). Breit knew of the "skip" distance radio propagation measurements of A. H. Taylor and E. O. Hulburt at the NRL and their results also convinced him of the reality of the ionosphere and of the practicality of measuring its height and characteristics.

As mentioned, Breit launched his attempt to measure the height of the upper conducting layer as soon as he arrived at DTM, and following two research paths:

> One method consists in directing a beam of Hertzian waves up to the layer by means of a parabolic reflector and looking for the reflection. The other is based on the different time taken by waves to travel over the earth's surface and by way of the layer. This difference in time should affect the modulation, and this effect should be different for aerials with different directional properties [LC, Tuve, 11].

Note that neither method clearly indicates pulse-echo. The initial idea for using the parabola was as a *beam* system, so that only a relatively narrow beamwidth sky wave would be transmitted and the receiving station could then estimate the downcoming angle of the received wave. The other method is similar to the method Appleton employed in England beginning in December 1924. That is, the direct ground wave, and the ionospherically reflected wave would interfere more or less constructively as a function of frequency. Appleton had the transmitting station alter frequency by several percent, more or less like the frequency-shift-keying data observed by de Forest and Fuller from 1912-1914. Breit, however, thought that minor (1.5 Khz) modulation changes of frequency could suffice to show the effect, especially if measured by different types of antennas. This comparison of signal received via, say, loop versus dipole antenna was also in Appleton's thinking and it was something he added early-on to his experiments between London and Cambridge, England.

Breit calculated that a paraboloid strung from eight 75-foot-tall oak poles could operate at about 3 meters wavelength with a 6 degree beamwidth. The paraboloid would be oriented 20 degrees off the vertical. The circular mouth of the antenna reflector surface would be about 67 feet wide at the top. From the bottom up through a cross-section width of 15 feet, the antenna would be solid sheet copper set in concrete. From there up to a cross-section width of about 30 feet, the antenna reflecting surface would be wood covered with copper-mesh window screen. The rest of the reflector surface would be covered with 2-inch wire mesh [LC, Tuve, 11]. Rough calculations by this author of the antenna pattern beamwidth to the half-power points for the proposed parabola give for a circular mouth diameter of 20 meters, a beamwidth of 8 degrees at 3 meters wavelength, 17 degrees at 6 meters wavelength and 29 degrees at 10 meters wavelength. An undated sheet of pencil sketches and calculations suggests that Breit did a preliminary calculation for a circular mouth diameter of 25 meters, which would result in a beamwidth of almost 7 degrees at 3 meters operating wavelength, since he notes "wavelength = 3 meters, . . . could have delta = 6 degrees, therefore alpha = 25 meters" [LC, Tuve, 11]. This would indeed suggest that Breit's original idea for operating frequency was in the 3 meter (100 Mhz) range (VHF). The antenna would be oriented toward Baltimore, 43 degrees East of North.

J. P. Ault, acting assistant director of DTM, approved Breit's plan and wrote a request on November 10, 1924 to L. A. Bauer, director of DTM, asking that $1625. be allocated for poles, wire, cable, mesh, etc., and that the cost be distributed from several DTM accounts for 1924 and 1925 [LC, Tuve, 11]. Breit followed Ault's letter immediately, on November 12, 1924, with a three-page "Memorandum regarding proposed experiments to determine height of conducting layer" [LC, Tuve, 11], accompanied by the blueprint of the proposed parabola and by a bibliography of over 25 references, including J. Larmor's very recent important Heaviside Layer communication in the *Nature* issue of the first of November [Larmor, 1924], which could only have been received a day or two before by Breit. Breit presented this memo to a planning meeting on November 12 at DTM with L. W. Austin, L. Cohen, J. H. Dellinger (BS), A. H. Taylor (NRL), Bauer, Tuve, and Breit attending.

This present essay does not concern developments in the central *theory* of ionospheric propagation, the magneto-ionic theory, as I have written about this at length elsewhere [Gillmor, 1982] but it is sufficient to note here that Breit's memorandum shows a complete understanding of the experimental and theoretical situation as of the fall of 1924. Breit argues that the evidence is very strongly on behalf of the existence of the upper conducting layer suggested by Kennelly and by Heaviside. The evidence comes, he says, from diurnal variations of geomagnetism, from long-distance radio transmission, from the aurorae, and from fading and swinging of radio waves in propagation. The parabola would be used to conduct experiments between Washington and Baltimore. The parabola would allow a sufficiently narrow beam for definite measurements to be made of angle of takeoff of the wave. The angle would be chosen to reflect the waves from a height of 100 kilometers into Baltimore. Breit proposed a portable receiving set in Baltimore, using a small receiving parabola, plus other cooperating receiving stations, since he supposed that the height of the conducting layer would vary and thus the maximum signal level would vary from place to place. He hoped that quantitative measurements of signal strength and of wave polarization could be made at Johns Hopkins. He cited the recent successes of F. W. Dunmore and F. H. Engel at the Bureau of Standards in Washington, and of G. Marconi, in the reflection of short radio waves by means of parabolic reflectors.

At first, Bauer was against the idea. Breit wrote Tuve on November 17, 1924, Bauer "is a damned nuisance . . . " [LC, Tuve, 2]. Perhaps Bauer was a nuisance, but as he explained to Breit, he had to do quite some juggling of DTM finances to accommodate Breit's ionospheric project [LC, Tuve, 11]. From Breit's correspondence with Tuve, it seems Breit soon lost any great affection he may have had for Bauer and for Fleming.

Breit was always ambivalent about his radio interests. Following his Ph.D. in 1921, he spent a year at Leiden with Kammerlingh Onnes and Paul Ehrenfest. He had a very broadly productive career, but Breit characterized his main interests as nuclear physics, quantum theory . . . considerably ahead of ionospheric physics. Breit's work with Tuve in ionospheric physics spanned five years. But he was mercurial; just while pushing the giant parabolic reflector, immediately before presenting the blueprints and the memo asking for building funds, Breit wrote to Tuve on November 5, 1924, just four months after arriving at DTM, to say that the atmosphere at DTM was stifling for quantum theory. Breit suggested that maybe Tuve could do all the experimental work on the ionosphere project. Breit wrote that he would like to resign from DTM in the near future [LC, Tuve, 2].

Tuve, too, had his problems with Bauer and Fleming, but contrast his feelings about DTM with those expressed by Breit—here is Tuve writing in November 1926 to Hilding, an old friend in Duluth, Minnesota:

> I am more happily situated here than I could ever have dreamed of being. Have a sort of free-lance research job here, working with Gregory Breit whom you may have known at Minneapolis. . . . We are working on problems of our own choice. [LC, Tuve, 2].

Throughout the fall and spring of 1924-1925, Breit tried experiments using local and regional transmitting setups. This was in connection with Breit's second approach: measuring the differential fading of radio signals of slightly different frequencies. Breit discussed his plans in a paper given on December 30, 1924 in Washington, D.C. at the meeting of the American section of URSI, the International Scientific Radio Union, and then in a paper given before the American Physical Society. Experiments in the fall between Washington and Baltimore were followed by tests in early 1925 with Professor W. G. Cady at Wesleyan University in Middletown, Connecticut, with Hartford, with Cornell University in Ithaca, and with Hamilton, Massachusetts. These were in turn followed by conferences Breit held on a visiting trip up the East Coast in February and March. Breit secured the help or promise of help from station WSC at Tuckerton, New Jersey on 600 and 670 meters and of WIM, Chatham, Massachusetts on 600 and 735 meters. E. F. Alexanderson and others at GE, G. W. Pickard, Western Electric, F. H. Krager of RCA, Kennelly (then at MIT), and C. H. Anderson of AT&T were helping [LC, Tuve, 11]. All were either planning to transmit and/or receive, advise on equipment or exchange data. Breit hoped to measure differential fading of frequencies differing by approximately 1.5 Khz.

Because Tuve had done his M.A. thesis on 80 cm. vacuum tube oscillators and had been assisted by Breit in this work, perhaps one would expect Tuve to be sold on the UHF frequency range. But, according to Tuve, it was Tuve who suggested that the experiment would have a better chance if the operating frequency were at several Mhz rather than the 90 MHz range, as the parabola would have produced. Tuve recalled that the "echo" idea as finally produced was decided at dinner at the *Columbia Bakery* in Chevy Chase, Maryland around Christmas, 1924 [Tuve interview, 1972]. Breit then agreed to write up the idea "Note on a Radio Method of Estimating the Height of the Conducting Layer," which he so did on January 4, 1925 and then submitted it to the journal *Terrestrial Magnetism and Atmospheric Electricity* [Tuve and Breit, 1925] on January 8th. The giant parabolic reflector never was built, even though lumber for the reflector arrived at the Broad Branch Road Campus of DTM in early December 1924.

Breit and Tuve's method differed slightly from the one unsuccessfully tried by Swann and Frayne at Minnesota. Instead of transmitting and receiving vertically, Breit and Tuve proposed to separate the stations and, while keeping the receiver from being overdriven, record both the *direct* wave from the transmitter to receiver, and also the indirect wave which travelled (supposedly) up and back from the ionosphere. To better discriminate between the two waves, they proposed not only to use a time base measurement, but to use also amplitude discrimination employing a directional loop antenna. The loop experiments had been conducted by Breit at least as early as November 1924 and continued. Also, if the conducting layer were about 80 miles in elevation, receiving stations from zero out to 320 miles from the transmitter could discriminate between the ground-wave and the sky-wave pulse travel time if the pulses occurred between 581 and 2461 cycles per second [LC, Tuve, 11].

If the upper conducting layer were 60 to 80 miles above the Earth and the radio signal traveled at the speed of light, then a signal of say 1/1000th second duration received over the ground 80 miles away would arrive sufficiently out of phase with the same signal from the transmitter reflected off the layer 80 miles up so that there would be considerable distortion. As Tuve was later to note [LC,

Tuve,11], this seemed a reasonable explanation for the "bad quality of modulation from certain broadcast transmitters which became apparent at distances of 100 to 150 miles, with perfect modulation at or near the transmitter, and at great distance from it." If, however, the pulse length were short enough, the difference in arrival time of the signal over the ground-wave versus the sky-wave pulse could be used to measure the height of the conducting layer.

Breit continued his propagation experiments with commercial, university, and government radio stations until Tuve became available to help directly at DTM in mid-June 1925 in a summer job at $170 per month. Breit's early tests in April with the RCA station WSC at Tuckerton on 600 meters wavelength, yielded indecisive results since there was sparking at the commutator which produced the pulses. Breit saw some "humps" (as they termed the signal returns) during the morning tests run three times a week. "Humps" were the rise in detector current at the receiver (and thus positive deflection of the galvanometer). But there were problems of instability of the transmitter and also the fact that the GE oscillograph had a natural period of the same order as the time delay between the ground- and sky-wave pulse and these problems also added to the inconclusive nature of the early tests [LC, Tuve, 2]. The GE oscillograph consisted primarily of a short period galvanometer and a suitable optical system for directing beams of light from a lamp to mirrors on the galvanometer vibrating elements and from there to a revolving film or to a visual screen. The standard vibrator in the instrument put out by GE had a current sensitivity of about 0.005 ampere per millimeter deflection and its *claimed* free period was approximately 1/5000th second, although the one used by Breit and Tuve had a period of about 1/600th second. The oscillograph was initially preferred over the new instrument, the cathode ray oscillograph (CRO). The CRO was expensive and the oscillograph seemed to provide for easier arrangements for photographing the data. Other early experiments with KDKA, Westinghouse's station in Pittsburgh, also produced no definite effects, although, again, some "humps" were photographed. Some of these experiments were done after midnight.

But Breit's spring of 1925 was not a waste. He lined up equipment (to be obtained or borrowed from the Navy, GE, Westinghouse, and the BS), fine-tuned his approach, and secured the use of NKF, a very stable, crystal-controlled, transmitter nearby at the NRL in Anacostia, a few miles from the DTM laboratory. Also, by summer, they had learned from a BS worker how to alter the oscillograph filament to decrease its period to 1/8000th second. Breit got Bauer in June officially to ask whether the NRL could cooperate with the DTM in further tests using the radio transmitter at NRL. Commander E. G. Oberlin assured Bauer of cooperation, noting however that the 71.3 and 41.7 meter operating frequencies were usually tied-up by the Navy for fleet communications at night [LC, Tuve, 11].

As Tuve later noted, the frequency output of non-crystal-controlled transmitters is always slightly dependent upon the plate voltage supplied. Tests with BS transmitter WWV confirmed this and showed how such instabilities could cause "apparent" echoes. Although Breit and Tuve were quite restricted in their use of the NKF transmitters, they considered themselves quite fortunate and received "all possible cooperation on the part of the Navy officials in the transmission of test signals." Tuve wrote, "Since the two transmitters at NKF are almost the only ones in this country which are operated with crystal control, it is seen that the transmitting facilities available for our experiments were distinctly limited . . . " [LC, Tuve, 388].

Definite results were not long in coming once they used the NKF transmitters. They had continued running tests with KDKA at 300 meters wavelength, even on July 23-25, getting some Humps. Meanwhile, they had made a new keying device which produced a 425 Hz pulse rate with each pulse considerably shorter than 1/1000th second. Their first successful tests at NKF were on July 28, 1925 at 10:30 am and 3:15 pm on 71.3 and 41.7 meters wavelength. They were unsuccessful in receiving echoes on shorter wavelengths of 20 meters and down. They again received clear echoes on August 6, where receivers and oscillographs were operated both at DTM and at the NRL transmitter site 7 miles southward at Bellevue, Anacostia. The returns suggested a layer height of 50 to 100 miles. They recorded some double and triple humps. Their first suggestions for these were possibly a "Wavy surface in the layer and successive reflections . . . " Tuve noted that successive reflections was not so likely as the answer; they suspected roughness in the layer. Later, they ran tests to rule out the possibility that the reflections were due to lateral deviation, such as from the Blue Ridge mountains.

A day after the August 6th tests, Breit and Tuve sent off a note to *Nature* which was published on September 6, 1925 [Breit and Tuve, 1925]. On the 19th of August, the Navy sent out a press release that NRL and DTM had confirmed the "theory of an ionized region on the higher levels of the earth's atmosphere." What happened next? They gained publicity, spoke at meetings, and continued their experiments. They constructed a more sophisticated pulse-modulating circuit and operated during daytime and nighttime. At the annual CIW exhibition, on December 11-14, 1925, Breit demonstrated the conducting layer experiment for visitors, showing "humps" both from the Navy's NKF station and the BS transmitter WWV [LC, Tuve, 11]. Breit got J. S. Ames at Hopkins to accept Tuve's conducting layer work as his Ph.D. thesis, which

Tuve wrote up in March 1926 and submitted later that Spring. The thesis is similar to the article with the same title published by Breit and Tuve in the *Physical Review* in September 1926. The article manuscript was submitted May 19, 1926 and accepted by Tate on 16 June. It is a beautiful article, carefully and clearly setting out the history, the methodology, and the results of the DTM conducting layer research from the fall of 1924 through spring of 1926 [Breit and Tuve, 1926].

There were still *lots* of questions about the ionosphere in 1926: Was the conducting layer of ions? of electrons? What was the particle density? Just how high was it and how thick? How did it vary in time and in different geographical regions? Did waves "bounce" from reflection, or were they "bent" through refraction, or both? What about collision frequency of the particles and absorption of the waves when within the ionosphere? What about "critical" frequencies, above which waves sharply disappeared?

Meanwhile, Appleton in England had somewhat scooped everybody [Gillmor, 1982]. He had been the "first" to get in print his earliest observations (of December 11, 1924 and February 17, 1925). Appleton, too, was using a "network" of helpers. Where Breit and Tuve had help from broadcasters, the BS, and especially, the NRL, Appleton had his own new research group at King's College in London, his former research group at Cambridge, the British Broadcasting Company, and a radio research group at the National Physical Laboratory near Slough, England. Appleton was incredibly energetic. His real genius was in recognizing the importance and promise of radio methods used to investigate the ionosphere and dropping nearly all of his other work to concentrate on the new research. Appleton could charm, cajole, or otherwise persuade. He devoted all his efforts for 15 years to building the science and the organizations devoted to the ionosphere, stepping down in a sense only to administer the British Department of Scientific and Industrial Research during World War II. He authored or co-authored 95 articles on the ionosphere. By 1927 Appleton was on the British Radio Research Board (the policy and funding board for British radio research), and on four of its seven committees.

Meanwhile, by 1928 Tuve heard again about Breit's wishes to leave the ionosphere, and even experimental hi-voltage physics. Breit wanted to take leave from DTM to study quantum physics in Europe. This leave was granted in 1928.

Tuve and Breit's first collaboration on radio was published in 1925, their last in 1928. Work on the ionosphere continued at DTM with Odd Dahl and L. R. Hafsted. But it was the arrival at DTM of Lloyd Berkner in 1933 that re-invigorated the program, developed a network of ionospheric sounders, and from 1935 organized at DTM Annual Conferences on "Ionospheric Research" for about 50 people from the U.S. and Canada representing academia, industry, and civil and military government bodies.

Tuve kept an interest in the ionosphere, and his temper could be aroused: Appleton created a stir at Washington in October 1927 when he read two short papers at an international URSI meeting, including one on the magneto-ionic theory [Gillmor, 1982]. In the other paper Appleton stated that he had discovered that there was more than one ionized layer in the upper atmosphere, and published a note in *Nature* concerning this [Appleton, 1927]. He managed later to have this layer called the "Appleton" layer, to be in symmetry with the previously known layer called by the British, the Heaviside layer. Tuve, Berkner, and others collided with Appleton over the years. Tuve was still smarting when he wrote to K. K. Darrow in 1940:

> There can be no question that the "echo-method" was developed here at our Department by Breit and myself (1925), and Appleton did not try using this method until 1928 or 1929, in spite of our urging him to try it because of the inherent mathematical ambiguity of his own interference method. . . . I might remark that Fuller carried out interference measurements between Honolulu and the West Coast in 1914, very similar to Appleton's later work . . .
>
> It is correct that Appleton was the first to announce in print that the ionosphere comprised two layers. My own impression would be that Appleton should be credited with having hazarded this proposal in print, with a 50-50 chance of being right, but that perhaps the echo method, rather than any individual was responsible for its discovery or proof. . . . The fact is that until observations were available on different wave lengths, no one could say which interpretation would be correct. . . . Appleton was lucky, perhaps, but more than anything else he was astute. Radio has been his life work, and as such it was entirely correct for him to hazard such guesses as he wished, in order to be first in the field [LC, Tuve, 4].

And there were other competitions as well: After the Bureau of Standards, the NRL, and the DTM supposedly agreed in the early 1930s who was going to do what in ionosphere work without stepping on other's toes, J. H. Dellinger wrote on April 18, 1940 to Lloyd Berkner, complaining about Berkner's bragging about DTM work and competing with the Bureau:

> . . . I have in mind our mutual understanding that your Washington ionosphere laboratory is a laboratory for

the development of equipment and not for carrying on of regular observation programs, and the need of avoiding the impression that it duplicates the Bureau's work. . . . visitors to your laboratory (Prof. Mitra, for example) have received the impression that you are carrying on a regular ionosphere observation program at Washington . . . This impression is reinforced by the publication in some of your papers of ionosphere data from your Washington laboratory. . . . On several occasions since you left here to join the Carnegie Laboratory you have told me that the latter would not engage in a regular ionosphere observation program at Washington or otherwise duplicate the ionosphere work being done by this Bureau . . . [NA-US, Dellinger, J. H., Papers, RG 167].

Still, the early radio work in Washington involved "personal" networks: Bureau of Standards, DTM, and NRL (or its predecessor), and Johns Hopkins were linked since before World War I. Austin and Cohen had worked both for the Navy and for the Bureau of Standards, almost simultaneously. E. O. Hulburt of NRL had studied at Hopkins and taught Breit there. Breit studied with Hulburt and Hulburt's father at Hopkins, worked at the Bureau, then met Swann and Tuve at Minnesota, then arranged for Tuve to come to Hopkins, and to DTM. Swann went from the DTM to Minnesota, met Tuve and Breit, later tried to get Tuve to work at Bartol. Berkner studied E.E. at Minnesota, worked at the Bureau of Standards, did radio electronics in the Navy, came to DTM with Tuve's assistance.

This early networking aided in the subsequent development of ionospheric research into "Big Science" following World War II. Radio studies at DTM in the 1920s allowed DTM to lead in the exciting new fields of radio applications to physics and geophysics. These studies also helped justify DTM's continuation and evolution of its more traditional research programs in terrestrial magnetism. In some ways this was similar to the metamorphosis of portions of DTM's traditional researches into atmospheric and geo-electricity into the new field of cosmic ray physics. And, DTM's experience, and network of geophysical field stations built in the 1920s and 1930s was to prove extremely valuable to Allied Forces in World War II.

ARCHIVAL SOURCES

To save space, the format is condensed for the numerous citations to the papers of M. A. Tuve in the Manuscript Division of the Library of Congress. References herein are to the individual containers. Thus [LC, Tuve, 11] indicates the Tuve papers, container number 11. References are to containers 2, 3, 4, 11, 384, 388, 391, 401. Also cited are the papers of J. H. Dellinger, at the U.S. National Archives, Record Group 167.

REFERENCES

Appleton, E. V., The existence of more than one ionized layer in the upper atmosphere, *Nature*, *120*, 330, September 3, 1927.

Breit G., and M. A. Tuve, A radio method of estimating the height of the conducting layer, *Nature*, *116*, 357, September 5, 1925.

Breit, G., and M. A. Tuve, A test of the existence of the conducting layer, *Phys Rev*, *26*, 554-75, September 1926.

de Forest, L., Recent developments in the work of the Federal Telegraph Company, *Proc IRE*, *1*, 37-51, 1913.

Dubin, A., and J. Burke, *Mechanical Man*, M. Witmark & Sons, New York, 1929.

Duddell, W., and J. E. Taylor, Wireless telegraphy measurements, *Electrician*, *55*, 258-61, 299-302, 349-351, issues of June 2, 9, and 16, 1905.

Eccles, W. H., On the diurnal variations of the electric waves occurring in nature and on the propagation of electric waves round the bend of the earth, *Proc R Soc London*, *A*, *87*, 79-99, 1912.

Eckersley, T. L., The effect of the Heaviside layer on the apparent direction of electromagnetic waves, *Radio Review*, *2*, Nos., 2, 5, 1921.

Fessenden, R., Wireless telegraphy, *Proc Am Inst Electr Eng*, *4*, July 1908.

Geddes, K., *Guglielmo Marconi: 1874-1937*, Science Museum, H. M. S. O., London, 1974.

Gillmor, C. S., The history of the term "ionosphere", *Nature*, *262*, 347-8, 1976.

Gillmor, C. S., Wilhelm Alter, Edward Appleton, and the Magneto-Ionic Theory, *Proc Am Philos Soc*, *126*, No. 5, 395-440, 1982.

Goldsmith, A. N., Editorial notes, *Proc IRE*, *1*, 51-53, 1913.

Heaviside, O., Telegraphy, in *Encyclopaedia Britannica*, 10th edition, *33*, 213-235, December 19, 1902.

Kennelly, A., On the elevation of the electrically-conducting strata of the Earth's atmosphere, *Electrical World*, *39*, New York, 473, March 15, 1902.

Kennelly, A. E., *Wireless Telegraphy*, 2d ed., Moffat, Yard and Co., New York, 1909.

Larmor, J., Why wireless electric rays can bend round the Earth, *Nature*, *114*, 650-1, November 1, 1924.

Pierce, G. W., *Principles of Wireless Telegraphy*, McGraw-Hill, New York, 1910.

St. Germain, A., The flowering of mass society: An historical overview of the 1920s, in *Dancing Fools And Weary Blues: The Great Escape Of The Twenties*, edited by L. R. Broer and J. D. Walther, pp. 14-18, Bowling Green State University Press, Ohio, 1990.

Swann, W. F. G., The penetrating radiation and its bearing upon the earth's magnetic field, *Eos*, *2*, 65-73, 1921.

Swann, W. F. G., Status, scope, and problems of the section of terrestrial magnetism and electricity, *Eos*, *4*, 60-8, 1923.

Tuve, M. A., and G. Breit, Note on a radio method of estimating the height of the conducting layer, *Terr Mag*, *30*, 15-16, March 1925.

Tuve, M. A., interview with C. S. Gillmor, June 30, 1972.

Villard, O. G. Jr., The ionospheric sounder and its place in the history of radio science, *Radio Science*, *11*, No. 11, 847-60, November 1976.

Building a Washington Network for Atmospheric Research

BRUCE HEVLY
Department of History (DP-20), University of Washington
Seattle, WA 98195

INTRODUCTION

In June, 1927, a brief letter was sent from the Department of Terrestrial Magnetism (DTM) of the Carnegie Institution of Washington (CIW) to spectroscopist J. A. Anderson at the Mt. Wilson Observatory, concerning a new research tool for investigations of atmospheric physics:

> Dear Dr. Anderson: After your visit in Washington at the time of the April meeting of the Physical Society we have been very much interested in the rocket for the purpose of exploring the upper atmosphere. It seems possible to try to excite the interest of some wealthy people here. . . . Commander Oberlin of the Naval Research Laboratory is doing his best to find out the possibilities in the Navy. We talked to Dr. Wright of the Geophysical Laboratory in connection with the possibilities of cooperation with the Army. Dr. Merriam [CIW president] seems interested; doubtless he would be glad to learn from you your ideas [NA-US, RG 19, Breit, Tuve, and Hulburt to Anderson, June 17, 1927].

The letter referred to Robert Goddard's development of liquid-fueled rockets, and suggested that it be brought under the aegis of a centralized, coordinated system for the systematic compilation of data in the realm of terrestrial physics, a system which extended from Washington, to Mt. Wilson, to the CIW's field stations and to the explorations of its research vessel, the *Carnegie*, mapping the terrestrial magnetism around the world. The letter goes on to request specifics concerning the capabilities of Goddard's rockets and the cost of supporting them, and concludes, "All of us feel that the experiment you wish to carry through will answer many questions in which we are interested. We should like to help you in every possible way in getting it started."

Merle Tuve and Gregory Breit were of course temporarily partners in the DTM; E. O. Hulburt was the civilian scientist in charge of the Optics Division of the four-year-old Naval Research Laboratory (NRL) in Anacostia, DC. Together they represented a substantial element of a network for atmospheric physics research in the interwar period in Washington, DC. It was a network with the CIW at its center, which included principally its DTM, at least through the 1920s; its atmospheric electricity division; the National Bureau of Standards; and elements of the Naval Research Laboratory. At least in the DTM-NRL axis, it was a relationship based on mutual interests, complementary needs and resources, and, again temporarily similar visions of how the physics of the Earth should be studied.

Breit, Tuve, and Hulburt collaborated in the letter to Anderson because they recognized the organizational realities which lay at the base of the atmospheric physics network. When the CIW began to support Goddard's rocket research, it did so through a system of supervisory committees and review procedures which aimed to integrate the new tool into the network's ongoing studies. The process transmits to the present a glimpse of how these organizations defined a newly-extended research field.

This process of definition is one of great interest to historians, philosophers, and sociologists of science, because of their interest in the way in which scientific communities arrive at standard, agreed-upon understandings. The term "network," then, at times carries a special meaning. In the history of science literature, it can denote the means by which consensus may be created and enforced [Latour, 1987, ch. 3, 4, 5]. A tradition in the history of technology established by the work of Thomas Hughes [Hughes, 1983; Bijker, Hughes, and Pinch, 1987], the systems approach, emphasizes that technologies are defined in part by their interlinkages within a network, and so by a surrounding social-technological system which draws its elements into a state of mutual consistency. In both cases, the scientific or technological network is of interest because of its supposed power to establish a type of conformity.

Thus studying the Washington network for atmospheric research affords the opportunity both to describe an important setting in the history of terrestrial physics, and to question whether the network functioned to enforce standard research techniques or scientific conclusions. In addition to describing briefly the cooperation on problems of upper atmospheric physics which took place in the

The Earth, the Heavens and the Carnegie Institution of Washington
History of Geophysics Volume 5

1920s, I want to make the point that this "network" existed without serving to impose total agreement about matters of research procedure or on the interpretation of results. Studying the relationship between the DTM and the NRL reveals a dynamic situation in which separate institutional identities remained. That is, despite the exchange of support for problems of mutual interest, those separate identities indeed conditioned the interpretations of data and the construction of theories, and they were not subordinated to any kind of enforced consensus. Thus it seems better to discuss the network as a site of action rather than as a more staid setting where alliances, once achieved, dictated a homogenization of views.

FOUNDATIONS OF RESEARCH AT NRL AND THE DTM

As we know thanks to the work of Nathan Reingold and others, Washington had an established tradition as a site for research careers in the physical sciences, particularly the varieties of terrestrial physics [Reingold, 1991, ch. 2, 5, 6, 10]. From the mid-nineteenth century on, those ambitious for positions as professional researchers came to Washington, often in order to gather and catalogue physical data relevant to America's westward expansion. The CIW and in particular the DTM followed this pattern through the 1920s, gathering data on a global scale in preference to laboratory research, and establishing a group of research bureaus for the conduct of research, rather than principally supporting research at universities [Geiger, 1986, pp. 60-67; Dupree, 1986, pp. 297-298].

In establishing the NRL in 1923, the U.S. Navy made a similar choice. While the Navy had long-established research facilities for ship design, ordnance design and testing, and power plant engineering, twentieth-century warfare presented new technological problems [Allison, 1981, ch. 2, 3]. One solution would have been to farm out research in such recondite areas as radio and underwater sound, probably to corporate laboratories, or even to schools such as MIT (which did some aeronautical research for the Navy) [Dupree, 1985, p. 334]. Instead, NRL was established as a civil service laboratory under the financial control of the Navy's technical bureaus, one of a generation of new laboratories including those of the National Advisory Committee for Aeronautics and a mature National Bureau of Standards which emerged in postwar Washington to deal with technological problems which seemed somehow science-based. Along with these new institutions came a postwar passion for coordinated, organized efforts in a myriad activities including scientific research, as lampooned in Sinclair Lewis's *Arrowsmith*, and boosted by America's scientific elite [Rosenberg, 1976, pp. 126-127, 130].

In fact, Breit, Hulburt, Tuve, and others shared an important, more personal postwar condition: as Johns Hopkins-trained scientists they benefitted from the entré provided by Joseph S. Ames to Washington's scientific community, a connection forged not just by proximity but also by Ames's service on the National Advisory Committee for Aeronautics. Thomas Cornell has remarked on this in his study of Tuve's work at the DTM, and it held for others as well [Cornell, 1986, pp. 123-124]. Ames could thus steer Hopkins physics graduates to the National Bureau of Standards, as in the case of Breit, or to NRL. Hulburt, for example, did not even know he had applied for his job there until he was informed by the Navy that he was the finalist [NRL, Hulburt history interview]. These men then opened up opportunities for others, as Breit did for Tuve, and as Hulburt did for his NRL protégé, Herbert Friedman.

THE RADIO AND THE SKY AT NRL

Upon arriving at NRL, Hulburt's task was to pursue general studies in optical science which would be useful to the Navy. Unlike the NRL's more technologically well-defined studies of radio communications and anti-submarine warfare, Hulburt had to devise his own research program, one which could prove worthy of financial support. He began by cooperating with A. Hoyt Taylor of the radio division on understanding short-wave radio propagation through the Kennelly-Heaviside layer, or the ionosphere, trying to bring order to the data returned from investigations made with NRL's crystal-controlled, shortwave transmitter. For Taylor and Hulburt, the crucial measurement was that of skip distance—that is, of the distances skipped over by a radio signal as it was reflected from the ionosphere [Taylor, 1925; Taylor and Hulburt, 1926; Hulburt, 1974]. From this data and some geometrical reasoning, one could determine the height at which the density of charged particles was sufficient to bend the radio wave back to Earth, that height depending on the frequency of the signal, the time of day, point in the year, sunspot activity, and so on. Gathering and analyzing this data, which came in the form of responses to NRL's test signals from all over the world, revealed information about a portion of the upper atmosphere beyond the reach of direct investigation, such as instruments sent aloft in sounding balloons.

Compared to his fellow alumni Tuve and Breit at the CIW, Hulburt found himself in a much different research setting. With no budget of his own, his small optics research branch had to satisfy potential funders within the uniformed navy's technical bureaus. Using NRL's shortwave radio as a scientific research tool with an eye to improving it as an operational technology satisfied both Hulburt's professional interests and his sponsors. This balance, then, helped shape the way ionospheric research problems were defined and solved at NRL.

This work established Hulburt's interest in ionospheric physics, and so brought him, and NRL, into a cooperative arrangement with the DTM. The DTM offered a professional outlet for Hulburt, and to the NRL ready cash for research the like of which was not available from the interwar Navy. The DTM offered a biweekly seminar series on problems of ionospheric physics, to which Hulburt presented his theories on the causes of the radio reflecting layer and its disruption. "It is a great pleasure to hear from your letter," John A. Fleming of the DTM wrote to Hulburt in September of 1928, " . . . that you have now worked out a detailed theory of the causes of the aurora and of magnetic storms which appears to explain many features and details of these phenomena" [NA-US, RG 19, Fleming to Hulburt, September 25, 1928]. Fleming went on to announce the initiation of a biweekly seminar series on terrestrial physics at the DTM; he invited Hulburt to give the opening presentation. The problems of correlated terrestrial magnetism, atmospheric electricity, and ionospheric physics provoked a series of theories from Hulburt and others, and a series of articles presenting coordinated work from the DTM, the Bureau of Standards, and Mt. Wilson observatory [Bradbury, 1937; McNish, 1937; Dellinger, 1937; Richardson, 1937].

Such contacts sustained Hulburt in his difficult balancing act at NRL, where he was charged with, and happily accepted, the responsibility of doing technologically useful research, but also was expected to produce "real research" as evidence of the quality of NRL's staff. The work by Hulburt and Taylor on ionospheric propagation was hailed by the naval officers supervising NRL as proof of the laboratory's growing scientific prestige. But this success depended on Hulburt's ability to keep alive both a long-term commitment to ionospheric physics, and the short-term, project orientation required by the navy's method of managing and funding research tasks. Here the contact with the CIW—a "university without any students"—cemented a sense of participation in the professional scientific community.

COOPERATION WITHIN THE WASHINGTON NETWORK

The CIW also had funds available on less strict terms than those available from the Navy's technical bureaus. For example, when the NRL and the DTM had a joint interest in placing an investigator in Alaska to collect radio transmission data during the International Polar Year 1932-1933, the CIW provided some of the funding, and NRL provided the observer, radio engineer Harry Maris [NA-US, RG 19, Carnegie file]. The results of Maris's trip (or ordeal, better said) were then shared between the two bodies, fuelling a discussion of the existence of the ionospheric D layer, which in 1934 seemed to exist over Britain but not over America [NA-US, RG 19, Fleming to Greenlee, March 27, 1934].

Maris's Alaska trip also points to some of the resources that the NRL could offer researchers at the CIW. NRL had personnel on the civil service rolls, who could be lent out to other organizations on a temporary basis. These included not only Maris, who was engaged in a joint project, but also a glass blower who served at the DTM for an extended period of time. NRL also had technical expertise which could serve the CIW's interests. When high-frequency radio signals became important for both communications and atmospheric research, the set for the research vessel *Carnegie* came from the NRL: the transmitter was especially built in the NRL's shops, and the receiver was loaned from the navy. NRL also relayed messages from the *Carnegie* to the DTM, and in return the CIW assisted with gathering radio transmission data for the NRL by listening on a fixed schedule for test messages.

NRL had capital sunk in equipment as well as personnel, and was even more willing to loan it. Fleming begged the loan of "your new Dufour oscillograph" for the use of Breit and Tuve in June 1928, and the DTM kept the instrument until 1930 [NA-US, RG 19, Carnegie file]. Even more important was NRL's loan of its crystal-controlled transmitter for Tuve and Breit's initial work to detect the ionosphere, an instrument being used by Taylor and Hulburt in a different way for the same purpose.

As described in this volume by C. S. Gillmor, Breit and Tuve, sworn to secrecy regarding anything of military importance that they might learn at NRL, worked from the Anacostia laboratory in order to send high-frequency signals off the ionosphere practically directly overhead and down to the DTM a few miles away. In contrast, Taylor and Hulburt depended not on local, more precise ionospheric sounding data, but rather on the results of round the world transmissions. In their first reductions of this data, the NRL scientists used it to produce charts of predicted reliable frequencies for long-range communication, but Hulburt continued to draw on it to support his theorizing about ionospheric physics into the late 1920s and the early 1930s [Maris and Hulburt, 1929; Hulburt, 1935]. Hulburt always cited the long-range skip distance data as most important, and then the more detailed local studies, such as Tuve and Breit's measurements over Washington, DC.

Here, then, was a cooperative network of researchers. Many of them shared a similar educational background, and even had physics professors in common. They were in the same place at the same time, shared research instruments and at least one project, and discussed their results. Yet in the end they were committed to quite different research techniques, and this difference came out clearly when they took up the subject of research rockets.

I should touch on one further thing these researchers held in common: what I think was a crucial approximation made by both the NRL researchers, who relied mainly on skip distance measurements, and the DTM researchers, who made more precise measurements over a short baseline. The approximation is a simple piece of geometrical reasoning which is used to justify treating the refraction of radio waves through the conduction layer as though it were a simple reflection, by arguing that the path length of the true, curved route is equal to that of a reflection of slightly greater height. Since the paths are essentially the same length, any argument based on travel time of the signal will hold, and the computed height reached by the signal is a very close approximation. From the point of view of the experimenter, the ionosphere appeared to be a mirror overhead.

Learning to "read" the state of the ionosphere at second hand, by interpreting the recordings made by radio sounding equipment, required the development of specialized insight, as Gillmor has pointed out [Gillmor, 1981]. The dependence on radio-propagation data of one kind or the other also fostered an inclination to see the ionosphere in operational terms, as it was tied to the function of the transmitter as the central research instrument. Thus, despite the change in terminology, active experimenters such as Hulburt were inclined to think of the ionosphere still as the "radio conducting layer" or the "radio reflection layer."

But the depth of that commitment varied, and here is where I think we can see interesting distinctions between work at the NRL and at the DTM—both within the Washington network. As a Navy scientist, Hulburt's purpose in investigating the ionosphere was to provide the Navy with a technical system for reliable, worldwide communications, and he continued to think in global terms; thus skip-distance data seemed more to the point for Hulburt than did information about transmission conditions between Anacostia and Chevy Chase. For Hulburt, the central problem of ionospheric physics—the creation and disruption of a radio reflection layer—was linked to the practical problems of the applied physicist, a role which Hulburt embraced without complaint in the interwar period [Hulburt, 1939].

While the researchers at the DTM, and especially in the atmospheric electricity group, used long-distance radio-propagation data, I think it was characteristic of Tuve and Breit to shrink down, in a sense, their experiment. Their measurements give a kind of local precision reminiscent of the DTM's worldwide grid of pinpoint geomagnetic measurements. Even in a research program carried out over vast distances, Carnegie researchers tended to make precise, localized experiments. But also, during this period the Carnegie and the DTM determined that the efficient coordination of research would be best served by doing more experiments on laboratory models of nature, and experiments to discover universal constants of the Earth, rather than in gathering local data from far and wide [Cornell, 1986, pp. 140-151].

This march from the field into the laboratory was the cause of some internal arguments at the DTM, which were faintly echoed in the Institution's annual reports. The 1924 annual report of the DTM emphasized the necessity for a "combination of facilities for observation, experiment, and investigation" [CIW, 1925, p. 171]. The 1932 annual report, which announced the death of Louis A. Bauer, the DTM's director from 1904 to 1930 and a proponent of wide-scale data gathering, also included an endorsement of the Department's policy of "effecting [a] laboratory approach to its problems" [CIW, 1932, pp. 223-224]. This is the period in which Tuve's research, beginning with the ionospheric sounding project, moved from the field into the laboratory.

THE UTILITY OF GODDARD'S ROCKET

The contrast between Tuve and Hulburt comes out even more strongly when each had the opportunity to comment on the possible research uses of the Goddard rocket in 1929. While the Carnegie Institution hoped to aid Goddard's rocket project, seeking support from Charles Lindberg and money from the Guggenheim family, its leadership required assurances that the project was technically sound and of potential scientific utility. Thus it turned to members of the Washington scientific network, including Tuve and Breit, for opinions as to the prospects of Goddard's device [CIW, "Goddard Rocket Project" file, 1929-1931]. Both were positive that the rocket would be useful, but they perceived its utility in quite different ways.

Hulburt expected Goddard's rocket to be limited in its flight to altitudes of 30 to 80 kilometers. This would put it, he believed, into an area entirely untouched by direct experiment, between the region well understood from balloon-carried instruments, and the higher region "beginning at about 70 km and going upward, to say, 200 km or higher," which had been explored by radio waves [NA-US, RG-19, Hulburt to Fleming, December 5, 1929]. Given the commitment to utility which helped to shape his research, Hulburt saw the ionosphere as first and foremost the radio reflecting layer, best explored by radio waves.

Tuve saw the rocket as a potentially revolutionary research tool. Goddard, he said, clearly was the extraordinary man who could develop the device, if only he had more help with "initiative, especially in the financial directions . . . he has worked on this idea of such fundamental importance for fifteen years, and has managed to obtain only $5,000 for the support of his experiments!" While Tuve admitted (in an emendation to his typescript) that "the radio method gives perfectly definite evidence on certain points," still it seemed to him that "the only way we shall ever know much [originally, "anything"] about the upper atmosphere is by sending instruments into it."

Thus for Tuve the rocket could correct a lack of basic data, despite information from radio-propagation studies. "The only data we have to date," Tuve wrote, "on almost every question about the upper atmosphere has been obtained by extrapolation . . . in most cases without foundation. Even indirect experimental attack, as by the study of radio-wave propagation, has required very exaggerated extrapolations and the use of assumptions whose truth is unsubstantiated by any other evidence." Before thinking better of it, he had originally written that the assumptions underlying radio-based research were "entirely unsubstantiated" [LC, "Report on the Goddard Rocket," Tuve papers]. Despite their close connection with the Washington network, by 1930 Tuve and Hulburt had moved to quite distinct philosophies of how to do research. Hulburt continued to rely on radio-wave measurements over long distances, while Tuve was moving inside the laboratory for more controlled experiments. Hulburt regarded the ionosphere, in its basic characteristics at least, to be well understood. For Tuve it was a region approached only indirectly and thus about which little was finally known. It appears that the network failed to enforce consensus, at least at this stage.

Even keeping in mind the fact that Tuve was trying to provide support for the idea of aiding Goddard, it is striking how much less confidence he had in radio investigations of the ionosphere than did Hulburt. Again, the difference lies in their very different institutional contexts. Hulburt had a certain level of confidence in his view of the ionosphere because it was functioning manageably as a radio reflecting layer for fleet communications. On the other hand, Tuve, concerned with the search for physical constants in a laboratory setting, appeared less certain of the state of the art in 1929.

The Optics Division at NRL and the DTM at the CIW were among several mutually-supporting institutions which participated in Washington's interwar network for upper atmospheric research. As was suggested in the first letter on Goddard's rockets, these groups were able to cooperate effectively because of common interests and complementary natures. While the CIW might serve as liaison to "wealthy people," it was also good to have connections to the resources of the Navy.

We are so often concerned with how it is that scientists arrive at agreed-upon knowledge, commonly-held knowledge, and authoritative knowledge that perhaps it is good to test to what extent consensus is actually achieved, even among cooperating scientists. In the case of this "Washington network," it seems that groups at separate institutions could hold to quite different interpretations of their research efforts, that such differences were determined by institutional context, and that they did not prevent extensive cooperation.

ARCHIVAL SOURCES

NRL's institutional records from before World War II are held by the National Archives in Record Group 19, abbreviated above as RG 19. Some materials are held in the laboratory historian's office, including the E. O. Hulburt oral history interview cited in the text. Within RG 19, materials are filed according to the U.S. Navy's filing manual. Most of the NRL-Carnegie correspondence is in files identified by the designator A3-2(5)/CI (abbreviated above as "Carnegie file"). The only exception cited here is the Breit-Tuve-Hulburt letter of 1927, which is filed in NRL's file S70-1(5).

REFERENCES

Allison, D. K., *New Eye for the Navy*, 228 pp., Naval Research Laboratory, Washington D.C., 1981.

Bijker, W. E., T. P. Hughes, and T. Pinch (Eds.), *The Social Construction of Technological Systems*, 405 pp., MIT, Cambridge, MA., 1987.

Bradbury, N. E., Fundamental mechanisms in the ionosphere, *J. App. Phys.*, *8*, 709-717, 1937.

Carnegie Institution of Washington, *Yearbook, 1924-1925*, 438 pp., CIW, Washington, D. C., 1925.

Carnegie Institution of Washington, *Yearbook, 1931-1932*, 465 pp., CIW, Washington, D. C., 1932.

Cornell, T. D., Merle A. Tuve and his program of nuclear studies at the Department of Terrestrial Magnetism, Ph.D. dissertation, 535 pp., Johns Hopkins University, Baltimore, 1986.

Dellinger, J. H., Sudden disturbances of the ionosphere, *J. App. Phys.*, *8*, 732-751, 1937.

Dupree, A. H., *Science in the Federal Government*, 460 pp., Johns Hopkins, Baltimore, 1986.

Geiger, R. L., *To Advance Knowledge*, 325 pp., Oxford, New York, 1986.

Gillmor, C. S., Threshold to space: Early studies of the ionosphere, in *Space Science Comes of Age*, edited by P. Hanle and V. D. Chamberlain, pp. 101-114, Smithsonian, Washington, D.C., 1981.

Hughes, T. P., *Networks of Power*, 474 pp., Johns Hopkins, Baltimore, 1983.

Hulburt, E. O., Theory of the ionosphere, *Terr. Mag.*, *40*, 193-200, 1935.

Hulburt, E. O., On the physicist in government service, *Am. Phys. Teacher*, *7*, 157-159, 1939.

Hulburt, E. O., Early theory of the ionosphere, *J. Atoms. Terr. Phys.*, *36*, 2137-2140, 1974.

Latour, B., *Science in Action*, 274 pp., Harvard, Cambridge, MA, 1987.

Maris H. B., and E. O. Hulburt, Wireless telegraphy and magnetic storms, *Proc. IRE*, *17*, 494-500, 1929.

McNish, A. G., Terrestrial magnetic variations and the ionosphere, *J. App. Phys.*, *8*, 718-731, 1937.

Reingold, N., *Science, American Style*, 429 pp., Rutgers, New Brunswick, 1991.

Richardson, The nature of bright chromospheric eruptions, *J. App. Phys.*, *8*, 752-756, 1937.

Rosenberg, C. E., *No Other Gods: On Science and American Social Thought*, 273 pp., Johns Hopkins, Baltimore, 1976.

Taylor, A. H., An investigation on the higher radio frequencies, *Proc. IRE*, *13*, 677-683, 1925.

Taylor, A. H., and E. O. Hulburt, Propagation of radio waves over the Earth, *Phys. Rev.*, *27*, 185-193, 1926.

To Watheroo and Back: The DTM in Australia, 1911-1947

R. W. HOME
History and Philosophy of Science Department
University of Melbourne
Parkville, Victoria 3052, Australia

As a confederation of British "settler" colonies, Australia traditionally looked to Britain for cultural leadership. Only in the aftermath of the Second World War, in the wake of the military alliance with the U.S. forged during the war years, did Australia's intellectual horizons expand to embrace America as well as Britain. In science as in culture generally, links were established with the U.S., at both an individual and an institutional level, that have endured [Home and Kohlstedt, 1991].

Such links were not, however, built entirely from nothing. This paper describes a remarkable instance of longstanding Australian-American cooperation in geophysical research that began a generation earlier and that continued after the war. The case is notable partly because it was both initiated and wholly sustained by a private organization, the Carnegie Institution of Washington (CIW), rather than by governments, and partly because it was not initially envisaged as a cooperative venture at all. Despite (or perhaps because of) these somewhat unusual features, it offered a clear instance of a linkage from which benefits flowed to both countries, one that other scientists could look to as they responded to Australia's new international environment.

*****

When the CIW established its Department of Terrestrial Magnetism (DTM) in 1904, its aim was declared to be "to investigate such problems of world-wide interest as relate to the magnetic and electric condition of the Earth and its atmosphere, not specifically the subject of inquiry of any one country, but of international concern and benefit" [Fleming et al., 1947a, p.iii]. As Tuve later noted [Tuve, 1947, p.43], "there can be little doubt that in the background there was hope of discovering the origin of the earth's magnetic and electric fields". An essential prerequisite to understanding was to determine the general structure of the field and the way it varied with time, and it was on this that the DTM focused during its early decades.

The Earth, the Heavens and the Carnegie Institution of Washington
History of Geophysics Volume 5

Given the DTM's stated aim, it was natural for it to concentrate initially on ocean surveys, at first using a chartered wooden vessel, the *Galilee*, and later employing, until its loss in a disastrous fire in 1929, the specially constructed non-magnetic yacht, the *Carnegie*. Not only had the Earth's field not been adequately mapped over large parts of the world's oceans, these areas also clearly lay outside the interest of "any one country." On land, matters were somewhat more complicated. Nevertheless, in countries where the local authorities had not undertaken magnetic surveys of their territories, the DTM under its founding director, Louis A. Bauer, did not hesitate to move in. As a result, reports of the DTM's early work [e.g. Bauer et al., 1921; Fisk and Sverdrup, 1927] are replete with romantic accounts of expeditions by its magneticians across more-or-less remote regions of the Earth's surface, recording the elements of the Earth's field as they went.

Australia was a principal target of this remarkable program of privately sponsored scientific imperialism, for despite more than a century of European colonization and settlement, much of its vast and often inhospitable land mass remained little known from a scientific point of view. So far as a magnetic survey was concerned, Georg Neumayer had covered much of Victoria between 1858 and 1864 [Neumayer, 1869], while a preliminary survey of Tasmania had been undertaken in 1901 [McAulay and Hogg, 1902]. Fixed stations had been maintained at Hobart from 1840 to 1854 [Savours and McConnell, 1982] and Melbourne since 1858 [Melbourne Observatory, 1872-1911], and some isolated determinations had been made along the coast by ship-borne observers. Otherwise, the map was effectively blank. Bauer visited Australia in 1911 to investigate possibilities. He persuaded officials of the smaller state astronomical observatories to take on some magnetic recording as secondary base stations, using "complete magnetic outfits" to be supplied by the DTM; the Melbourne Observatory, where records had been kept continuously since Neumayer's day, became the principal base station.

All these observatories were, however, small and declining institutions which had suffered severely during the economic depression that ravaged Australia in the 1890s. More recently they had seen the principal basis of their public support, the meteorological service they provided, transferred to a new federal authority. Their leadership had been weakened by the retirement of several long-serving directors. They were in no state to take on a major new task like a magnetic survey of the continent. Bauer, with the agreement of the Australians, committed the DTM to filling the breach [Bauer, 1911]. The DTM would measure the three elements of the magnetic field at locations 100 miles apart, on average, throughout the continent. Bauer's report makes it clear that he hoped that, with a grid of observations thus established, the Australians would gradually fill in the intervening gaps.

During the next few years, the senior officer assigned by the DTM to the Australian survey, the New Zealander Edward Kidson, travelled widely through many of the more accessible parts of the country and also undertook two major transcontinental expeditions by camel, one from south to north through the center, more or less along the line of the overland telegraph, the other along the recently opened up Canning Stock Route through the interior of Western Australia. Several other DTM officers (F. Brown, A. L. Kennedy, W. C. Parkinson) also covered large distances in difficult and sometimes arduous circumstances, taking observations at regular intervals. Kidson in his report [Kidson, 1921, p.172] acknowledged the "great assistance and encouragement" that he and his colleagues had received from government officials and settlers in all parts of the country. From a scientific point of view, however, the only significant Australian input came from G. F. Dodwell, Government Astronomer for South Australia, who made systematic magnetic observations on behalf of the DTM during an expedition to the Musgrave Ranges, in the remote center of the continent, and later in other parts of South Australia, and from staff of the Melbourne Observatory, who helped with the intercomparison of instruments. On some of his journeys, Dodwell was accompanied by Kerr Grant, the professor of physics at the University of Adelaide, and his student R. S. Burdon [Dodwell, 1921, p. 154].

Surveys such as these delineated the general shape of the Earth's magnetic field within the region concerned, but they revealed nothing about the time variations to which the field is liable. To determine these, a program of regular observations at fixed sites is required. To this end, in 1915 the decision was taken at the DTM to establish a number of fixed magnetic observatories—initially, nine were envisaged [Tuve, 1947, p. 45], though in the end only two were erected—to fill the most pressing gaps in the existing international network of observatories. Of the two that were eventually commissioned, one was at Huancayo, Peru, close to the magnetic equator. The other was located near Watheroo, Western Australia, in a position almost antipodal to the magnetic observatory operated by the U.S. Coast and Geodetic Survey at Cheltenham, Maryland, and bridging the huge gap between the existing magnetic observatories in South Africa and Melbourne.

The decision to locate an observatory in southwestern Australia was thus taken for reasons of general global strategy within the DTM's overall research program. This was a wholly American project, conceived in and entirely managed from Washington, where all significant data analysis and publication would be undertaken. The larger program required an observatory in Western Australia, so one would be established there. No thought was given to seeking the views of Australian scientists on what was proposed, or to involving them in the work. So far as local arrangements were concerned, it was simply assumed -- correctly, as it turned out -- that Australian officials would cooperate, once the Americans let them know what they wanted. Such an approach is inconceivable today. That it went unquestioned in 1915 reflects both Australia's generally subservient political position and the weakness of the Australian (and especially the Western Australian) scientific community at that time. Nobody thought to consult local scientists about the project at least partly because, as Bauer would have been well aware as a result of his 1911 visit, there were almost none to consult.

In fact, when DTM magnetician W. F. Wallis joined W. C. Parkinson in Perth in mid-1916 to seek out a suitable site for the observatory, he brought with him from Washington very definite specifications as to what was required. A site was to be selected between 28° and 36° south latitude and 114° and 118° east longitude. It was to be flat, free from present or future magnetic disturbances, remote from towns with large industrial establishments but accessible enough for communication and supplies, close to sea level but at least 80 km from the coast, with a dry climate, an adequate water supply and no tall trees. For earth-current determinations, space was also needed for installing two cables, each at least 15 or 16 km long, at right angles to each other, with one preferably in the magnetic meridian and the other in the prime vertical [Fleming et al., 1947a, p. 1]. Several possible locations were investigated before the site at Watheroo was selected. One particularly promising site near Pindar, on the railway line from Geraldton to Meekathara and Wiluna, was finally rejected because of fears that this little used line to one of the more remote parts of the country might one day be electrified!

The site selected was on a level section of the vast undulating sand plain that stretches north along the coast from Perth. It lay in uninhabited and very poor and sparsely vegetated country, 17 km west of the village of Watheroo on the Midland Railway. Winter rainfall was reasonable but ground water extremely scarce. The sandy surface of the surrounding plain made access difficult in the early years; until a road was built in the mid-1920s, a round trip by horse-drawn vehicle took a whole day. In the absence of competing claims to this unattractive piece of land, it was granted to the CIW by the Western Australian government for the cost of having it surveyed.

The buildings erected for the observatory were made of local materials but to a standard design developed in Washington for use in all the observatories then projected. Among other things, this meant that space was provided in the living quarters for an extensive heating system, and that the roof was designed to carry a heavy load of snow—an unlikely eventuality in the mild Western Australian winters where snow is never seen! Apart from non-magnetic sand-lime bricks made from local sand for the magnetometer piers, all materials were carted in. Construction was delayed by war-time shortages of men and materials, and it was January 1, 1919 before the observatory officially commenced operations.

The core of the observing program was of course the regular measurement of the Earth's magnetic field. One of the DTM's standard magnetometers (No. 7) was used for absolute determinations of declination and horizontal intensity, while its No. 2 earth inductor was used to measure the inclination. Variations in the field were recorded using the DTM's Eschenhagen magnetograph No. 1, supplemented from 1933 by data from a La Cour rapid-running magnetograph [Fleming et al., 1947a, pp. 5-9]. The observing program also, however, included measuring other geophysical quantities. In 1922, continuous recording began of both positive and negative electrical conductivity of the atmosphere, while atmospheric potential gradients and earth-current potentials began to be recorded in January 1924. Later on, the program was expanded to include systematic probing of the ionosphere as well. The data obtained were reduced on site, but were then sent to DTM headquarters for analysis. Barometric pressure, temperature and humidity were recorded continuously from the beginning, together with wind speed and direction and daily rainfall.

Monthly accumulations of meteorological data were passed to the Australian Meteorological Bureau but in the absence of a telegraph link, observations could not be passed on in time to be included in the daily synoptic weather charts issued by the Bureau. Moreover, at least in the early years, no effort seems to have been made to render the observatory's data compatible with records from the Bureau's own stations. The rain gauge installed at Watheroo in 1918, for example, was U.S. rather than Australian standard issue [Fleming et al., 1947a, p. 4]. It was therefore perhaps symbolic of a more general change in attitude on the part of the DTM authorities when, in 1932, after the original rain gauge rusted out, it was replaced by an Australian-standard instrument [National Climate Center, Bureau of Meteorology, Melbourne, Station file 90/008132 (Watheroo Magnetic Observatory)]. There is no mention of this in the brief account of the observatory's meteorological instruments published in 1947, which suggests that the account was actually written in the 1920s. This account likewise doesn't mention the thermometers being housed in a weather screen of any kind, let alone in a Stevenson screen as was standard practice at Australian Meteorological stations; yet in fact a Stevenson screen appears to have been in place from the early 1930s at the latest [W. D. Parkinson, personal communication, 1992].

The engagement of the observatory staff in the research program was thus in its original conception limited to ensuring a steady flow of reliable data to Washington. The strict subservience of the observatory to Washington is also evident from the thick files of letters, now preserved at the DTM, that passed between successive observers-in-charge and the Department. With two short-term exceptions to which we shall return, observers-in-charge were all experienced, long-serving DTM officers.

Having selected the site for the observatory, both W. F. Wallis and W. C. Parkinson remained in Western Australia to oversee its construction. In due course, Wallis became the first observer-in-charge, with Parkinson as his assistant. Later in 1919, however, Wallis returned to Washington, his place at Watheroo being taken by Edward Kidson who had recently been released from war service in the British Army. Two other senior DTM magneticians, G. R. Wait and H. F. Johnston, also served as observers-in-charge during the 1920s, each for a period of several years. Then the Englishman Parkinson, having spent most of the previous decade on DTM projects in other parts of the world after a brief stint as observer-in-charge following Kidson's resignation, returned to take charge again in April 1930. Thereafter, except for a three-month break during the summer of 1933-1934, and a much longer break of almost three years in 1936-1938 when he was relieved by another senior DTM magnetician, J. W. Green, Parkinson remained as observer-in-charge until the CIW relinquished control of the observatory in 1947 [Fleming et al., 1947a, p. 71].

Kidson later developed links with the wider Australian scientific community, beyond those depending directly upon his work for the DTM. In January 1922 he resigned from the CIW to join the Commonwealth Bureau of Meteorology at its headquarters in Melbourne. Soon afterwards, he became head of a newly created (and short-lived) Research Division within the Bureau. From this vantage point he played a significant role in the formation of an Australian

Branch of the British-based Institute of Physics, an important stage in the development of a community of physicists in Australia [Home, 1987]. Later, in 1927, he returned to his native New Zealand to become director of the meteorological service there [Kidson, 1941].

Kidson was, however, an exception. The Watheroo Observatory was clearly conceived of, by those responsible for both its establishment and its continued operation, as a wholly American institution, an outpost of the DTM in the southeast Indian Ocean region. Despite this, the observatory soon took on something of a life of its own. Links that had not been anticipated by Bauer and his colleagues gradually developed with the Australian scientific community, and these steadily expanded over the years with benefit to both sides.

Some of these were opportunistic and without long-term consequences for the work of the observatory. For example, following Eddington's 1919 announcement that his observation of the total solar eclipse visible in that year from the island of St Helena had confirmed Einstein's theory of general relativity, there was much interest in observing the next total eclipse, due in September 1922 and visible from sites in a band running east-west through the center of Australia [Crelinsten, 1983]. Bauer had DTM personnel make special magnetic and atmospheric-electric observations in connection with the eclipse, both at Watheroo (which lay outside the belt of totality) and in cooperation with Australian and other observing parties within that belt [CIW, DTM Miscellaneous 1917-1928 (IV), "Magnetic and Allied Observations During the Solar Eclipse of September 21, 1922"]. At a more mundane level, in 1923, in an effort to reduce the observatory's running costs, the then observer-in-charge, G. R. Wait, sought advice from Western Australia's Director of Agriculture as to the possibility of cultivating part of the observatory site, and this led to a proposal that experiments be carried out by the Department of Agriculture at the observatory into the use of superphosphate to improve the fertility of the Western Australian sand plains generally [CIW, DTM Miscellaneous 1917-1928 (III), Bauer to John C. Merriam, May 28, 1923].

Of much greater significance for the observatory in the longer term was the emergence of a new site of local scientific expertise with the opening of the University of Western Australia in 1913. To be sure, this long remained a tiny institution. Nevertheless, the appointment of the first professors constituted a dramatic advance in a society where people with scientific training had until then been few and far between. Among the new arrivals was the young Scottish professor of mathematics and physics, Alexander David Ross (1883-1966), a student in Lord Kelvin's laboratory at Glasgow during the master's final years who had subsequently worked as a research student at Göttingen and then lectured at Glasgow before being appointed to the chair in Western Australia. In Perth, he found himself without facilities of any kind for continuing his research, with no advanced students, and with only one other person within over a thousand miles with any extensive knowledge of physics, a lecturer at the local technical college. The unexpected prospect of having a major geophysical laboratory erected close by was from Ross' point of view an extraordinary boon, and from the moment Wallis and Parkinson arrived to search for a site, he did everything he could to advance the undertaking. "It is obviously most important that the University should be kept in the closest possible touch with an institution of such an international character," he told the Vice-Chancellor, "and that it should offer what assistance it can" [UWA, 0981, Ross to Vice-Chancellor, June 17, 1916]. The University Senate concurred, and Ross accompanied Parkinson on several surveying expeditions as the search for a site proceeded [UWA, 0981, Ross to Members of Senate, October 14, 1916].

Once the observatory was functioning, new links with Ross and, through him, with the University, quickly developed. Ross, it appears, with no facilities whatsoever for advanced scientific training, thought of using Watheroo as a training ground for his more promising students. To this end, when the *Carnegie* visited Perth in September 1920, he arranged for her captain, J. P. Ault, to attend a meeting of the University's Student Physical Society at which one of the students, John Shearer, read a paper on the *Carnegie*'s work. Ault was sufficiently impressed to write to Bauer about it: "It was a remarkable paper in many ways and showed a keen interest in the subject on the part of the speaker." "From various indications," Ault went on, "I am sure he is anxious to get into the work of the Dept. and I feel that he would make a good addition to the Dept. Prof. Ross is anxious to recommend him, as an Observatory assistant. . . . I am writing this just to let you know that there is a man here available for Obs'y assistant in case you decide to appoint a local man" [DTM, Shearer personal file, Ault to Bauer, September 30, 1920].

This was a clever idea, because the proposal suited both sides. For Shearer, there was no other avenue open in Western Australia for pursuing a career as a physicist, and scholarships to take him elsewhere were virtually nonexistent. For the DTM, appointing a competent local man offered substantial savings over sending someone to Western Australia. Hence it comes as no surprise that when a vacancy arose a couple of months later, consequent upon Kidson's departure for the Meteorological Bureau and Parkinson's promotion to observer-in-charge, Shearer was appointed. He stayed at Watheroo for over two years, then used the money he saved from his salary to support his graduate study at Australia's then premier physics school at the University of Melbourne. He subsequently taught physics for many years at the University of Western Australia [Home, 1990, p. 182].

Shearer was the first of a steady stream of students from Ross' department to spend time at Watheroo. Indeed, from the time of Shearer's appointment in late 1920 until the arrival of military personnel on the site during World War II, all but four of the steadily increasing number of assistant observers employed at Watheroo were Ross' students. (Two of the "outsiders" were specialists sent from Washington, namely O. W. Torreson, 1924-1926, and S. L. Seaton, 1935-1938; R. G. Curedale was a ship's wireless operator recruited by Parkinson in 1930 to operate the observatory's recently installed radio link with Washington; and the fourth, A. A. J. Crowne, was a more or less untrained young Englishman, also hired by Parkinson, who spent something over a year at Watheroo in 1930-1931 [AustAS, AustIP papers, MS 86/1/10, Crowne to Ross, August 10, 1931].) Most of the students who were appointed stayed for several years. During the late 1920s, Ross also arranged for up to three senior undergraduates at a time to spend their long summer vacations gaining "work experience" at Watheroo. Full lists of the observatory staff over the years are given in [Fleming et al., 1947a, p. 71; 1947b, p. 25].

Several of the students appointed as assistants in due course pursued postgraduate qualifications at the University of Western Australia, based on research they had done at Watheroo. In effect, the observatory became Ross' substitute for a graduate school. It gave his students training in research that was simply not available, given the want of laboratory space and equipment, in Perth. For many of them, Watheroo opened a path toward a continuing career in physics.

The work of the observatory was, as we have seen, closely controlled from Washington, with all significant data analysis being done there. In the early years in particular, when the work at Watheroo consisted merely of maintaining the routine magnetometer and meteorological readings, the opportunities for research were very limited. It was a lonely life, and rather tedious. Later in the 1920s, however, the work diversified, and there was more scope for the young Australian assistants to undertake original investigations.

For example, as noted already, routine measurements of atmospheric electrical conductivity began in 1922 and of earth current potentials and atmospheric potential gradients in 1924. For these latter determinations, the DTM sent out a specially trained staff member, O. W. Torreson, to install the equipment and operate it for an initial period [CIW, recorded interview with O. W. Torreson]. Later, in order to clarify the earth-current results, W. J. Rooney from DTM spent seven months at Watheroo from December 1925, making an earth-resistivity survey of the whole area [Fleming et al., 1947a, p. 4]. Problems arose, however, with the atmospheric potential-gradient data, subsequent to Torreson's departure, and this gave two outstanding young graduates of Ross' department who were then on site, J. E. I. Cairns and G. Builder, an opportunity to do more challenging work. Analysis of the first three years' records had led to the suspicion that insulation leaks in the collector were producing systematic errors in the data. Cairns and Builder were given the far from straightforward task of determining whether the leakage was small enough to keep the data accurate within 1%, as required.

From the mid-1920s, there were also increasing opportunities for work on radio propagation, which the DTM encouraged because of the obvious links with its traditional field of investigation. At Watheroo, radio research was first taken up in 1926 as a spare-time activity by Cairns, who wrote up his results for a Master's thesis that he submitted to the University of Western Australia in 1927. He also presented two papers on his research at the 1926 Perth meeting of the Australasian Association for the Advancement of Science—the first meeting of the Association to be held in Western Australia—and the DTM subsequently arranged for these, together with a third paper by Cairns arising out of this work, to be published in America [Home, 1990, p. 44]. Cairns' project was well conceived to use work already going on at Watheroo. The core of the investigation was a systematic study of fading in broadcast radio signals received at Watheroo. Cairns showed a clear relationship between his data on fading and the observatory's atmospheric potential-gradient records. In addition, he used the results of Rooney's ground-resistivity survey to link a forward inclination of the radio waves with a low-resistivity subsoil. He also recorded the strength, direction, and waveform of a large number of cases of the interference then generally referred to as "atmospherics."

For his efforts, the University of Western Australia in 1928 awarded Cairns its recently endowed Hackett Research Scholarship, which enabled him to travel to England for further research training as a Ph.D. student at the University of Cambridge. Following his degree, he was reengaged by the DTM as observer-in-charge of the Huancayo Magnetic Observatory, 1931-1934, after which he returned to England to make a career as a research physicist with the electronics company, E.M.I. Ltd.

The meeting of the Australasian Association in Perth in August 1926 in fact marked the beginning of what appears to have been a policy initiative of the then observer-in-charge, H. F. Johnston, to establish more regular contacts between Watheroo staff and the wider community of physicists in Australia. Johnston attended the Perth meeting with Cairns and also presented a paper [Johnston, 1926]. He likewise attended the next meeting of the Association, at Hobart in January 1928, and later reported [CIW, DTM Directors (1904-1934) (IV), Fleming to Merriam, February 29, 1928] presenting a paper on the observatory's potential-gradient work (though for some reason this was not mentioned in the published proceedings of the meeting). The journey to Hobart also gave him an opportunity to visit the observatories in Adelaide and Melbourne and advise them on their magnetic work. There was evidently some dissatisfaction among the physicists present at the Hobart conference, for during the meeting they resolved to hold a separate conference of their own, in Canberra, later in the same year. Johnston was unable to attend this, but he and another of Ross' former students who was now employed at Watheroo, F. W. Wood, submitted a paper on the calibration of the atmospheric potential-gradient records, while Cairns and Builder reported on their investigation of the leakage from the collector [Conference of Australian Physicists, 1928, pp. 31-36].

Meanwhile Ross was also working to strengthen the links being Watheroo and Australian science. Early in 1928, the Australian and British governments decided to establish the Imperial Geophysical Experimental Survey, the aim of which was to submit various new geophysical techniques for prospecting for minerals to extensive field testing in Australia, while simultaneously training Australian workers in the use of these techniques [Butcher, 1984]. Ross immediately lobbied the Australian scientific authorities, using the presence of the Watheroo observatory and the availability of its records as an argument for conducting some of the Survey's work in Western Australia:

> I do not know whether he [i.e. A. B. Broughton Edge, director of the Survey] proposes to make any tests [b]y magnetometric methods, but if so, the presence in this State of a magnetic observatory with the most modern equipment and obtaining continuous registrations of all the magnetic elements, places W.A. in a uniquely favourable position. Even for the electrical methods which he will most certainly employ, I think the continuous records of earth currents kept at Watheroo would be of considerable value. [AustAS, AustIP papers, MS 86/1/7, Ross to A. C. D. Rivett, February 11, 1928]

Ross' arguments were, however, in vain, and Broughton Edge and his team spent comparatively little time in Western Australia.

Johnston apparently suggested that Australia's newly founded Council for Scientific and Industrial Research (CSIR), through its Radio Research Board, should follow up Cairns' work by installing a directional atmospherics recorder on loan at Watheroo [Evans, 1973, p. 186]. The recorder, of the Watson-Watt type, was operational by November 1928, but by then Cairns had departed for England. F. W. Wood was therefore placed in charge of it with a view to his submitting an M.Sc. thesis to the University of Western Australia, based on the results obtained [*Year Books*, 1929, p. 236 and 1930, p. 281]. However, there were persistent problems with the apparatus, and the thesis was never completed. In August 1929, Johnston attended the second Australian conference of physicists, mathematicians and astronomers, held in Melbourne, and presented three papers by Watheroo staff, one by Wood reporting results obtained with the atmospherics recorder and two by Builder arising from the atmospheric potential-gradient work. Johnston subsequently assured Fleming in Washington that the contacts he had made at the meeting and the feedback he had obtained would "be of material assistance in the work of the Observatory" [CIW, DTM Directors (1904-1934) (III), Fleming to Merriam, October 2, 1929]. Builder's papers were published soon afterward in the DTM's journal, *Terrestrial Magnetism and Atmospheric Electricity* [Home, 1990, p. 38].

Earlier in 1929, Johnston had been struck down by a sudden attack of appendicitis that led to his being hospitalized for a time. There was no time to arrange a replacement, and the running of the observatory had had, perforce, to be left in the hands of the most senior of the young Australian assistants then working at Watheroo, F. W. Wood. Under his direction, its affairs, Johnston reported, went on "splendidly" [CIW, DTM Directors (1904-1934) (III), Fleming to Merriam, May 18 and May 31, 1929]. Later that year, W. C. Parkinson's scheduled arrival to replace Johnston as observer-in-charge was delayed for several months by his involvement in the salvage operations following the destruction of the *Carnegie* [CIW, DTM Miscellaneous 1934-35]. A decade earlier, the nearest available DTM magnetician would undoubtedly have received instructions to proceed to Watheroo with all speed to take over in his absence. Now, however, with Wood's successful management earlier in the year still fresh in mind, the DTM authorities in Washington did not hesitate to reappoint him as acting observer-in-charge until Parkinson arrived. Wood was later designated to represent the observatory at the third conference of Australian physicists, held at Sydney in August 1931 [CIW, Wood personal file, Fleming to Merriam, March 5, 1931]. He stayed at Watheroo until the end of 1932 before going to King's College, London, to work under E. V. Appleton.

He returned to Australia later in the 1930s to become a research officer with the Radio Research Board. In 1946, pending the handing over of the Watheroo Observatory to the Australian government, Wood returned there as transitional observer-in-charge to see in the new arrangements.

In relying on Wood, the DTM authorities were of course relying on one of their own officers, albeit a young man who was more or less unknown to them. A much more remarkable illustration of the close links that the DTM had by then established in Australia, and its confidence in the people involved, came during the summer of 1933-1934, when Parkinson took leave from the observatory for several months. Only one of the young assistants then on the staff, R. G. Curedale, had been there for more than a few months, and he did not occupy an established position on the observatory payroll but was classified instead as a temporary employee "engaged locally on a more or less continuous basis" [CIW, DTM Miscellaneous 1932-33, Fleming to W. M. Gilbert, November 26, 1932]. Hence, rather than appointing him acting observer-in-charge, the then DTM Director, J. A. Fleming, arranged for Professor Ross himself to be seconded from the University to take over.

This was, in fact, but a public manifestation of how the DTM had come to rely on Ross for help and advice, as well as for a steady supply of observatory assistants. He had visited the observatory regularly from its foundation, and on at least one occasion had undertaken a formal inspection on Washington's behalf [*CIW Year Book*, 1931, p. 308]. In Perth, he provided hospitality and introductions for DTM staff, and lobbied officials and government ministers on the DTM's behalf whenever problems arose—as they often did with Customs authorities in particular, who were prone to assess heavy customs charges on the specialized equipment being brought in for the observatory, despite a standing agreement that no such charges should be levied. The Department's annual report invariably included an effusive acknowledgement of the support he had provided during the year. Later in the 1930s, when personal problems threatened for a time to disrupt the work of the observatory, it was once again to Ross that Fleming turned for an independent assessment of the situation [DTM, Parkinson personal file]. DTM staff visiting or transferring to Watheroo were invariably ordered to make contact with him at an early stage. For example, the official instructions issued to L. V. Berkner for his extended visit to Watheroo in 1938 included the exhortation, "Please take the first opportunity you have to call on Professor A. D. Ross at Perth, who has been of so much help in our work at Watheroo" [LC, Berkner papers, Box 4, Fleming to Berkner, February 8, 1938]. Likewise W. C. Parkinson was told, when he was reassigned to Watheroo later in the same year,

> I have no doubt that while you have been in Perth you will have called on Prof. Ross to renew your old acquaintance and to assure him again of the appreciation of the DTM of the valuable assistance he has given us throughout the life of the Observatory. In view of the many years of contact with Prof. Ross we should spare no effort to maintain and keep alive his interest which has been so helpful to us in Western Australia" [DTM, Parkinson personal file, Fleming to Parkinson, September 12, 1938].

Ross was the person principally responsible for the formation of a body of professional physicists in Australia under the aegis of the British-based Institute of Physics [Home, 1987], and he worked constantly to bring the observers at Watheroo into the fold. Not only did he encourage their participation in AustAAS congresses and in the national conferences of physicists that were held regularly from 1928, he also encouraged those whom he regarded as qualified to join the Institute of Physics. When an inquiry was set up into Australia's observatories, he made sure that cognizance was taken of Watheroo, even though it was with the provision of facilities for astronomical work that the inquiry was supposed to be concerned [AustAS, AustIP papers, MS 86/1/10-11, Ross to Parkinson, November 28, 1931, and Fleming to Ross, January 22, 1932]. And it was doubtless also he who arranged Parkinson's nomination as a foundation member of the ambitiously styled but shortlived Royal Society of Australia at its formation in 1932 [CIW, Parkinson personal file, Merriam to Fleming, April 11, 1932; Lewis, 1978]. Some years later, when a Western Australian Division of the by then officially constituted Australian Branch of the Institute of Physics was formed, Ross arranged for Parkinson to be elected its first Chairman [AustAS, AustIP papers, MS 86/2/42, Ross to Secretary, Institute of Physics, April 26, 1940].

During 1929 and 1930, Geoffrey Builder was engaged on establishing a shortwave radio link between Watheroo and Washington. Eventually, he built for this purpose a 250 W transmitter with crystal frequency control [*Year Book*, 1930, p. 282]. He left the observatory, however, in August 1930, preceding Wood to King's College, London, where he did a Ph.D. under Appleton's supervision. At Watheroo, the radio work was taken over, as noted, by Curedale. There were obvious practical benefits in establishing this radio link. Yet there was also a scientific rationale for the observatory's growing engagement with radio research. As Cairns' study had demonstrated from yet another perspective, there were clear connections between the propagation of radio signals and the Earth's magnetic and electric fields. In fact, DTM headquarters in Washington was already at the forefront of research into this question, with DTM researchers Gregory Breit and Merle Tuve developing in the mid-1920s their well known pulse-echo technique for probing the ionosphere.

The investigation of the ionosphere had also quickly become a major focus of research for the Australian physics community following the establishment of the Radio Research Board in late 1926 [Evans, 1973]. The DTM's decision in 1931 to install ionospheric recorders at its Huancayo and Watheroo observatories therefore inevitably led to new links developing between Australian researchers and Watheroo. Because of its favorable position close to the magnetic equator, Huancayo received its equipment first, and then there were problems with the power supply for Watheroo's transmitter. As a result, it was not until January 1935, following the arrival of S. L. Seaton from Washington, that regular recording began at Watheroo, using manually operated equipment. A year later, a second generation of equipment was installed, embodying continuous recording on a single frequency supplemented with manually adjusted multifrequency observations at regular intervals. The instrumentation evolved rapidly, however, and in 1938, Berkner spent several months visiting Watheroo from DTM headquarters in Washington, installing new equipment for continuous automatic multifrequency recordings.

Meanwhile, Australian Radio Research Board workers based at the University of Sydney had been making rapid strides in developing ionospheric recording equipment of their own. A manually controlled variable-frequency pulse-echo apparatus designed by O. O. Pulley was operating by mid-1935 and generating the data that underpinned an important paper by Pulley and D. F. Martyn on the constitution and temperatures of the upper atmosphere [Martyn and Pulley, 1936]. By May 1936, locally developed fully automatic swept-frequency ionosondes were operating at both the University of Sydney and Mount Stromlo Observatory outside Canberra. Ionospheric research was a major item on the agenda of the 1936 conference of Australian physicists, held in Sydney that month, and the DTM's interest is evident from the fact that Watheroo's ionosphere specialist, Seaton, was nominated to represent the observatory, and that the papers he presented at the meeting had the names of Berkner and another of the DTM's senior ionospheric physicists, H. W. Wells, on them as co-authors. Parkinson, who had left Watheroo shortly before, also attended the meeting. Cooperation was very much in the air. One of the Board's researchers, A. L. Green, had visited Watheroo a year earlier and expressed the hope that the two groups could work closely together [CIW, DTM Miscellaneous, 1934-35, Fleming to D. C. W. Florance, August 5, 1935], while from the Watheroo side, the new observer-in-charge, J. W. Green, told Ross that at the Sydney meeting, over which Ross was to preside, "we would also like to participate in a discussion of Co-operative Ionosphere Research. Mr. Seaton will be prepared to present the DTM's program in this matter, including Washington, Huancayo, Alaska and Watheroo" [AustAS, AustIP papers, MS 86/1/15, Green to Ross, April 8, 1936]. While in Sydney, Seaton completed arrangements with the Australian group for regular exchanges of ionospheric data [Evans, 1973, p. 241].

Berkner's visit to Australia in 1938 brought the DTM researchers and the Australians still closer together. As J. P. V. Madsen, Chairman of the Radio Research Board, told Fleming, the Australians welcomed the prospect of real cooperation rather than just exchanges of information:

> From time to time I have looked into the question of closer cooperation, but as we have been concentrating on different objectives and both Watheroo and our own observations have been limited to special periods daily, we have not found it possible to do much effective cooperative work. With the installation of the new equipment I think it is most important that we should try to set out clearly some definite plan of coordinate effort. Berkner's visit to us will give an excellent opportunity of elaborating an effective scheme. [LC, Berkner papers, Box 4, Madsen to Fleming, May 2, 1938]

Moreover, in sharp contrast to the situation twenty years earlier, in this work, the Americans now treated the Australians as equals. Berkner's visit to Australia had another objective beyond installing the ionosonde at Watheroo, as the instructions he received from Fleming before his departure make clear:

> You will also communicate with the research workers in Australia, especially those associated with the Radio Research Board and the University of Sydney. Upon completion of your work at Watheroo you will proceed to Melbourne and Sydney and, if it [be] agreeable to those in charge, work for a limited period with Australian radio research men associated with the organizations indicated above in order that the Department may derive the greatest possible benefit from the researches they are conducting. [LC, Berkner papers, Box 4, Fleming to Berkner, February 8, 1938]

Along with these instructions, Berkner was given letters

of introduction to a list of influential people connected with radio science in Australia whom he was to arrange to visit if possible: "It is important that these men have an understanding of the work we intend to carry on at the Observatory in order that we may have their cooperation in it." The list is comprehensive: the DTM's officers had clearly done their homework. (A similar list had been drawn up for Seaton before his departure for Australia three years earlier.)

Ross of course appeared on the list, and Berkner quickly established cordial relations with him. Ross was sufficiently enthused by Berkner's project that for virtually the first time since arriving in Perth in 1913, he thought to begin some research of his own. Furnished with a small grant from CSIR for equipment, he sought Berkner's advice, prior to his departure, as to what he might do. Berkner suggested some useful observations that Ross could make on the reception in Perth of the ionosonde signals emitted at Watheroo [LC, Berkner papers, Box 2, Ross to Berkner, December 5, 1938 and Berkner to Ross, 17 February 1939]. In the event, however, Ross was unable to devote enough uninterrupted time to the work [AustAS, AustIP papers, MS 86/1/17, Ross to Parkinson, July 5, 1939], and it did not lead to any publications.

While Berkner was in Australia, a controversy arose in print between him and his English collaborator Henry G. Booker, on the one hand, and the Australians D. F. Martyn and G. H. Munro, on the other, over the validity of the Lorentz polarization term when considering the reflection of waves from the ionosphere. Booker and Berkner had included the term in their analysis of the problem, but Martyn and Munro argued that, on their interpretation of the Sydney ionosonde records, it should not be [Booker and Berkner, 1938; Martyn and Munro, 1938a, 1938b, 1939]. Berkner and Martyn engaged in a stimulating correspondence while Berkner was in Western Australia. Then after Berkner had finished his work at Watheroo, he spent several weeks in the eastern Australian states in accordance with the instructions he had received from Fleming.

At the congress of the successor organization to AustAAS, the Australian and New Zealand Association for the Advancement of Science, held at Canberra in January 1939, Berkner took an active part in the "very lively" discussions on the ionosphere, including an airing of the controversy over the Lorentz term, that were a major feature of the physics program [LC, Berkner papers, Box 2, Berkner to Fleming, January 23, 1939]. Afterwards, he spent a month in Sydney, working chiefly with Martyn. This time spent together was extremely fruitful. So far as the Lorentz correction was concerned, the two men reached complete agreement as to what the experimental data were—a considerable advance in itself—but were unable to resolve their differences as to how the data were to be interpreted. They did, however, suggest some experiments that might resolve the disagreement. They also had long discussions about the nature of radio fadeouts and their connections with chromospheric eruptions on the Sun and changes in the F_2 region of the ionosphere. Here, they succeeded in completely resolving the apparent differences between them, leading Berkner to make some last-minute alterations to a paper of his that was then in press [LC, Berkner papers, Berkner to Fleming, February 16, 1939].

Martyn and Berkner also drew up recommendations concerning future collaboration between the DTM and the Australian Radio Research Board [LC, Berkner papers, Box 2, Berkner to Fleming, February 27, 1939]. They urged the Australian authorities to follow the DTM's lead in publishing regularly the monthly average hourly values of the penetration frequency and the equivalent heights of the various regions in the ionosphere, that blueprints of the actual hourly values for each day be exchanged between the two groups, and that back runs of data be interchanged immediately. They recommended, in addition, systematic exchange of drafts of papers, and in some cases collaboration leading to joint publication. Berkner presented these recommendations to a meeting of the Radio Research Board that he attended prior to leaving Australia. The Board's Chairman, J. P. V. Madsen, and CSIR's Chief Executive Officer, Sir David Rivett, who was in attendance, also discussed with him the possibility of one of the Board's research workers spending some time at the DTM as they knew Booker had recently done. Rivett, Berkner reported, "especially stressed the isolation from other research units and mentioned his feeling that while they had been sending most of their workers to England, he felt it most desirable that some be sent to America to make contact with the work there."

Not long afterwards, F. W. Wood, who was by now in Sydney working for the Radio Research Board, was sent to Watheroo to obtain details of the new ionospheric recorder. Much of the momentum generated by the Martyn-Berkner collaboration was dissipated, however, by the gathering storm clouds in Europe. Within days of Berkner's departure, Martyn was sent without warning to England in response to an urgent telegram from the British government, to be given highly secret information about Britain's vital new electronic weapon, radar, to carry to Australia. Then, on his return, he was given the task of setting up a top-priority secret new laboratory within CSIR, the Radiophysics Laboratory, to do further research on this. To staff the Laboratory, he hired most of the leading workers from the Radio Research Board, deflecting them at one stroke from the fundamental ionospheric physics research on which they had previously been engaged, and into a situation where collaboration with the DTM was no longer feasible. The Board itself was so shorn of staff, indeed, that it was only with great difficulty that the continuity of the ionosonde records from Mount Stromlo was maintained.

The U.S. not at first being a belligerent, the observatory at Watheroo was for a time little affected by the onset of war, and the steady program of recording proceeded as before. Trained physicists were in great demand, however, and as time passed it became increasingly difficult for the observatory to find and retain the assistants who until then had continued to be supplied in a steady stream by Ross' department. Then, commencing with a request from the Department of Air in December 1940 for regular information on the day-to-day state of the ionosphere, the observatory became engaged in a new form of collaboration. A similar request was made at the same time to Mount Stromlo, in the hope that predictions could be made that would permit the Australian Air Force to maintain its long-distance radio communications network without interruption. Data from these two observatories, together with data from the Radio Research Board's recorder at Liverpool, near Sydney, became critical ingredients in the worldwide wartime ionospheric prediction service that was later established by the Allied nations [Evans, 1973]. By 1943, however, the scientific staff at Watheroo had shrunk to just Parkinson and his son, W. D. Parkinson, who had completed his degree at the University of Western Australia shortly before. To keep the meteorological and ionospheric observations going, Australian Army and Air Force staff moved on to the observatory site, and W. D. Parkinson was employed partly on this work and on only a part-time basis by the CIW to help maintain its traditional program of work [DTM, Parkinson personal file, Parkinson to Fleming, November 24, 1941, Fleming to Parkinson, January 5, 1942, and H. F. Johnston to Fleming, June 12, 1944].

Shortly after the war ended, J. A. Fleming retired as Director of the DTM and was replaced by Merle A. Tuve, who promptly undertook a comprehensive review of all the Department's activities, including the observatory program. The latter had initially been planned to continue through one sunspot cycle, but had been extended in the 1930s because of the new possibilities for investigation of the ionosphere [Tuve, 1947, pp. 46-52]. Now, in line with plans made when the program was first conceived, the decision was taken to offer the DTM's observatories at Watheroo and Huancayo, together with all their equipment, to the Australian and Peruvian governments, if they continued the observing program.

In the Peruvian case, the transfer to local control was not at all straightforward because of a lack of local scientists with the skills necessary to take over the work. In Australia, however, there was by this time a considerable cadre of expertise in geophysical research, and the transfer was able to go ahead without difficulty, as from July 1, 1947. The observatory at Watheroo was maintained under Australian government control until 1959, at which point its functions were transferred to other sites closer to Perth and the Watheroo site was abandoned.

We have seen that when the Watheroo Observatory was established a generation earlier, Australia did not possess the scientific manpower or skills required to operate a geophysical research station on this scale. That it now did testifies to the remarkable expansion in expertise in this branch of science that had occurred in the interim, to which the Watheroo Observatory had itself contributed significantly.

Reference has been made already to the Imperial Geophysical Experimental Survey that operated in Australia, 1928-1930, with which several Australian workers gained valuable experience in geophysical techniques before government funding was withdrawn with the onset of the economic Depression. In 1935, the Australian government joined forces with the governments of Queensland and Western Australia to establish the Aerial, Geological and Geophysical Survey of Northern Australia—a longterm survey, using chiefly geophysical methods, of mineral resources in the northern regions of the country. The DTM agreed to lend one of its magnetometer-inductors to the new Survey, and L. A. Richardson of the Survey's staff was trained at Watheroo in the use of this and other magnetometers for survey purposes. The instrument was on loan for many years. Collaboration continued into the war years with systematic exchanges of data between the DTM and the Survey [CIW, DTM Miscellaneous 1941-45, Fleming to J. M. Rayner, October 13, 1941].

Early in 1942, with Australia's links with most of its traditional suppliers cut off, the federal government extended the search for mineral resources nationwide [Mellor, 1958, pp. 101-102]. Ultimately this led to the formation in 1946 of a new agency, the Commonwealth Bureau of Mineral Resources, Geology and Geophysics, which recruited on to its staff a strong group of geophysicists, most of whom had previously had experience on the Northern Australian and wartime surveys and some of whom had also worked, before that, on the Imperial Geophysical Experimental Survey [Butcher, 1984, p. 39]. It was this new Bureau that took over operational control of the Watheroo Observatory when it was transferred to the Australian government in 1947.

From the point of view of the present paper, what is

especially remarkable is how disproportionate a number of the geophysicists who joined the Bureau in its early years had gained their introduction to the subject at Watheroo. Most in due course rose to senior positions within the Bureau. One of these was L. A. Richardson, whose being sent by the Northern Australian Survey to Watheroo for training has already been noted. The rest had worked for the DTM as assistant observers at Watheroo. These included F. W. Wood, R. G. Curedale, L. S. Prior (who eventually became Chief of the Bureau's Geophysical Branch), N. G. Chamberlain and E. McCarthy. W. D. Parkinson, who worked at Huancayo and elsewhere for the DTM after leaving Watheroo, also later worked for the Bureau for many years, including a stint back at Watheroo as officer-in-charge, before taking up a position at the University of Tasmania. When he joined the Bureau in the 1950s, four of the five most senior staff of the Geophysical Branch were ex-Watheroo men [W. D. Parkinson, personal communication]. Under Australian administration, the observatory continued to provide an entry path into the Bureau for young graduates of the University of Western Australia.

The impact on the local scientific community of the DTM's activities in Australia is made particularly clear by the statistics just set out. Though the Watheroo Observatory was initially conceived as operating entirely independently of Australian science, the policy that was adopted from an early stage of recruiting observatory assistants locally meant that it in fact became an important training ground for Australian geophysicists. Its impact in this regard was all the greater because, except for the Northern Australian Survey, there was virtually nowhere else in the country where young scientists could gain experience in geophysical work.

The influence of the Watheroo Observatory on Australian science spread further, however, than this. Locally in Western Australia, as we have seen, it long provided almost the only avenue available to young graduates in physics for advanced training, or for employment outside schoolteaching. Elsewhere in the country, it functioned as a point of reference and source of expertise for the small number of physicists, chiefly at the observatories in Melbourne and Adelaide, who were doing similar work. The regular participation of Watheroo staff in national scientific conferences that was pioneered by H. F. Johnston in 1926 not only doubtless provided a valuable stimulus for the work of the staff themselves, it gave a slightly different focus to the meetings, both because of the observatory's orientation towards field rather than laboratory research and because of its links with American investigators in an environment where for many years the links had been almost exclusively with Britain. And then as radio propagation and the behavior of the ionosphere became principal foci of Australian physics research, the Watheroo Observatory became an integral part of a continent-wide research program in which Australian investigators no longer looked up to the Americans for advice and assistance but worked with them as equals. Australian and American physicists thus established a pattern of friendly cooperation that presaged an era of steadily expanding linkages between Australian and American scientists in the postwar world.

ARCHIVAL SOURCES AND ACKNOWLEDGEMENTS:

My interest in Watheroo was aroused by an excellent class paper prepared some years ago by my student Hugh Hamersley. I am grateful also to Dudley Parkinson, who kindly shared with me his memories of life at the observatory and his knowledge of its work, and, in Washington, to James D. Ebert, then President of the CIW, L. Thomas Aldrich at the DTM, and Adrienne Powell and Raymond Bowers at CIW headquarters, who provided helpful access to the relevant archival records under their control. Citations of materials in the archives of the University of Western Australia are abbreviated UWA. References to the Australian Institute of Physics papers at the Australian Academy of Science are abbreviated AustAS, AustIP.

REFERENCES:

Bauer, L. A., The magnetic survey of Australia, *Terr. Mag.*, *16*, 215-217, 1911.

Bauer, L. A., et al. (Eds.), *Land Magnetic Observations, 1914-1920*, edited by L. A. Bauer et al., Researches of the DTM, vol. IV, CIW, Washington, D.C., 1921.

Booker, H. G., and L. V. Berkner, Constitution of the ionosphere and the Lorentz polarization correction, *Nature*, *141*, 562-563, 1938.

Butcher, B. W., Science and the imperial vision: the Imperial Geophysical Experimental Survey, 1928-1930, *Hist. Rec. Aust. Sci.*, *6*, 31-43, 1984.

CIW, *Year Book*, 1902-.

Conference of Australian Physicists, Canberra, 15th to 18th August, 1928: Proceedings and Abstracts of Papers, Canberra, 1928.

Crelinsten, J., William Wallace Campbell and the "Einstein problem": an observational astronomer confronts the theory of relativity, *Hist. Stud. Phys. Sci.*, *14*, 1-91, 1983.

Dodwell G. F., On magnetic work in South Australia, 1914 to 1918, in *Land Magnetic Observations, 1914-1920*, edited by L. A. Bauer et al., pp. 152-154, Researches of the DTM, vol. IV. CIW, Washington, D.C., 1921.

Evans, W. F., *History of the Radio Research Board, 1926-1945*, CSIRO, Melbourne, 1973.

Fisk, H. W., and H. U. Sverdrup (Eds.), *Land Magnetic and Electric Observations, 1918-1926*, Researches of the DTM, vol. VI, CIW, Washington, D.C., 1927.

Fleming, J. A., et al. (Eds.), *Magnetic Results from Watheroo Observatory, Western Australia, 1919-1935*, Researches of the DTM, vol. VII-A, CIW, Washington, D.C., 1947.

Fleming, J. A., et al. (Eds.), *Magnetic Results from Watheroo Observatory, Western Australia, 1936-1944*, Researches of the DTM, vol. VII-B, CIW, Washington, D.C., 1947.

Home, R. W., The beginnings of an Australian physics community, in *Scientific Colonialism: A Cross-cultural Comparison*, edited by Nathan Reingold and Marc Rothenberg, pp. 3-34, Smithsonian, Washington,

D.C., 1987.
Home, R. W., *Physics in Australia to 1945: Bibliography and Biographical Register*, University of Melbourne, Department of History and Philosophy of Science, Melbourne, 1990.
Home, R. W., and S. G. Kohlstedt (Eds.), *International Science and National Scientific Identity: Australia between Britain and America*, Kluwer, Dordrecht, 1991.
Johnston, H. F., Determination of the atmospheric potential-gradient reduction-factor at the Watheroo Magnetic Observatory, Western Australia, *Rept. 18th Mtg. Aust.A.A.S.*, 128, 1926.
Kidson, E., On the General Magnetic Survey of Australia, and On an Expedition over the Canning Stock-route, Western Australia, 1914, in *Land Magnetic Observations, 1914-1920*, edited by L. A. Bauer et al., pp. 164-172, Researches of the DTM, vol. IV, CIW, Washington, D.C., 1921.
Kidson, I. M. (Ed.), *Edward Kidson*, Whitcombe & Tombs, Christchurch, 1941(?).
Lewis, M., The Royal Society of Australia: an attempt to establish a national academy of science, *Rec. Aust. Acad. Sci.*, *4* (pt 1), 51-62, 1978.
McAulay, A., and E. G. Hogg, A preliminary magnetic survey of Tasmania, 1901, *Rept. 9th Mtg. Aust.A.A.S.*, 81-94, 1902.
Martyn, D. F., and O. O. Pulley, The temperatures and constituents of the upper atmosphere, *Proc. Roy. Soc. Lond.*, *A*, *154*, 455-486, 1936.
Martyn, D. F., and G. H. Munro, The Lorentz polarization term and the Earth's magnetic field in the ionosphere, *Nature*, *141*, 159-161, 1938.
Martyn, D. F., and G. H. Munro, The Lorentz "polarization" correction and the behaviour of radio echoes from the ionosphere at frequencies near the gyro-frequency, *Nature*, *142*, 1159-1160, 1938.
Martyn, D. F., and G. H. Munro, The Lorentz "polarization" correction in the ionosphere, *Terr. Mag. & Atmos. Elec.*, *44*, 1-6, 1939.
[Melbourne Observatory], *Monthly Record of Results of Observations in Meteorology, Terrestrial Magnetism etc. etc. taken at the Melbourne Observatory*, Government Printer, Melbourne, 1872-1911. (Subsequently titled *Record of Results of Observations* . . . and later *Results of Observations* . . .)
Mellor, D. P., *The Role of Science and Industry*, in series Australia in the War of 1939-1945, Series 4 [Civil], vol. V, Australian War Memorial, Canberra, 1958.
Neumayer, G., *Results of the Magnetic Survey of the Colony of Victoria Executed during the Years 1858-1864*, J. Schneider, Mannheim, 1869.
Savours, A., and A. McConnell, The history of the Rossbank Observatory, Tasmania, *Ann. Sci.*, *39*, 527-564, 1982.
Tuve, Merle A., Review of magnetic survey and observatory program of the Department of Terrestrial Magnetism, 1904-1946, in *CIW Year Book No. 46, 1946-1947*, pp. 43-53, CIW, Washington, D.C., 1947.

Some Memories of the Watheroo Magnetic Observatory

W. D. Parkinson
68 Risdon Rd., New Town, Tasmania 7008, Australia

These are rather random memories of the Watheroo Magnetic Observatory. Perhaps I should start by stating my connection with DTM. My father, W. C. Parkinson, joined DTM in about 1913 and spent the rest of his working life there, mostly at its observatories. He was sent to Australia soon after joining DTM and was at first engaged in magnetic field work.

I might say a word about magnetic field-work as it was then performed. The object was to determine the general global structure of the geomagnetic field. A location was chosen, as far as could be determined visually, free from interfering structures, such as buildings, railways, etc. and a site where the observations could be repeated years later without interference. Observations of three components of the field were made, with an attempted precision of one gamma, the units then in vogue. Observations took all day. They were spaced so that the diurnal variation was averaged out. This diurnal variation has an amplitude of from 20 to 50 gammas and so observations covering only part of a day could well be biased. But what was not realized was that in many places changing the position by a few meters changed the field by hundreds of gammas.

My father completed surveys in Western Australia, the South Pacific, and New Zealand before starting the search for a site for an Australian observatory. During a sojourn in Sydney he met and married my mother. This is worth mentioning because otherwise someone else would have the pleasure of addressing you today.

Originally, I believe, DTM intended to establish many magnetic observatories, but the budget stretched to only two, Watheroo and Huancayo, Peru. This sounds very familiar, but apparently things were much the same in the early days of this century.

Regarding Huancayo, the magnetic equator was an obvious site, and western South America a conspicuous gap in the existing observatory network in the early 1920s. As for Watheroo, observatories existed at Batavia, Cape Town, and Melbourne. The nearest land to the middle of this gap was the south west of Australia. My father was already in Australia when he was joined by Mr. Wallis in 1916. Although my father worked with Mr. Wallis for several years, I never heard him refer to him by his first name, so I have no idea what it was. Together they sought a suitable site: (1) free of magnetic anomalies, (2) with sufficient water, (3) accessible, and (4) sufficiently far from habitation that it would not be interfered with by some human activity. For Wallis, the senior of the pair, the last was the overriding factor.

It's worth mentioning that the encroachment of buildings, railways, etc. has always been the worst enemy of magnetic observatories. Many, in fact most, observatories have started life some distance from a city, only to find themselves in an inner suburb a few years later. Wallis was determined this would not happen to Watheroo, and it didn't. In fact it is the only case in history of an observatory moving closer to town, but that is a later story. The country south of Perth was too closely settled for Wallis' liking, so they surveyed to the north.

Probably few of you have seen the instruments that were used in magnetic field work into the 1950s. They consisted of an oscillation-deflection magnetometer (also used as a declinometer), a dip needle, theodolite, and chronometer. So Mr. Wallis and my father selected random sites in what is called the Midlands, about 100 miles north of Perth and 60 miles from the coast and spent a day at each, measuring declination, inclination, and horizontal intensity.

I should say something about the geology of the region. The western edge of Australia is fringed with a narrow deep sedimentary basin, mainly of Jurassic and Cretaceous age. Its eastern boundary is a sharp, probably faulted, contact with the Archean shield. The sediments are undisturbed and non-magnetic, but the shield is laced with mafic and ultramafic intrusions. The contact is obvious. To the east are red soil, eucalyptus trees, and farms; to the west white sand, low scrub, and no farms. The culture follows the geology. The railway, and later the road, ran north, with a swathe of farms beside them. The Western Australian wheat belt was confined to the thin strip between the start of rich soil and the failure of rainfall further inland. Today, new breeds of wheat can grow in the sandy soil of the coastal region. When I visited the area in 1991 I was surprised to find towns where there had been no habitation when I lived there.

Most of the DTM's survey sites were on the east side of

The Earth, the Heavens and the Carnegie Institution of Washington
History of Geophysics Volume 5

the contact, and their results showed an abundance of magnetic anomalies. They searched the area for a full year before finally choosing a site about 12 miles west of the town of Watheroo. I call it a town because it is not sufficiently independent to deserve the name village. It is really only a stop on the railway, with the appurtenances that this brings, such as a wheat silo, general store, church and town hall.

I am often amazed by the year of surveying that went into site selection for the observatory. Today a survey plane could cover the same area in about an hour. In fact, this has been done. The irony is that immediately east of the observatory is one of the largest and most intense anomalies in the district. The Observatory was just off this anomaly and although the gradient is very small, the declination at Watheroo was always differed somewhat from the general pattern. Of course there were no airborne magnetometers in those days, but a half an hour's discussion with a geologist would have saved half or more of their work.

The chosen site was free of magnetic contamination, certainly never likely to be interfered with by development, inhabitable (just), and within reach of the township of Watheroo (just). The surrounding vegetation was low scrubby bushes, but a grove of trees in a hollow suggested water, and the water table was shallow and its quality was good.

The site certainly satisfied the criteria. Initially there was no access, but the country was flat and it was not difficult to establish a track sufficient for the horse-drawn cart. It was certainly sufficiently far from habitation, and this led to its eventual downfall. My father had favored a site further south, near the much larger town of Moora, but Wallis feared that the railway might become electric. Electric trains have always been the terror of magnetic observatories, and Wallis was not satisfied with the remark of the local who said "It's only just steam!"

Construction of the wooden observatory buildings began in 1917. My parents lived in tents beside the well until the first building was inhabitable. All the building material was carted from Watheroo, four hours away. First constructed were the variometer building, to house the recording instruments, and the absolute house, where standardization measurements were made. The recording variometers are sensitive to temperature so their building consisted of several concentric structures, with sawdust insulation between the walls. It was very effective. There was a seasonal change in temperature, but no diurnal change. The office was attached to the variometer building. These buildings were designed in Washington and identical ones were constructed at Huancayo.

Next the main dwelling house was built. This also was designed in Washington and an identical building was built at Huancayo. They were designed to be built of pine, but the one at Watheroo was actually built of jarrah, a much heavier and stronger wood. The carpenter said there was sufficient timber in the house for three houses of that size. The house was extended, first with three rooms on one end, then with a 10 foot wide veranda all the way around, encased with fly-wire. This made it suitable for the Western Australian climate.

Regular recording started in 1919, the year I joined the staff at the age of five weeks. Wallis left late that year and was replaced as observer-in-charge by Kidson. After a little over a year Kidson left to become Government Meteorologist in New Zealand. I believe he was the first person to detect the sun-spot cycle in meteorological variables. When he left, my father took charge of the observatory. He held this position, with two breaks, for the next 26 years.

We left Watheroo late in 1921 and Ray Wait took charge. I am told that soon after they arrived Mrs. Wait looked out at the rather dull arid scene and said plaintively, "Ray, how long do we have to stay here?" During his time, in 1923, two additions were made to the research program: measurement of the conductivity of the air and of the electric field at the Earth's surface. It is interesting to trace their origins and subsequent developments.

I must divert into a little physical theory at this point. The best iso-magnetic charts available at the beginning of this century indicated that if you took a line integral of the magnetic field around a closed curve on the surface of the earth, the result was not zero. According to Maxwell's theory of electromagnetism this indicated a large electric current flowing through the surface of the Earth. If such a current existed, it could be measured. Such air-earth currents had indeed been measured, mainly in Germany, but they were orders of magnitude smaller than indicated by magnetic results. Louis Bauer, first Director of DTM, was convinced of the reality of the so-called "non-potential field" and so became interested in measuring the air-earth currents at the observatories. The discrepancy between the two methods bothered him and so he encouraged the more accurate measurement of this air-earth current in various parts of the world. Hence Ray Wait at Watheroo installed instruments for measuring both the conductivity of the air and the electric field at the surface of the earth. Once you start measuring a geophysical variable it becomes of interest in itself, and so, although the atmospheric electric measurements at Watheroo did little to solve the non-potential problem, DTM was for some time the leader in atmospheric electric research. This subject, however, has declined in interest in the geophysical community, perhaps mainly because it was the first victim of atmospheric pollution.

A so-called earth-current recorder was installed about the same time. It did not measure earth currents, but the potential difference between two electrodes buried in the ground. We actually had five electrodes, two north and two east of the reference electrode. It was considered a different method of measuring magnetic variations. So for years we had measures of the variations in the magnetic field and simultaneous measures of the electric field in the earth. A quarter of a century later Tichonov in Russia pointed out that the ratio of these two variables gives information about the resistivity structure of the earth beneath. Surely one of the lost opportunities in geophysics!

Ray Wait was followed by Freeborn Johnson in 1924. The normal staff at the observatory consisted of the observer-in-charge, two or three geophysicists (called observers), a mechanic and various domestic staff.

Living conditions improved throughout the 1920s and 1930s. The road was upgraded as vehicles improved. When I visited the observatory in 1928 they had a Citroen with caterpillar tracks, in an attempt to beat the soft dry sand. In 1930 we were back at Watheroo with my father again in charge. One of my mother's first remarks on returning was "One day there will be lawns all around this house." She was as good as her word. I should explain that the average rainfall is about 15 inches per year. The stunted vegetation is mainly due to the poor soil and hot dry summers. My mother set about creating lawns and a garden. After a wet winter we were able to keep sprinklers on for most of the summer. Otherwise, recycled bathwater kept most plants alive. She had a lawn on two sides of the house, with gardens and aviaries of finches and canaries. We had a standing reward for any visitor who did NOT say "It's like an oasis in the desert." None claimed the reward.

One of the improvements during Freeborn Johnson's time was the addition of a screened veranda all the way around the house, which almost doubled the floor area, and gave us room for table tennis. In the attic we found an old quarter-sized billiard table. We replaced the cloth and Frank Wood's Christmas present to everyone was new cues and balls. Frank Wood was also a keen marksman and he made a small rifle range. By this time rabbits had become a plague in the country and we had 22-calibre rifles. Soon afterwards the first golf course was laid down. The only grass that would grow there was not at all suitable for greens, which were oiled sand. Later the golf course was moved to a different site and extended to 9 holes. There had been a tennis court there for some time, and this was refurbished with gravel.

I don't want you to think that the scientific side of life was abandoned. The Hale spectrohelioscope was added and we took our turn in a world-wide watch on solar activity. Perhaps the most important addition to the geophysical program in the 1930s was the introduction of ionospheric sounding, in 1938. Stuart L. Seaton was in charge of these investigations. Ever since Schuster showed in 1889 that the magnetic diurnal variation was due to some external source it was obvious that exploration of the upper atmosphere was pertinent to geomagnetism. There were no ionosondes when we started and Stuart had to make eye readings of images on a cathode ray oscilloscope. Also, the amateur radio station VK6MO started about this time with Stuart as the chief operator. We kept regular schedules with an amateur in Pennsylvania who relayed our messages on to Washington.

Early in 1936 we left Watheroo for a three year break during which my father did more magnetic field work in Australia, the South Pacific Islands, and South East Asia. Wilbur Green took charge during this time. He was a keen golfer, so the golf course was kept in good shape. The garden and aviaries did not fare quite so well, but recovered soon after we returned in 1939. Lloyd Berkner was there then installing the CIW ionosonde. So we then had the facilities to see solar activity, monitor the magnetic field, and determine its effect on the ionosphere. As well, the atmospheric electric and telluric equipment were still in operation. Meteorological observations had been going on since the beginning of the observatory.

Providing fresh water was always a problem at Watheroo. The shallow aquifer simply held accumulated rain water and was subject to the variable rainfall. In 1939 we had a bore sunk. At 400 feet we struck sub-artesian water, but it was salty. The obvious thing to do was to build a swimming pool. It was nothing fancy; simply a corrugated iron tank sunk in the ground, but it was a great boon during the hot summers.

Then came World War II. At first there was little impact, but gradually the staff drifted, or were drafted, away. My father was left with only one assistant. I was an undergraduate at that time and had become interested in ionospheric research. The Air Force approached my father asking if he could supply information to help with planning short wave radio communication. In exchange, two technicians were assigned to the observatory from the Air Force. I was employed to predict communication conditions such as maximum usable frequencies. If you think weather forecasting is difficult, you should try ionospheric forecasting! The earth-current recorder proved to be a great boon because it printed onto paper, whereas the magnetographs were photographic recording. It gave us instant warning of the beginning of a magnetic storm; the worst ionosheric effects occur some hours later.

At last the war ended. The observatory was then 27 years old. The original intention of DTM was for the observatories to run for two sunspot cycles or about 22 years. So by 1946 they had exceeded their expected life. What to do with them? In Peru a new department of the government had to be created to absorb the Huancayo Observatory. There was no such problem in Australia. The Bureau of Mineral Resources had been formed some years earlier and had collaborated somewhat with the Watheroo observatory. The BMR was then becoming interested in fundamental geophysics and had taken over the magnetic observatory at Toolangi, near Melbourne. They were willing to continue the operation of Watheroo, too.

Frank Wood, chief assistant in the 1920s and 1930s, was appointed observer-in-charge for a period of joint control before the Bureau took over completely. After a transition period the observatory passed completely to the Bureau. But the difficulty of operating at such a remote site proved insurmountable for a government instrumentality, and by 1956 a substitute magnetic observatory was in operation at Gnangarra, near Perth. Both observatories operated during the IGY and Watheroo was closed down at the end of 1958.

My father left in 1946 for work at the DTM in Washington. I lost touch with the observatory at this stage, and, except for a brief visit, did not see it until 1956. The observer-in-charge then was Alf Tillett. By then I was working for the BMR and when Alf Tillett was due for long service leave I was asked to take charge for a couple of months.

The difference between a Carnegie-operated observatory and a Government-operated one was something of a shock. There were two full time clerks to attend to the paper work, a luxury my father never asked for nor expected. The scientific program was much as it had been, except that the atmospheric-electric and telluric recorders had been shut down, and a more modern ionosonde was in operation.

Merle Tuve and his wife visited us during this time. He thought there should be a house for visitors to the observatory, and Carnegie supplied the money to build one. It came to be known as "Carnegie Hall."

My short stay at Watheroo produced a profound change in my scientific life. As a result of doodling with some substorms recorded on the magnetograms I stumbled onto the idea of induction vectors and have been in the induction business ever since.

But the strain of trying to operate such a place under government regulations proved too much. Some time in 1959 the observatory site and buildings were sold to a farmer. By then strains of wheat that would grow in the sandy soil had been developed. So again my contact with Watheroo stopped until recently. In early 1991 I visited Perth and made a point of driving to Watheroo to see what was left of the observatory. I was heartbroken to see the buildings in ruins. The farmer, who had another property, was unable to keep the buildings in repair. The veranda had almost fallen off the main house, the roof had large holes, tanks had fallen down. The swimming pool was a rubbish tip. But the saddest sight was the interior of the variometer house, the most sacred place of all, where the Eschenhagen magnetometer had operated faithfully for decades, where no one dared bring in a knife or watch or even switch on a light. It had been used as a shearing shed. The concentric walls were gone and old sheep hides were lying around the floor.

But one thing remained. During the war my mother and father planted an avenue of gum trees on each side of the driveway. I remember them carrying buckets of water to keep the saplings alive during the summer. They are now a magnificent avenue of mature trees, giving some hint of the glory that once was the Watheroo Magnetic Observatory.

So that is the story of the Watheroo Magnetic Observatory. I might add one postscript. In 1991 I also visited the Tucson Magnetic Observatory, a place that I had read about for many years. I was surprised to find that it was in a suburb of Tucson, rather than in the wilderness that I had come to regard as essential for magnetic observatories. Also the staff totaled one, and he said that the observatory was to be upgraded so that it would operate entirely automatically. I realize that this kind of automation is inevitable and has many advantages. But I would hate to see the day when magnetic recording is confined to the machine production of a lot of figures. Scaling magnetograms is a tiresome job, but I believe there are still interesting phenomena to be found in them and that there is no substitute for human beings actually looking at the raw data.

A Canadian Life with Geomagnetism: The Research of Frank T. Davies

J. E. KENNEDY
Professor Emeritus of Physics, University of Saskatchewan

W. O. KUPSCH
Professor Emeritus of Geological Sciences
University of Saskatchewan
Saskatoon, Sask., S7N 0W0, Canada

While it might be claimed that Space Science started with the launching of Sputnik in 1957, Peter A. Forsyth has stated alternatively that the birth of Canadian Space Science was much earlier and originated in Saskatchewan:

> I think it started when a completely frustrated Saskatchewan farmer picked up a pitchfork and chased an incompetent foreign labourer once around the cowbarn and then off into the sunset. I am uncertain about the date but it must have been about sixty years ago. The incompetent Welsh labourer . . . was Frank Davies and the experience convinced him that he had better find a university and continue his studies in physics. The university that he found was [the University of Saskatchewan] in Saskatoon and the rest is history [Sask., Kennedy Fonds, Forsyth Speech, "How did we get here from there?" 1987].

Frank Davies was born in 1904 in Merthyr Tydfil, a coal mining town in Wales (Fig. 1). His father was a schoolmaster. Frank graduated with a B.Sc. (Honours) in physics from the University of Wales, University College Aberystwyth [Pounder, 1983; Anon., 1981; Sask., Faculty Biographies Collection, vol. 7]. In 1925 he came to Canada and journeyed to the West on a harvest excursion. Proving unsatisfactory on the farm, Frank joined the staff at the University of Saskatchewan in Saskatoon as a demonstrator-instructor in physics for the 1925-26 academic year. Here, Frank fortunately met another physics student, Balfour Currie [1976, pp. 32-34]. A close association developed and was maintained throughout their long and distinguished careers in geomagnetism.

From Saskatoon, Frank went to Montreal and McGill University where he earned in 1928 the M.Sc. (Physics) [Davies, 1928]. During part of his residency, Frank shared a room with Currie. Long winter evenings of study and discussion likely shaped their future research careers.

After graduation Frank Davies, sometime in 1928, joined the staff of the Department of Terrestrial Magnetism (DTM) of the Carnegie Institution of Washington (CIW), which was then involved in the scientific preparations for the first Byrd Antarctic Expedition. The DTM loaned both Davies and air-potential apparatus to the expedition [Bauer and Fleming, 1929, p. 218]. On August 25, 1928 the sailing vessel *The City of New York* departed for the Antarctic with Davies on board. They returned from "Little America" in February 1930, after 14 months at their exploration and observation base [Byrd, 1930, p. 225; Davies, 1930].

Among the scientific studies undertaken by the Byrd Expedition was the gathering of data on terrestrial magnetism. This task was assigned to the expedition's physicist, Frank Davies. A non-magnetic hut, in which only brass and copper nails were used, was built well away from the radio towers. As the magnetic recording apparatus was photographic, the hut had to be dark. A hole deep enough to contain the hut was dug and snow was packed around it [Byrd, 1930, p. 179].

During the winter of 1929 in Antarctica Frank Davies " . . . continued magnetic observations with his many sensitive recording instruments". He must have spent countless hours as " . . . a night watchman to record that weird spectacle, the aurora australias . . . " [Byrd, 1930, p. 173]. After his return to the Carnegie Institution Davies put forward this illuminating guidance:

> Based upon Antarctic experience, Davies assisted in the preparation of suggestions on methods of observing auroral displays, stressing (1) as simple and regular a program as possible, (2) reliable standard of intensity, (3) use of simple devices for measuring approximate azimuth and altitude, and (4) importance of recording conditions of cloud, moonlight or daylight at time of observation and particularly visibility of stars [Bauer

The Earth, the Heavens and the Carnegie Institution of Washington
History of Geophysics Volume 5

and Fleming, 1930, p. 316].

The year 1930 was busy for Davies. He reduced the Byrd Expedition's magnetic data, presented a seminar at the DTM about the auroral work, and delivered a series of lectures at McGill University on the range of scientific studies undertaken at "Little America" [Davies, 1931a, b, c, and d]. Along with other members of the DTM, he participated in the Department's land magnetic survey, building up observations "for obtaining more precise knowledge of the characteristics of secular variation" [Bauer and Fleming, 1931, p. 298].

During this extremely busy period of his life as a physicist at CIW, Frank found time in 1931 to marry Ada Eleanor (Nell) Bennett of Saskatoon. In 1992, Mrs. Davies still resides in Ottawa.

Davies was granted leave from the DTM in 1932-33 to lead the Canadian scientific team for the Second International Polar Year, to be based at Chesterfield Inlet, Northwest Territories [NAC, Davies Papers; Currie, 1987; Collins, 1934; Davies, 1986]. Currie, on leave from the University of Saskatchewan, also was a member of this group. Davies and Currie spent a most productive year in the Arctic [Kupsch and Roots, 1986].

Two articles by G. G. Shepherd [1987] and E. J. Llewellyn [1987] provide excellent accounts of the pioneering studies on aurora borealis carried out at Chesterfield Inlet. The other members of this party were Stuart McVeigh (meteorologist) and John Rea (general assistant). Shepherd stated that: "Their natural ingenuity, good judgement and willingness to perform hard work would have made any program successful" [1987, p. 40].

The field experience acquired by Davies in the Antarctic enabled him to maintain, as leader, reasonable harmony among this small but diverse group of individuals in the isolation and close confines of a northern outpost. His additional training at CIW gave him the benefit of being able to decide exactly what the group could accomplish during a year at Chesterfield Inlet.

At Chesterfield physicists Davies and Currie were brought together for the third time within the short span of seven years. A mutual respect developed that each carried for the scientific pursuits of the other throughout the remainder of their lengthy careers.

The diaries and letters of Frank and Balfour illustrate that as well as being serious and dedicated scientists, traces of humour crept into their isolation in the Arctic. Currie built the magnetic hut at Chesterfield (Fig. 2) and described it as " . . . the best insulated place in the N.W.T. It has no windows and only small doors so as to prevent much cold getting in around them, and as a result is an object of much curiosity among the natives" [Sask., Currie Fonds, Currie to E. L. Harrington, Sept. 26, 1932]. In his diary Davies offers a different viewpoint: "Find Balfour has made the door of his E.C. [earth current] shack just large enough for him and damn tight for me" [NAC, Davies papers, 27a]. In profile, Frank could be described at best as corpulent. Did Balfour intend to keep Frank out of his domain?

The descriptive entries in the diaries and the drawings in the notebook of auroral displays observed at Chesterfield Inlet are outstanding evidence of their scientific studies of this phenomenon (Fig. 3). In December 1933, Davies and Currie published a paper on the audibility of the aurora [Davies and Currie, 1933]. In a recent review paper by Keay [1990], many references indicate that studies have been made on the audibility of aurora throughout the present century. A better understanding of the cause of these sounds, along with a possible explanation, appears close at hand. Davies and Currie contributed significantly to this body of knowledge.

Davies returned to CIW in 1934, completing and publishing an analysis of auroral data from the Arctic and elsewhere [Davies, 1935a, b, and c; 1936a and b; Davies et al., 1937]. Two years later, on December 6, 1936, Davies became Observer-in-Charge of the Huancayo Magnetic Observatory in the Peruvian Andes, where he stayed for almost three years [Davies and Thomas, 1968]. Throughout this time, reports prepared by the Officer-in-Charge were published in *Terrestrial Magnetism* on a quarterly basis covering in detail those days on which unusual magnetic disturbances took place. Davies resigned from the Carnegie Institution in 1939 for active service with the British Forces [Fleming and Gish, 1940, p. 98]. The contributions of Frank Davies to the British forces during the first years of World War II, and in the later years to the Canadian forces in Operational Research, are omitted from this paper, as much of this material is still classified.

During the later part of the war, it became increasingly apparent in Canada that a new support branch for the Armed Forces was required [Goodspeed, 1958]. This was designed to look after the growing needs in research of the Navy, Army, and Air Force. After the termination of hostilities, plans were formulated for a "civilian arm" of the three services under the Defence Research Board (DRB) of Canada. In 1947, Davies joined DRB and continued as a vital member of this organization until his retirement in 1969. Initially, Frank was Head of the Radio Propagation Laboratory (RPL) and shortly after he became Acting Head of the Electronics Laboratory (EL) as well, both sections located within the city of Ottawa [Personal communication, Colin O. Hines to Kennedy, Apr. 14, 1992; Davies, 1967]. The EL coped with the urgent need for communications equipment. Northern Canada had become a zone separating two major world powers. The need to better understand the

Fig. 1. Frank T. Davies 1904-1981. (From *Pounder* 1983:74).

Fig. 2. Davies and Rea (left) at the magnetic hut, Chesterfield Inlet 1932-33. (University of Saskatchewan Archives).

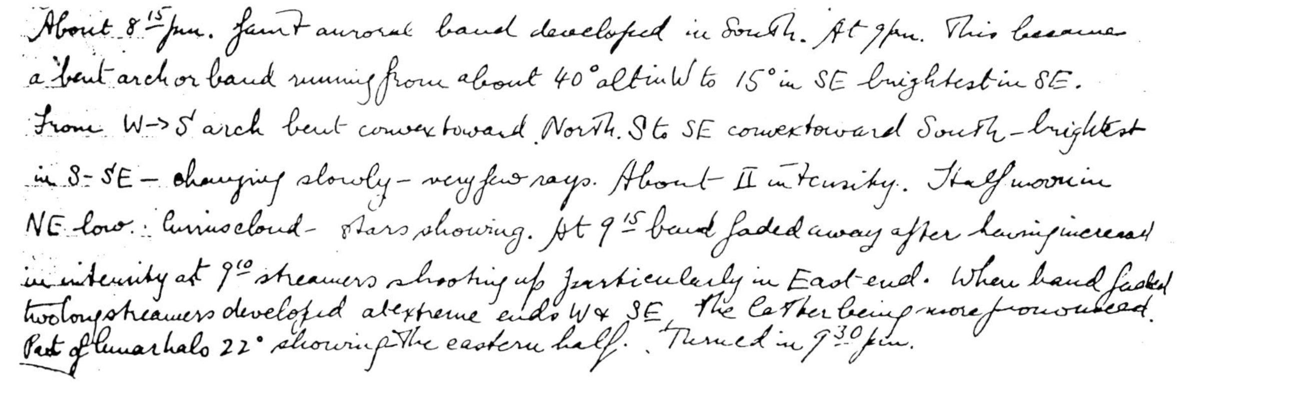

About 8 15 pm. faint auroral band developed in South. At 9pm. This became a bent arch or band running from about 40° alt in W to 15° in SE brightest in SE. From W-> S arch bent convex toward North. S to SE convex toward South - brightest in S-SE - changing slowly - very few rays. About II intensity. Half moon in NE low. Cirrus cloud - stars showing. At 9 15 band faded away after having increased in intensity at 9 10 streamers shooting up particularly in East end. When band faded two long streamers developed at extreme ends W & SE, the latter being more pronounced. Part of lunar halo 22° showing the eastern half. Turned in 9 30 pm.

Fig. 3. Excerpt from Frank Davies' Log Book and Diary showing entry on observation of the aurora. Chesterfield Inlet, September 21, 1932. (University of Saskatchewan Archives, Kennedy Fonds).

problems of radio communications in this region was imperative. The number of ground-based ionospheric stations operating across this vast expanse of territory was increased. Frank's detailed knowledge of the aurora and its effect on radio communication was essential to the program [Davies, 1950a and b; 1954].

RPL and EL both needed more laboratory space and an area where man-made radio interference would be somewhat less than within a city. A site was chosen at Shirley Bay, a few miles out of Ottawa, and a new building was constructed. Frank Davies was appointed Chief Superintendent of this new facility, known as the Defence Research Telecommunications Establishment (DRTE). Under his capable leadership, a dedicated group of scientists was assembled. The *Annual Reports* of DRTE illustrate clearly that the basic research conducted there had as one of its goals the improvement in radio transmission across Canada's North [Davies, 1966].

A few years after becoming Chief Superintendent of DRTE, resisting the transfer as much as was humanly possible, Frank found himself in the Headquarters of the Defence Research Board, as Assistant Chief Scientist and Director of Physical Research. It was by no means a posting to his liking, but he was able to influence decisions related to research projects in northern Canada, essential at the time to maintain a visible Canadian presence in these regions. However, after a relatively short period, Frank was able to return to Shirley Bay and his post as Chief Superintendent of DRTE [Hines, personal communication; Davies, 1966].

During his absence from DRTE, the group of scientists at this establishment had charted new courses for their research. On his return, Frank was concerned that their resources were spread too thinly and that the envisaged program of an Alouette satellite could not be adequately supported with the available personnel and budget. In retrospect, it is easy to see how his leadership brought about the launching of Alouette I in September 1962, a highly successful venture. Canada's first satellite sounded the ionosphere from above and produced an incredible amount of data during its lifetime. An article by Davies [1965] ends on this note:

> It is not unreasonable to state that the scientific value of Alouette I during the past two years is at least equal to that of the world ground network of ionospheric stations, numbering about 170.

With the experience and scientific data garnered from the performance in orbit of the Alouette I satellite, the antenna and orbit of the Alouette II satellite were changed. Alouette I and II were followed by the ISIS satellites (International Satellite for Ionospheric Studies). This extensive program occupied much of the manpower and other resources of DRTE during the 1960s, acting as a necessary prelude to the launching of Canadian communications satellites in the early 1970s, the first being the Anik satellite.

For the last decade before his retirement in 1969, Frank Davies became Director General of DRTE. Near the end of his tenure, the name was changed to the Communications Research Centre, as DRTE became the research arm of the newly created Department of Communications [Melody, 1985].

Subsequent to his retirement, Frank was the recipient of Honorary Doctorate of Science degrees from York University in North York, Toronto, and McGill University in Montreal. Along with medals and his earlier elections as a Fellow of the Royal Society of Canada and as a Fellow of the Arctic Institute of North America, these honors provided a fitting recognition of Frank's contributions to Canadian science, with his major efforts centered in geomagnetism.

By providing an opportunity for many years of basic training in geomagnetism and its field observations, the Carnegie Institution deserves credit for starting a young and enthusiastic scientist along his road to success. From "Little America" to Chesterfield Inlet, from Washington to Huancayo, then after an interruption to RPL, DRTE, Headquarters of DRB, and terminating at the Communications Research Centre, Frank Davies was always the pioneer in the scientific tasks he undertook.

Frank always displayed a genial personality along with a keen sense of humour. His interest in people was genuine, particularly for his staff at all levels within the establishment. To many of his close friends and associates, Frank will always be remembered as a "character". He was a solid individual, ranking high among the top level of scientists in Canada, a man who in retrospect we truly believe it was our good fortune to have known.

ACKNOWLEDGMENTS

Our special thanks to Peter Forsyth and Colin Hines for providing material cited in this paper and to Cheryl Avery, Assistant Archivist, University of Saskatchewan, for her assistance with locating archival material and references.

ARCHIVAL SOURCES

National Archives of Canada, Davies Papers, MG 30 B 73, [cited as NAC, Davies Papers].

University of Saskatchewan Archives, B.W. Currie Fonds (unprocessed collection) [cited as Sask., Currie Fonds].

University of Saskatchewan Archives, Faculty Biographies Collection.

University of Saskatchewan Archives, J. E. Kennedy Fonds, MG 102, [cited as Sask., Kennedy Fonds].

REFERENCES

Anonymous, Dr. Frank Davies, *The Times* (London) p. 1, Oct. 14, 1981.

Bauer, L. A. and J. A. Fleming, Department of Terrestrial Magnetism, in *CIW Year Book [for] 1928-29*, No. 28, 209-276, CIW, Washington, D.C., 1929.

_________, Department of Terrestrial Magnetism, in *CIW Year Book [for] 1929-30*, No. 29, 277-278, CIW, Washington, D.C., 1930.

_________, Department of Terrestrial Magnetism, in *CIW Year Book [for] 1930-31*, No. 30, 316, 335-336, CIW, Washington, D.C., 1931.

Byrd, R. E., The Conquest of Antarctica by Air, *Nat. Geog. Mag.*, *58*, 127-227, 1930.

Collins, J. R., Notes on lecture, International Polar Year and the Canadian Expedition to Chesterfield Inlet delivered by F. T. Davies, *J. Roy. Astr. Soc. Canada*, *28*, 93-94, 1934.

Currie, B. W., *The Physics Department 1910-1976*. 226 pp., University of Saskatchewan, Saskatoon, 1976.

_________, The Second International Polar Year, 1932-33, Canadian Participation, *Musk-Ox*, *35*, 3-6, 1987.

Davies, F. T., I. The Measurement of Surface Temperature with Special Reference to Surfaces Undergoing Evaporation. II. The Construction and Testing of a Hygrostat. III. The Relation of Indoor to Outdoor Humidities and the Determination of a Correlation Factor Between them for Certain Periods of the year. M.Sc. thesis, 60 pp., McGill Univ., Montreal, 1928.

_________, Davies Tells of Byrd Expedition, *Montreal Gazette*, p. 5, Nov. 15, 1930.

_________, Aurora australis observed on the Byrd Antarctic Expedition (abstract), *CIW Year Book [for] 1930-31*, No. 30, 335-336, 1931a.

_________, Aurora australis observed on the Byrd Antarctic Expedition, *Wash. Acad. Sc. J.* 280-283, 1931b.

_________, Notes on Green Flash observed August 29, 1928 and October 16, 1929, *Terr. Mag.*, *36*, 61-62, 1931c.

_________, Observations of the Aurora Australis, Byrd Antarctic Expedition, 1929, *Terr. Mag.*, *36*, 199-230, 1931d.

_________, The diurnal variation in magnetic and auroral activity at three high-latitude stations, *Terr. Mag.*, *40*, 173-182, 1935a.

_________, Review of I. M. Stagg - Some general characteristics of aurora of Fort Rae, N.W.T., Canada 1932-33. London, 1935, British National Committee for the Polar Year 1932-33, *Terr. Mag.*, *40*, 400, 1935b.

_________, Correlation of auroral and magnetic activities for different periods of the night at Chesterfield, Canada, 1932-33, *Terr. Mag.*, *40*, 456, 1935c.

_________, Auroral display at Forsyth, Montana, *Terr. Mag.*, *41*, 211-212, 1936a.

_________, Auroral display at Fortine, Montana, *Terr. Mag.*, *41*, 317-318, 1936b.

_________, Visual Auroral Observations in Canada 1943-7, *Atmospheric and Terrestrial Magnetism and Electricity Bulletin*, *13*, 255-273, 1950a.

_________, Progress Report on Atmospheric Ionization in Canada, 1939-48, *Atmospheric and Terrestrial Magnetism and Electricity Bulletin*, *13*, 88-91, 1950b.

_________, The Ionosphere Over Northern Canada, *Arctic*, *7*, 188-190, 1954.

_________, Alouette I and II, *Physics in Canada*, *21*, 5-14, 1965.

_________, Defence Research Telecommunications Establishment Annual Report for 1966, *DSIS document no. 68-05094*, [Defence Research Telecommunications Establishment], 33 pp., Ottawa, [1966].

_________, The Second International Polar Year Expedition to Chesterfield Inlet 1932-33, in *Living Explorers*, edited by S. Milligan and W. O. Kupsch, 109-122, Outcrop, the Northern Publishers, Yellowknife, NWT, 1986. Reprinted in *Musk-Ox*, *35*, 26-39, 1987.

Davies, F. T. and B. W. Currie, Audibility of the Aurora and Low Aurora, *Nature*, 856-857, Dec. 2, 1933.

Davies, F. T., W. E. Scott, O. W. Torreson, and H. E. Stanton, A solar eruption of November 27, 1936, and simultaneous disturbances in earth's magnetism, earth-currents, and the ionospheric regions, *Terr. Mag.*, *42*, 93, 1937.

Davies, F. T. and D. G. Thomas, *Semper Revolvuntur Res (History Repeats Itself)*, [Defence Research Telecommunications Establishment], Ottawa, 10 pp. unnumbered, 1968.

Fleming, J. A. and O. H. Gish, Department of Terrestrial Magnetism, in *CIW Yearbook [for] 1939-40*, No. 39, 98-99, 1940.

Goodspeed, D. J., *History of the Defence Research Board of Canada*, 259 pp., Queen's Printer, Ottawa, 1958.

Keay, C. S. L., C. A. Chant and the Mystery of Auroral Sounds, *J. Roy. Ast. Soc. Canada*, *84*, 373-382, 1990.

Kupsch, W. O. and E. F. Roots, The International Polar years, *Musk-Ox*, *34*, 1-8, 1986.

Llewellyn, E. J., Auroral Spectra and the Second International Polar Year: The Saskatchewan Contribution, *Musk-Ox*, *35*, 42-46, 1987.

Melody, W. H., Satellite Communications, Canadian Developments, *The Canadian Encyclopedia*, *3*, 1645, 1985.

Pounder, E. R., Frank T. Davies 1904-1981, *Proc. and Trans. Roy. Soc. Canada*, Series IV, *21*, 74-76, 1983.

Shepherd, G. G., The Legacy of Auroral Observations at Chesterfield Inlet, *Musk-Ox*, *35*, 40-42, 1987.

Dr. C. T. Kwei and the Carnegie in China in 1930s and 1940s

WANG SHEN, LIANG BAIXIAN, AND HU XINRU
Department of Space Physics, Wuhan University
Wuhan, Hubei 430072
People's Republic of China

Dr. C. T. Kwei (1895-1961) was a professor of physics at Wuhan University, China. He was also one of the pioneers and founders of modern geophysical research in China. As a lifetime interest, his dedication to and contribution to this important branch of natural science was closely related to his early work in the Department of Terrestrial Magnetism (DTM), Carnegie Institution of Washington (CIW), in the 1930s and 1940s.

From 1931 to 1935, supported by the Institute of Physics, China, and the CIW, Dr. Kwei and Dr. F. C. Brown had made patrolling geomagnetic measurements in western, southern, and northern China during the summer or winter vacations. The elements measured were longitude, latitude, dip, declination, and horizontal force. For example, from July to September 1932, they had made measurements at 10 sites in northern China. The data they collected and the analysis they made comprised valuable information [Brown, 1933a and 1933b].

In 1936, when he was on sabbatical leave from the Central China College, Wuchang, China, Dr. Kwei visited the USA and worked for a period of time at DTM CIW. Upon his analysis of the geomagnetic data observed by Amundsen's Arctic Expedition from 1903 to 1905, he noted "some evidences of the dependence of diurnal variation on magnetic disturbance in the polar latitudes on longitude." A paper was published in the same year [Kwei, 1936].

During his stay in the DTM CIW in those years, under the influence of L. V. Berkner, J. A. Fleming, and M. A. Tuve, Dr. Kwei was keenly interested in ionospheric research, and some preparations for the work had been made. After he returned to Wuchang, a manual ionosonde of Berkner's design was constructed, and noon-hour observations were made from October 1937 to June 1938. He and one of his students, P. L. Sung, later reported, among others, two phenomena of interest. One is that the "critical frequencies of E- and F2-layers are higher for Wuchang than for Washington"—now known to be due to latitudinal effect and geomagnetic control. The other is "a large train of echoes closely packed together, the train itself varying in amplitude with time"—now called spread-F [Sung and Kwei, 1938; Wang, 1992].

In the spring of 1943, Dr. Kwei went to the USA again. In the winter of 1944, through his effort, a cooperative research project on the ionosphere was agreed upon between Wuhan University and CIW. A whole set of manual ionosonde was provided by DTM CIW. With assistance in transportation from the US Navy, the equipment arrived in Wuhan University, Loshan, China, on November 23, 1945. Through a strenuous effort of installation, hourly observations were started at the right beginning of the New Year's Day, i.e., 00 hr LMT of 1946. The Wuhan University Ionospheric Laboratory was then established with Dr. Kwei as the director. That was one of the earliest laboratories of this kind.

As a development of the Laboratory, the Institute of Radiowave Propagation and Space Physics was established in 1978. The Institute has now a staff of about 70 members. About 20 groups of researchers pursue about 10 projects annually. The main interest is upper atmospheric phenomena and their dynamics over mid- and low-latitudes. The existing experiments include: ionospheric vertical and oblique backscatter soundings, HF and VHF Doppler measurements, HF absorption measurements, satellite beacon measurements, low-latitude whistler observations, geomagnetic micropulsation observations, solar radio emission measurements, space plasma computer simulation experiments, and antenna measurement and design.

* * * *

Flower and fruit come from but a tiny seed. The seed sowed by Dr. Kwei and CIW grows well in China. Another spring turns up. Further prosperity should then be anticipated.

REFERENCES

Brown, F. C., The Magnetic Survey of China, *Lingnan Science Journal*, *12*, 101-104, 1933a.

Brown, F. C., and C. T. Kwei, Results of magnetic observations in North China, *Chinese Journal of Physics*, *1*, 91ff., 1933b.

Kwei, C. T., Some evidence of the dependence of diurnal variation of

The Earth, the Heavens and the Carnegie Institution of Washington
History of Geophysics Volume 5

China, *Chinese Journal of Physics*, *1*, 91ff., 1933b.
Kwei, C. T., Some evidence of the dependence of diurnal variation of magnetic disturbance in the polar latitudes on longitude, *Terr. Mag.*, *41*, 57ff., 1936.
Sung, P. L., and C. T. Kwei, Ionospheric measurements at Central China College, Wuchang, China, October 1937 to June 1938, *Terr. Mag.*, *43*, 453-461, 1938.
Wang, Shen, Dr. C. T. Kwei and the Spread-F, in *Exploring the Earth: Progress in geophysics since the 17th century*, edited by Wilfried Schroeder, Michele Colacino, and Giovanni Gregori, pp. 125-127, Interdivisional Commission on History of the International Association of Geomagnetism and Aeronomy, Bremen-Roennebeck, Germany, 1992.

Chopping and Changing at the DTM 1946-1958: M. A. Tuve, Rock Magnetism, and Isotope Dating

H. E. Le Grand
History and Philosophy of Science Department
University of Melbourne
Parkville, Victoria 3052, Australia

Merle A. Tuve assumed the directorship of the DTM in July 1946 after long service as a physicist there and secondment during the war to work on radar development and the proximity fuse. As director he reoriented research at the DTM toward laboratory physics and the application of physics to other disciplines [DTM *Annual Report*, 1946-1947, p. 55]. He described himself as "an opportunist," whose strategy was to try to spot newly-breaking fields of research, while "just letting people be interested in things that were striking." For him, since the DTM had no students to teach nor shareholders to please, one could take the risk of plunging into speculative fields. However, risk-taking in choice of research topics was to be balanced by periodic reviews:

> Every program or effort by the Institution should be *scrutinized at regular intervals* and pruned or reoriented to "front lines" emphasis as needed [LC, Tuve, Box 120, memo to self, August 19, 1945].

A comparative history of two programs in "the geophysics of the earth's crust," in Tuve's words, illustrates the positive—and some of the negative—aspects of these two prongs of Tuve's research policy. He was initially well disposed toward the program in rock magnetism which he had inherited. But, when his high expectations for it seemed ill-founded, he decided to prune it root and branch. In contrast, he played the major role in the establishment of the joint DTM-Geophysical Laboratory (GL) program in isotopic dating. Though the dating group met many technical problems in its early years, he carefully nurtured it through the 1950s.

Planning for the Future

Tuve had reflected on redirecting research at the DTM before succeeding John A. Fleming as director. This had been promoted by Vannevar Bush, CIW president, who, in June 1945 as the war neared its conclusion, formally requested statements of future research directions and objectives from both the DTM and the GL [CIW, DTM files, Terrestrial Magnetism, folder "Future Program"]. Fleming thought that some change of emphasis was needed; for example, gradually to phase out the observatory program in favor of some expansion of laboratory-based geophysics. However, he does not seem to have foreseen any major reorientation. This inference is supported by the minutes of a staff meeting held November 21, 1945 to discuss a response to Bush's request ["Future Program"]. Others in the DTM were concerned with more practical matters: salary increases, who might be the new director, and administrative structures. With respect to the last, there was significant support for the proposition that the director be in effect a chairman elected by the staff to implement policies determined by a steering committee elected by the staff ["Future Program"]. This proposition was not accepted, a decision with notable consequences. The new director would indeed *direct* the post-war program of the DTM.

Tuve's thoughts on the future of the DTM were rather more radical. In a letter to Fleming [October 31, 1945, "Future Program"] he cautioned that although no sudden decisions should be made and that "the 'new' program *should grow out of the present program*," nonetheless the activities and program "should be *expected to change* in certain major ways during the next several years, looking toward a reduction of routine observational and computational work and an emphasis on exploratory efforts in new directions." Tuve pressed for consideration of three possibilities: biophysics, nuclear physics, or a broad program in geophysics. Indeed, Tuve thought that this change of emphasis would soon entail a change of name for the DTM. By March 1946 Bush had decided upon Tuve as Fleming's successor, although he acknowledged that he would "have to use skill to keep Tuve in the path of best accomplishment, for he is a man of positive convictions and of unusual approaches to his problems" [CIW, Vannevar Bush files, Bush to Members of the Committee on Terrestrial Sciences, March 11, 1946]. The appointment

The Earth, the Heavens and the Carnegie Institution of Washington
History of Geophysics Volume 5

meant in effect that it was largely Tuve's agenda and his judgments which guided the DTM over the next two decades. As one former senior scientist there recalls, "The kind of freedom we had here was the freedom to do anything the director thought was a good idea" [No attribution by request].

In June 1946 Tuve responded to Bush's request for a statement of the DTM's future program [CIW, DTM files, Director's File, "Statement Concerning the Scientific Program of the Department for the Immediate Future," June 22, 1946]. Much of Tuve's reply was later published, almost *verbatim* [DTM *Annual Report*, 1946-1947]. Tuve took the view "that we must make every possible effort to emphasize creative work, work with new potentialities, and work which lies on the front lines of knowledge." These objectives may seem platitudinous, but Tuve sought to achieve them. In particular, he aimed to phase out the observatory program and projects related to analysis of the data gathered by it in favor of exploratory (*not* exploration) geophysics and theoretical and laboratory geophysics [p. 2]. Like many other scientists—especially physicists—and the public at the beginning of the nuclear age, he believed " . . . that physicists could do anything better than anybody" [Wetherill, i1992, T 1 S 1]. For him the DTM should be a physics department the research of which did not depend on government funding and the staff of which were encouraged to extend themselves into new fields [Wetherill, i1992, T 1 S 1]: "If a nuclear physicist wanted to become a biologist or a geochronologist and it was O.K. with the director, that was all there was to it . . . there wasn't any problem of being qualified as a biologist to get funding to do biology." Within this context, he sketched out his expectations of present and future staff:

> . . . individuals should be judged by their creative research contributions; steady or devoted work is almost irrelevant as a criterion of accomplishment or virtue. Since individuals differ in their capacity to contribute creatively, however, they will be expected to recognize this and to invest their energies willingly in directions which are pointed out. . . . [DTM *Annual Report*, 1946-1947, p. 2]

On these grounds Tuve was critical of the observational programs on earth currents, and cosmic rays, and terrestrial magnetism—"there wasn't anybody except for very dull kinds of things" [M.A. Tuve, i1982]. This view was shared by other staff: McNish, a senior physicist, in a 1946 memo to Tuve [LC, Tuve, Box 121] setting out his preferred future directions for the DTM urged the "complete and immediate abandonment of our present program of collecting and reducing data . . . " on secular variation and the emphasis shifted to interpretation. Moreover, the manning and maintenance of such observatories as Watheroo in West Australia [see articles in this volume by Home and Parkinson] absorbed both staff and funds which he was keen to channel into laboratory physics [CIW, DTM files, folder "Huancayo and Watheroo Transfer," Tuve to Rear Admiral L.O. Colbert (Director, U.S. Coast and Geodetic Survey), November 29, 1946]. This field of research was sharply pruned almost immediately after Tuve's appointment and reoriented toward work by a small group on interpretation and publication of data previously gathered.

NURTURING AND PRUNING: PALEOMAGNETISM

Tuve, in contrast, was enthusiastic about the resumption of the paleomagnetism program which had been begun before the war by A.G. McNish and E.A. Johnson [DTM *Annual Report*, 1946-1947, pp. 56-57; Le Grand, 1989]. Tuve singled out as the "most fundamental" deficiency for understanding the origin and behavior of the Earth's main field the lack "of quantitative data on the behavior of the earth's field with time." This might be met by paleomagnetic studies of the history of the Earth's field. He forecast that as these studies were pushed farther back in time, "important limits will be set on theories of the earth's main field." He hoped in particular that such studies could decide between the rival "internal" and "fundamental" theories of geomagnetism and characterized the rock magnetism program as one of the two main foci of "the new explanatory approach of the Department to the geophysics of the solid earth" [DTM *Annual Report*, 1946-1947, p. 57].

Although McNish left in 1946 and Johnson in 1948, their departures did not seem to jeopardize the program. Later in 1948 Tuve privately rated "Varves and rock magnetism" as "AA," one of his highest scores, with reference to probability of results and quality of work to date [LC, Tuve, Box 136]. John Graham had only joined the DTM in 1947 as a doctoral student in geology from Johns Hopkins, but he soon began to play a prominent role in the reduced group. He completed his Ph.D. in 1949 and by 1951 had achieved the status of a regular "Investigator" and *de facto* chief researcher in paleomagnetism. Graham at first continued to trace back the history of the Earth's magnetic field. His Ph.D. thesis reported measurements on sedimentary rocks dating back approximately 200 MYA, suggested two field tests for magnetic stability—the so-called "Graham tests" which were quickly accepted by others, and proposed it might be possible not only to discriminate between theories of the Earth's main field but also to answer "questions regarding large-scale movements of the crust, such as continental drift and polar wandering" [Graham, 1949, p. 160]. These positive results and the promise of important theoretical contributions were well in accord with Tuve's goals for the DTM.

Graham, however, soon came across samples the magnetic poles of which were the reverse of the Earth's present ones. Tuve requested funding for further sampling to explore whether "the crust of the whole earth had shifted relative to the poles, or that the continental masses have migrated, or that totally unexpected shifts of the magnetic axis have occurred . . . " [LC, Tuve, Box 136, Tuve to Bush, December 28, 1948, p. 1]. This was "clearly a gamble," but "a gamble of the type which always occurs when research men encounter a new discovery" [p. 3]. Tuve also requested a fellowship to keep Graham on staff to lead the project [LC, Tuve, Box 136, Tuve to Scherer, December 14, 1948]. In an interview with Charles Dollard, president of the Carnegie Corporation, he lobbied forcefully and successfully for the project. After the interview, Dollard commented to Scherer in a humorous vein that:

> Talking with Merle Tuve is like drinking Scotch before breakfast. The project which he outlined to me is at least a million miles off the line of our program, but . . . we are going to recommend it to our trustees. . . .
>
> The next time that fellow comes in here, I am going to have our Counsel on one side of me and the Treasurer . . . on the other [CIW, Carnegie Corporation Grants, folder "Earth Magnetic Polarization," February 7, 1949].

The Corporation made a grant of $30,000 early in 1949 to cover a fellowship for Graham, equipment, and field expenses ["Earth Magnetic Polarization," Tuve to Bush, March 29, 1949].

The gamble was unsuccessful. Further sampling led Graham to attribute rocks with reversed magnetism not to past crustal motion nor to reversals of the Earth's magnetic poles but to the physical and chemical constitution and history of the rocks themselves. Graham subsequently shifted his attention to rock magnetism itself rather than its application to theories of geomagnetism or to drift and expressed growing skepticism about the value of paleomagnetism as a tool to solve those debates. Tuve's initial enthusiasm as to the potential value of paleomagnetic studies declined considerably but even so he still remained more optimistic than Graham. Tuve reported October 31, 1950 on the special grant:

> As I remarked to you before the grant was made, we would begin our extensive studies with a simple picture and a simple question, but after a first regional survey everything would appear extremely complicated and much effort and study would be spent before a simple picture would again emerge.
>
> We are in the complicated phase, and very much aware of being there. As usual, since I am not involved in all the details, I am somewhat more sanguine than Graham, for example, that the final answers will be simple ["Earth Magnetic Polarization"].

We might conjecture that Tuve was concerned that the lack of immediate success of the project for which he had argued so strongly might tarnish his credibility as director and his subsequent annual reports may reflect some disenchantment with either Graham or rock magnetism. The rock magnetism section had in 1947-1949 come first in the annual report of the DTM, now it was placed after "upper atmosphere" and "seismology" [DTM *Annual Reports*, 1949-1953]. In the report for 1953-1954 there was no entry whatever for rock magnetism and Graham was listed in the isotope-dating group. Meanwhile, Graham moved farther and farther from the original aims of paleomagnetic research at the DTM toward, on the one hand, geology, which was not an area of emphasis at the DTM and, on the other hand, problems in physics which were far removed from his original competencies in geology. Aldrich [i1987] recollects that Graham seemed to come up with a new explanation for each new phenomenon or problem which cropped up and that this "was unsatisfactory for a physicist such as Tuve."

Graham's progress and interpretations did not reassure Tuve, who had hoped for reliable and significant results in paleomagnetism. Tuve reviewed Graham's position at the DTM in 1953. In his opinion, Graham's "very determined or even stubborn" choice of research emphasis was directed too much toward, and his abilities lay in, geology rather than physics. Moreover, Graham was working largely in isolation from other DTM staff and had not heeded Tuve's suggestion that he develop additional research interests to enable him to contribute to and collaborate with another group. Graham was thus breaking with Tuve's explicit policies. First, Graham's geological focus ran counter to Tuve's conviction, expressed in almost every Annual Report, that the DTM was and should be focussed on physics and its applications in new research fields. Second, Graham had not followed Tuve's suggested directions. Third, both Graham's single-minded devotion to rock magnetism and his preference for solitary research were undesirable: Tuve had frequently criticized over-specialization as opposed to broad and cross-disciplinary research [*E.g.*, DTM *Annual Report*, 1949-1950, p. 63; 1951-1952, pp. 65-66] and had stressed the importance of working in small research groups "so that . . . [the researcher] encounters the fresh stimulus of new ideas or research progress in a whole range of scientific fields" [DTM *Annual Report* 1952-1953, p. 97]. At a personal level, there was likely considerable friction generated by a clash of two strong wills. Many of Graham's contemporaries refer to his stubbornness. Of the director, a former senior scientist at the DTM recollects: "Tuve was an absolute tyrant. By the time I had been there seven years or so, I began to get tired of Tuve screaming at me" [No attribution by request]. To cap it off, Graham was obtaining mostly negative results and reaching negative conclusions: sufficient reasons in themselves for Tuve to prune or reorient this speculative program. His hopes of seven years ago now seemed dim indeed. Graham's work may have been steady and devoted but for Tuve this was "irrelevant as a criterion of accomplishment or virtue."

Tuve advised Graham that he should plan to leave the DTM within five years. He also *directed* him to devote half his time to working with Aldrich's group in isotopic dating to enable him to become more employable [LC, Tuve, Box 176, Tuve to Bush, September 24, 1953]. This proved unsatisfactory to both sides: Graham evinced little interest or enthusiasm for isotopic dating and he failed to master the mass spectrometer. Bush seems to have agreed with Tuve's assessment but thought that Tuve should not insist that Graham leave [LC, Tuve, Box 176, memo to self, January 28, 1957].

Graham was soon granted a temporary reprieve. Beginning in 1954, a stream of publications from British physicists and geophysicists who applied paleomagnetic studies to polar-wander and to crustal motion sparked world-wide interest. Tuve regained his hope that at last rock magnetism might settle debates over geomagnetism and drift. Tuve once more gave prominence to paleomagnetism in his annual report and a separate entry for Graham's research section reappeared [DTM *Annual Report*, 1954-1955, pp. 42-43 and 54-55]. Graham received support to collect samples in New Mexico and Arizona of roughly the same ages as those used in the British studies. If their magnetism corresponded, this would constitute evidence for a general relative motion between the magnetic axis (as well as the axis of rotation if one accepted that their near-coincidence was no coincidence) and the crust; if they diverged in a systematic fashion, this would constitute evidence for relative motion between Britain and North America. He concluded that there was a rough correspondence and "thus it follows that the Wegener hypothesis of continental drift finds no support. . . ." If observations from other continents fit this same pattern, "then the Wegener hypothesis will have ceased to be a concept worth serious consideration" [Graham, 1955, p. 345]. However, the polar positions he calculated were widely scattered and "there seems to be no lack of mechanisms by which it is possible to explain away the polarizations that fail to fall within a group" [p. 339]. To overcome this difficulty he suggested a systematic survey aimed at finding "in rocks of various ages, on a world-wide basis, the persistent and simple patterns of magnetization that have withstood all remagnetizing influences" [p. 337].

Tuve again interceded with Bush on behalf of the rock magnetism program, requesting special funding to enable Graham to visit the Bernard Price Institute to collaborate on a study of South African rocks. Tuve also foreshadowed a request for a much larger sum to enable Graham to collect samples in Brazil. Tuve expected the work supported by these funds to confirm a general shift of the Earth's magnetic and rotational axes [LC, Tuve, Box 171, Tuve to Bush, September 27, 1955]. Once again, however, Tuve's hopes were dashed. The results were "disappointing." Graham related that "no magnetization pattern impressively consistent over a large area was obtained . . . [and] no obviously sound explanations for the internal inconsistencies and the lack of confirmation of the earlier deductions were adduced." Despite these difficulties, however, he judged that "some of the available magnetic data . . . can be most easily interpreted in terms of continental drift" and that other data, including some he had collected, "are most simply interpreted on the basis of polar wandering" [DTM *Annual Report*, 1955-1956, pp.85-87]. On his return to the DTM, he put forward yet another mechanism to explain the "grossly scattered magnetization directions" he had encountered: magnetostriction [Graham, 1956, p. 735]. If rocks were magnetized under stress and then the stress removed, the magnetic orientation of the rock might change as a consequence. The very act of removing a sample for study could, by releasing stress, alter the magnetization in unpredictable ways so that there might be no relation between the magnetism measured in the magnetometer and the "original" magnetization of the strata. He drew the obvious conclusion: if magnetostriction were a common phenomenon, then little confidence could be placed in paleomagnetic measurements [pp. 735 & 739].

Tuve's burst of enthusiasm quickly waned when Graham returned with a collection of wildly scattered pole positions and the new complication of magnetostriction. Graham summarized in a memo to Tuve [LC, Tuve, Box 171] his thoughts on the state of the field: "the easy cream has been skimmed" in paleomagnetism, "it now requires the application of the most sophisticated knowledge and techniques of solid state physics." Tuve was probably not reassured by the memo: there was no promise of an early breakthrough and Graham lacked the expertise in solid-state physics which Graham himself thought requisite for success. Tuve reviewed the ambiguities in paleomagnetism which Graham himself had pointed out: "soft components, reversals, oxidations, magnetostriction, reheating, etc." [LC, Tuve, Box 172, memo to self, March 17, 1956]. When one adds to magnetostriction all the other complications and ambiguities,

> the prospect is hardly encouraging . . . and they certainly do not foster the hope that has prevailed for decades that by way of the techniques of rock magnetism it will be possible to deal effectively with such major geophysical questions as continental drift and polar wandering [DTM *Annual Report*, 1956-1957, p. 99].

Given the uncertain future of paleomagnetism, he listed possible research areas into which Graham might move. He was unwilling to develop geology as a major field at the DTM but thought Graham unlikely to take up physics *per se*. He concluded that Graham should devote no more than 40% of his time to paleomagnetism and the remainder to isotopic dating or geochemistry in which he had worked previously and which could provide some long-term employment [LC, Tuve, Box 171, memos to self of March 10 & 17, 1956]. Curiously, neither Tuve nor Graham nor the members of Aldrich's isotope group seem to have explored the possibility of making paleomagnetic measurements on igneous rocks which could have been dated by the new methods being developed by that group.

It was in this context that Tuve received a request from Graham that he be allowed to pursue paleomagnetism full-time. He wished to leave aside the issues of polar wandering and drift which had attracted support from Tuve and instead explore the applications of rock magnetism to traditional geological problems. Graham also requested an additional staff member "to serve as a sparring partner," a technician, more equipment including a new astatic magnetometer, and funds for field work. He closed by emphasizing that his own wish was to push rock magnetism "more toward geology and away from physics and geophysics" [LC, Tuve, Box 176, Graham to Tuve, January 11, 1957]. Tuve's reaction was predictable: Graham had "clearly tossed overboard" the earlier agreement to pursue rock magnetism on only a part-time basis. Tuve set out two reasons for a director refusing to support a staff member's research program: either it "is tending toward either unsound or trivial conclusions" or the director lacked expertise to offer intelligent criticism and support. Tuve was agreeable to a program in geophysics but he had "no basis or desire to support pure geology" and no expertise in that field. He decided to refuse Graham's request and to pressure him to leave the DTM within two years [LC, Tuve, Box 176, memo to Director's Personal Files, January 28, 1957]. On February 8, 1958 Graham left to take up a post at Woods Hole Oceanographic Institution. His departure was quickly followed by the discontinuance of paleomagnetic research at the DTM.

Tuve by the mid-1950s seemed to have good reasons to wind down paleomagnetism at the DTM. He sought to promote research which was directed at major problems, which might yield significant results, which promised to make rapid progress, and which would keep the DTM to the fore in physics and applied physics. From this perspective, Tuve initially welcomed research which might decide such major questions as the origin of the geomagnetic field or drift, provided that research rapidly produced decisive results. However, when that promise was unfulfilled, then, given the limited resources at his disposal, it was reasonable to prune and reorient that program toward some other promising research ventures or to launch another program. Tuve's strategy was not without merit. A good initial strategy in a new and fast-breaking field is to push forward as rapidly as possible in the belief that apparent difficulties and objections will be resolved or removed later. Only after "the easy cream has been skimmed," as Graham put it, should one begin to worry about clearing up ambiguities and answering objections. This view was, in fact, explicitly put to Tuve by P. M. S. Blackett, head of one of the major British rock magnetism groups:

> In the early stage of a subject it is often wise to go rather rapidly ahead and skate over difficulties and anomalous results. Now clearly we must spend several years being very critical and looking into all possible objections . . . [CIW, DTM files, Director 1954-1959, folder 1, Blackett to Tuve, October 27, 1958].

There are both advantages and disadvantages in being a path-breaker, but if one stops to question the direction or to investigate side-trails, then one's colleagues and competitors may take the lead.

OPPORTUNISM AND COLLABORATION: ISOTOPIC DATING

The genesis and early history of the isotopic dating program at the DTM illustrates well Tuve's willingness to take up and nurture promising new research topics which were consistent with his overall research policy. (For an overview of isotopic methods [Dalrymple, 1991, chaps. 3 and 7] and for K-Ar dating [Glen, 1982, pp. 22-47].) In response to Bush's request for a statement of future directions, neither Tuve at the DTM nor L. H. Adams at the Geophysical Lab identified isotope dating as an area of priority in 1946. Although Adams mentioned "the measurement of radium in rocks for the purpose . . . of making further estimations of the age of the earth," as a topic "of vital concern," he did not think it could be taken up by the GL "without diluting our efforts to an unprofitable extent" and that other institutions should be encouraged to pursue it [CIW, GL files, "Geophysical Laboratory: Elements of a Proposed Program for Future Research," p. 31]. In preparing the report, Adams circularized leading geologists for their suggestions on future studies. Of the 31 who replied, only Harry Hess mentioned as one of seven areas "determination of isotope ratios, Pb-U ratios, Ra-He ratios or other ratios which could give an age determination" [Appendix A, p. A-16)]. By 1949 Tuve and Adams had changed their opinion and had decided to initiate a cooperative program.

Tuve and Adam's decision to develop research in isotope dating had as its proximate cause a seminar series, "Milestones in the Pre-Cambrian," held at the DTM between March 4 and May 13, 1949 in association with the GL and the USGS. The range of topics was extraordinarily broad: speculations about the origin of the universe, the Earth, the atmosphere, the oceans, and life, as well as discussions of specific techniques relevant to investigating the Pre-Cambrian. Tuve himself spoke on the mass spectrometer and on the origin of life. Philip Abelson, whom Tuve had selected to head the biophysics section at the DTM [LC, Tuve, Box 129, Tuve to self] talked about isotope separation techniques [DTM *Annual Report*, 1948-1949, pp. 76-77; LC, Tuve, Box 137]. These seminars brought to a focus for Tuve—and for others at the DTM and the GL—several converging institutional, scientific, and technical interests.

Tuve was an opportunist with respect to pursuing emergent research fields. These seminars gave him a vision of new opportunities which the application of new techniques in physics and chemistry to the Pre-Cambrian might offer. He described his and Adams's reactions in a letter to Bush as follows: " . . . the vague outlines of a whole new and significant field of investigation have come into view in our seminars." In particular, the exploitation of recent developments in trace-element and isotopic analysis might lead to the finding of "'fossil' evidence of life in terms . . . of the pattern of trace elements necessary for metabolism," determining the composition of the Pre-Cambrian atmosphere through trace element and isotopic studies, and explaining the origin of granites and thus inferentially the formation of continents and orogeny. As with paleomagnetism, to embark on research in this field and with this approach was risky, but he assured Bush "some exploratory thrusts are clearly justifiable" [CIW, DTM Files, Director's File, Tuve to Bush, May 6, 1949].

Such a program also met other of Tuve's criteria. He was particularly interested in developing collaborative research with groups outside the DTM and, as the seminars had shown, there were possibilities for co-operation with the GL, the USGS, and university groups at Minnesota, Chicago, MIT, and elsewhere. The approach was based squarely on physics and physical chemistry and their application to broad questions, consistent with Tuve's view of the DTM as "a physics department with special emphasis on experiments in geophysics and biophysics, making intensive use of the techniques and ideas of modern physics" [DTM *Annual Report*, 1946-1947, p. 54]. Moreover, it was a program which should promote collaboration among most of the existing research groups in the DTM, especially those supervised by Howard Tatel on the Earth's crust and those by Abelson on biophysics. Indeed, only four days after Tuve had written to Bush, Tatel and Abelson sent a memo to Tuve urging that the DTM and the GL develop research projects applying "tracer element and isotope measurements" to geophysical problems and pointing out that "at DTM there exist the tools and people capable of making measurements of this type" and that there was considerable enthusiasm, while at the GL, though there was less enthusiasm, there were "potentialities for high-class chemical analysis and . . . persons with excellent judgement in geology." They implied, however, that the appointment to the GL of someone with expertise in the latest techniques was needed [LC, Tuve, Box 136, Abelson and Tatel to Tuve, May 10, 1949]. Finally, in contrast with paleomagnetism, Tuve had some expertise in the relevant techniques and instrumentation. For example, the mass spectrometer, which had been recently developed by A. O. Nier at Minnesota, was opening up the field of trace-element and isotope analysis. Nier's refinements of the older mass spectrograph, especially his replacement of photographic plates as recording devices by electronic circuits in what then became known as the mass spectrometer, greatly facilitated rapid, accurate determinations of the presence of minute quantities of isotopes [Glen, 1982, pp. 22-23]. Tuve had at the DTM a mass spectrometer fabricated by Nier's machinist and set up by Tuve himself to try to extend Harold Urey's work on isotopic oxygen in rocks [Aldrich, i1992, T 1 S 1].

Tuve had most of the resources needed to launch the program he had outlined to Bush: the key piece of apparatus, researchers with expertise in nuclear physics and geochemistry, interested staff and other resources at the GL, contacts with relevant organizations, and, of course, funds freed by recent pruning of other activities with which to hire (in Tuve's usual phrasing) some "good young men" to be the nucleus of the program. Even prior to the seminar series, Tuve—a graduate of Minnesota—had asked Nier for advice on a possible appointment at the DTM in mass spectrometry and isotope analysis. Nier proposed one of his former students and collaborator, Tom Aldrich, who had just taken up a job at Missouri. Aldrich and Nier [1948] had determined that ^{40}K decayed to ^{40}Ar by electron capture and also that some ^{40}K decayed to ^{40}Ca. However, the half-life of ^{40}K was not then known precisely enough for dating purposes [Glen, 1982, p. 23]. Aldrich, when approached by Tuve, was interested in the offer to come to the DTM on full-time research and asked him to call again after he had completed his stint at Missouri [Aldrich, i1992, T1 S1].

Tuve in early 1950 launched the program with a renewed offer of an appointment to Aldrich [CIW, DTM Files, Director's File, Tuve to Aldrich, February 15, 1950]. Tuve had his sights on hiring one or two other researchers for the group but these arrangements seem to have fallen through [CIW, DTM, Director's File, Tuve to Bush, February 3, 1950; Aldrich, i1992, T1 S1]. He described to Aldrich as follows the research he was to undertake:

> . . . the intensive application of the "micro-techniques" of modern physics and chemistry, including isotope and radioactivity analyses, to ancient Paleozoic and pre-Cambrian rocks for information on such broad questions as the origins and dating of various granitic masses, dating and sequences of rocks in the Canadian shield, and the constituents of the earth's atmosphere in early pre-Cambrian time [Tuve to Aldrich, February 15, 1950].

This description makes clear that the aims of isotopic studies as originally conceived by Tuve, Tatel, Abelson, Adams, and others and communicated to Bush had undergone some modification. The dating of ancient rocks *per se* figures more prominently and less stress was placed upon such broader questions as the nature of Pre-Cambrian life and its environment. Over the next decade, isotopic dating and the development of techniques and methods associated with it was the dominant immediate concern of Aldrich and of George Tilton, George Wetherill, and Gordon Davis, who were to form the nucleus of the research team. This orientation was signified by the relevant headings of the group's entries in the annual reports of the DTM and the GL. Between 1951 and 1958 these always highlighted "Age" or "Dating." Glen has suggested both social and cognitive reasons why isotope daters might have been particularly interested in relatively ancient rocks:

> . . . the old Precambrian terranes, such as the region of central Canada called the Canadian shield, were so complex that only a radiometrist could say something definitive about age and correlation; thus the early isotope dates, most of them geologically untutored, could operate in old terranes with near impunity. Those vast regions of the continents comprising the oldest rocks represented the most exciting frontier: because so little was known of them, they offered hope of discoveries beyond the limits of other methodologies [1982, pp. 26-27].

Moreover, there was more than a touch of competitiveness in trying to find the oldest rock in the world and thus to fix the Earth's approximate age.

Reasons for a shift of emphasis to dating also included a growing recognition, perhaps informed by discussions with Aldrich, Nier, and others in the field, that the "micro-techniques" still needed considerable refinement, even for such an aim as the dating and sequencing of rocks in the Canadian shield. Some of these included techniques for separating elements with similar chemical properties, for example, Sr and Rb, from samples; isotope dilution methods, which had only been initiated in the mid-1940s and still needed further development; determining the reliability and applicability of different dating methods for different minerals with sometimes complex geological histories; and establishing precise half-lifes for some of the parent-daughter decay sequences [DTM *Annual Report*, 1956-1957, p. 100]. Indeed, the research agenda for the next few years largely comprised solving these and other problems connected with the isotope dating techniques themselves in order to turn the promise sensed in 1949 into, in Tuve's words, a "powerfully analytical probe" [DTM *Annual Report*, 1954-1955, p. 43]. The first goal was to use a range of isotope methods on the same sample—and of course, to secure agreement amongst them regarding date. Were this to be achieved, then one could have confidence in the absolute ages so measured and could proceed to the application of these techniques to such larger geological problems as the evolution of continental structures [DTM *Annual Report*, 1950-1951, p. 74].

Aldrich began at the DTM in September 1950. His special interest was developing dating techniques using the Rb-Sr and K-Ca decay systems, research closely related to that he had been pursuing at Missouri [Aldrich, i1992, T1 S1; DTM *Annual Report*, 1950-1951, pp. 74-76]. He was supplied with excellent equipment. Tuve provided as requested a new, improved mass spectrometer constructed at the DTM [CIW, DTM, Director's File, Tuve to Aldrich, February 15, 1950]. Davis [i1992, T1 S1] comments that the steady improvements in mass spectrometers engineered at the DTM considerably aided the work in isotope dating well into the 1970s. Tuve also supplied other specialized apparatus including a flame photometer acquired by the GL for potassium studies [DTM *Annual Report*, 1950-1951, p. 76]. Tuve, in preparing a submission to Bush for special funding [LC, Tuve, Box 144, "Aldrich re Apparatus Funds," April 5, 1951] listed among other items a vibrating reed electrometer @ $1500, special isotopes @ $500, electronics and magnets for a 9-inch mass spectrometer @ $2700, and other items and consumables @ $800. In 1951 those sums represented a sizeable investment in the new program. No less important, he arranged for technical assistance in the form of collaboration with Gordon Davis of the GL and J. B. Doak. Davis had expertise in geochemistry and some experience in chemical analysis of trace elements. Doak was a research assistant who also worked with Tuve in mass spectroscopy [Davis, i1992, T1 S1]. Almost immediately Tuve's promotion of interaction among research groups in different fields bore fruit. An early problem that Aldrich tackled was to separate the Rb and Sr found together in mineral samples. He and Tuve visited L. H. Ahrens at MIT, who was trying to effect chemical separations but this line did not seem promising: even very small residues of Rb could invalidate the Sr measurement [Aldrich, i1992, T1 S1; DTM *Annual Report* 1950-1951, p. 75]. The solution was much closer to home. Downstairs at the DTM, Abelson's biophysics group had recently started using the new technique of ion-exchange resin columns to effect biochemical separations. They suggested to Aldrich that he give it a try and he found that this method gave rapid, thorough Sr-Rb separations. This, coupled with improved isotope dilution procedures developed with Davis, promised "a highly effective procedure for age measurements on igneous intrusive rocks" [Aldrich, i1992, T1 S1; Davis, i1992, T1 S1; quote from DTM *Annual Report* 1950-1951, p. 76; Aldrich et al., 1953].

Tuve was doubtless encouraged by the headway being made. Not only had several technical problems been resolved but also effective collaboration had been established within the DTM, with the GL, with academics, and with the National Bureau of Standards (NBS, which permitted use of its mass spectrometer), the USGS (which assisted in the location and identification of outcrops for sampling), and the Atomic Energy Commission (which provided some isotopes). Moreover, Tuve believed that not only was the body of data produced by other investigators "skimpy" but also that the DTM-GL group had no real competitors with respect to the approach of targeting different isotope methods for different parent-daughter pairs in the same rock [LC, Tuve, Box 146, Tuve to Aldrich, November 26, 1951]. This may have reflected Tuve's impressions gained at the "Symposium on Mass Spectroscopy in Physics Research" at the NBS held earlier that year at which Nier and others spoke [LC, Tuve, Box 146, NBS covering letter and program]. Adams also seems to have been pleased. Late in 1951 he proposed to Bush "formal inclusion in our program of the radioactive measurement matter." Bush's initial response was that such a move did not "ring the bell" with him for he was concerned that it might spread resources "a bit thin" and, further, that he needed assurance that the program would occupy a position of leadership in the field [CIW, GL files, Director's File, Bush to Adams, November 14, 1951]. Bush's reservations seem to have been answered in subsequent discussions. Over the next few years, Tuve with Adams, and from 1952, with Abelson, the new director of the GL, steadily built up the isotope dating research group. This came in part through fellowships to such younger researchers as L. O. Nicolaysen (1951-1952), a South African graduate student who returned to the Bernard Price Institute, and Paul Gast (1954) from Columbia University but, more important, from well-planned additions to the regular staff.

Early in 1951 Tuve approached G. R. Tilton, a Ph.D. student in physical chemistry and radiochemistry under Harrison Brown at Chicago. Brown had visited the DTM and discussed Tilton's work with Tuve, hoping to find a position for his student [Tilton, i1992, T1 S1]. Tuve must have been delighted with the proposition. Tilton knew isotope dilution techniques and was completing a thesis on abundances of uranium in meteorites. Given the aim of using different isotopic methods on the same rock, and given that there was no expert at the DTM or the GL on the latest techniques of U-Pb dating, the classic isotopic dating system, Tilton was an ideal recruit. Tuve apparently mentioned in his letter offering a post to Tilton that he was expected to continue his U-Pb work at the DTM. Tilton replied that his own plans were not confined to those techniques:

> My own long term interests encompass much more than uranium-lead age determinations. I am interested in all phases of the age problem. For example, in the next several years, . . . I hope to collaborate with [P. M.] Hurley and others to attempt simultaneous age determinations on a suitable specimen by all possible methods: U-Pb, He, K-A[r], K-Ca, Rb-Sr, Rn, Tn. This would produce some idea of the relative reliability of the different age methods.
>
> I have greatest interest in further study of meteorites . . . . I believe the age of meteorites is best to be approached by the potassium-argon method . . . [CIW, DTM Personnel files, "Tilton," Tilton to Tuve, March 14, 1951].

The response must have delighted Tuve: Tilton had the

broader interests which Tuve valued, his own goal paralleled the immediate goal of the DTM program, he had expertise in U-Pb, and he wished to work on the K-Ar method. He took up a temporary appointment at the DTM in September and became a member of the regular staff in 1953. Tilton's appointment also helped cement relations between the DTM and the GL. The latter had the better facilities for chemical studies and Tilton spent most of his time working in the GL labs. Indeed, in July 1956 Tilton formally transferred from the DTM to the GL. Tuve probably did not discourage such a move. He regarded Tilton as a chemist rather than a physicist and his move to the GL opened a position at the DTM for a *real* physicist (see Le Grand, 1989).

By mid-1952 four major sets of isotopic methods were being actively pursued: Rb-Sr and K-Ca mainly at the DTM and K-Ar and U-Pb mainly at the GL. The immediate aim was unchanged: "to develop techniques and modify equipment for determining the age of several common minerals, in the same rock, using different methods which will give completely independent ages" [GL *Annual Report*, 1951-1952, p. 53]. Tuve was clearly sanguine about the progress that had been made in two years' time: already it seemed reasonable to begin a program of systematic sampling to investigate the theory that North America had developed by accretion on the margins of the Canadian shield [DTM Annual Report, 1951-1952, pp. 71-76]. Early results showed considerable disparities among dates yielded by several of the methods. Further refinements in the methods and in sample selection ensued [*E.g.*, GL *Annual Report*, 1952-1953, pp. 79-84; DTM Annual Report, 1952-1953, pp. 114-116].

George Wetherill joined the group in 1953. He had completed his Ph.D. in physics under M. G. Inghram at Chicago, a nuclear physicist well known to Tuve through his interest in mass spectroscopy. His thesis was on the measurement of the inert gases xenon and krypton produced by the spontaneous fission of uranium and thorium. He was familiar with isotope dilution techniques. What recommended him to Tuve was that he did "not have the restricted range of enthusiasms which we usually find among nuclear physicists . . . no doubt because he worked with rocks and his imagination was constantly stimulated by nature and the broad sweep of the past" [CIW, DTM Files, Director's File, Tuve to Bush, June 29, 1953]. Moreover, Tilton [i1992, T1 S1] had actually taught and been much impressed by Wetherill in a freshman chemistry lab at Chicago. Wetherill, who had also worked with radiogenic gases, began to work up the K-Ar method and also to be a resident theoretician for the group [Tilton, i1992, T1 S1]. With Wetherill the isotope dating group had assumed the form it was to retain through the end of the 1950s. The "team" was well-balanced, included a broad spectrum of abilities and interests and appears to have collaborated well on specific research projects. Although each specialized in one or two of the four major methods pursued, each had a working knowledge of all four. Wetherill comments: "I taught the others to do argon and they taught me rubidium and strontium which I had not done" [Wetherill, i1992, T1 S1].

Beginning in 1953 the group applied their methods to minerals in an igneous intrusive in Colorado and to the cross-dating of samples that had been gathered elsewhere. In collecting samples, the dating group received considerable help from university geologists and especially from the USGS. The group had no trained geologist so it was usually assisted by a member of the Survey [Tilton, i1992, T1 S1; Aldrich, i1992, T1 S1]. Davis [i1992: T1 S1] recalls: "Every summer, a number of people from the GL would talk to the people at the Survey and say 'Where are you going to be in the field this summer?' and . . . then we'd get into a couple of cars and tour the country, visit their places, and if we needed samples, we asked them for advice on where to get them." They were optimistic that "it can be safely stated that the initial goals of the program are within reach of the group" [DTM *Annual Report*, 1953-1954, p. 55]. Again, however, there were disparities between the two U-Pb systems and the Rb-Sr system [p. 62]. Further work over the next year, including the use now of K-Ar dating, did not resolve these differences. Further complexities cropped up. The group had "based our hopes of determining uranium-lead ages for granites almost entirely on the mineral zircon." Davis, Aldrich, and Wetherill [i1992, T1 S1] had pressed to get "some really good uranium-lead ages on single-crystal uraninites" to use as a baseline for the other methods. However, they encountered "gross discrepancies" among ^{238}U-^{206}Pb, ^{235}U-^{207}Pb and ^{207}Pb-^{206}Pb. This they attributed to either a loss of lead or a gain of uranium [DTM *Annual Report*, 1954-1955, pp. 57, 61-62]. In a succinct statement of their progress to date and the problems which they now faced, they argued as follows [p. 56]:

> The age of a mineral can be measured if (a) the relative amounts of parent and daughter isotopes can be determined, (b) the amount of daughter isotope initially present in the mineral at the time the mineral formed can be determined, (c) the decay scheme and decay constants of the parent isotope are accurately known, and (d) there has been no loss or gain of parent or daughter isotopes since the mineral was formed.

They were confident of (a) and (b) for all their methods and of (c) for U-Pb methods. However, (c) was not yet secure for K-Ar and Rb-Sr and there could be no certainty about (d) for any of the methods. Much work remained to be done before isotopic dating could be applied to any of the grand problems Tuve had sketched five years earlier. However, considerable progress had been made and, in the process, much collaborative research in both chemistry and physics had been conducted and published.

1956 witnessed a major turning point in the isotope program. Wetherill invented and began to apply the Concordia Scale for U-Pb dating. The term "Concordia" does not refer, as might be supposed to "concordance" of ages, but rather was chosen by Wetherill during negotiations for a chair at Concordia College [citation for Hess Medal, *Eos*, 1991]. The U-Pb method was the most widely-known and earliest method of absolute, radiometric dating. It had been pioneered by Arthur Holmes in the 1910s and further developed by him and others. However, still in the early 1950s geologists were reluctant to place full confidence in U-Pb dates. Their reservation was well-founded: how could one be sure that in particular rocks the proportions of U and Pb had not been altered over millions and hundreds of millions of years by various physico-chemical processes? This had been the problem that the isotope group had encountered with the discordant U-Pb ages measured in most zircons. The genesis of the scale lay, however, not in the zircon work at the DTM itself, but arose from papers by Ahrens [1955a; 1955b] reporting and depicting on a graph U-Pb results on uraninites from Rhodesia and Witwatersrand. Wetherill was struck by the graphical representation of convergent ages for ^{235}U-^{207}Pb and ^{238}U-^{206}Pb: "They looked like railroad tracks converging in the distance" [Wetherill, personal communication, June 15, 1992]. Within three days in the process of composing a reply to Ahrens's paper, he had worked out the scale [Wetherill, 1956a]. The Concordia Scale [Wetherill, 1956a; 1956b] provided an elegant and simple solution and indeed could be regarded as the most important single contribution to the field of the isotope group [Dalrymple, 1991, 115-119].

In brief, the scale uses the fact that the two different U-Pb systems have different half-lives. If one plots the values of the two U-Pb systems with respect to time, a smooth "concordia" curve results. If the rock be "closed," the ratios of the two U-Pb systems as measured in a sample will fall on the concordia curve and give the age of formation. If the rocks have been disturbed, resulting in some loss of the Pb, the ratios do not fall on the curve and the intercept of a line connecting the points generated by the sample with the concordia gives the age at which that disturbance occurred. Concordia put U-Pb dating on a secure footing: not only could it be applied to a wide variety of minerals, it could be applied to both "closed" and "open" systems; that is, rocks which had remained relatively unaltered since their formation or—far more commonly for the rocks of the Canadian shield, older rocks and metamorphic rocks generally—rocks which had been heated since formation or which had undergone loss of Pb or gain or loss of U [DTM *Annual Report*, 1955-1956, pp. 94-97]. Further, the isotope group could now use reliable, self-checking U-Pb dates to calibrate the other methods and to calculate, for example, the relevant decay constant for ^{87}Rb and cross-check the value for ^{40}K determined by Wetherill through the gamma-counting method [GL *Annual Report*, 1955-1956, pp. 161-63; Aldrich, et al., 1956; Wetherill et al., 1956].

This "breakthrough," as it has been often described, successfully concluded the first phase of the work of the isotope dating group. Tuve's overview in the *Annual Report* described it as a "milestone" which made possible investigating the age of the Earth and the geological evolution of the Precambrian Earth [1955-1956, p. 71]. Similarly, the GL *Annual Report* summarized these developments as follows: "The initial goal of the age group, discovery of a widely applicable means of obtaining trustworthy determinations of the age of rocks and minerals, has now been substantially achieved" [1955-1956, p. 161].

The next decade marked the isotope group's second phase: the application of their refined methods to large-scale dating projects aimed at chronological mapping of the evolution of continental crust, particularly on the margins of the Canadian shield, and periodizing the major, early orogenic episodes. Collaboration with government bodies and university-based researchers extended and expanded as more and more geologists and geophysicists adopted the isotope dating methods that the DTM-GL program had helped to develop.

Field-oriented geologists, including Survey geologists, had been very skeptical in the 1950s of the whole approach. As Tilton [i1992, T1 S1] recollects: "They said that if the geochronology was done properly it would agree with the geology! We could just as well say the reverse. . . ." Wetherill is even more forceful:

> Among the geological community they either didn't believe any of it [isotopic dating] or they felt it was something you give to a technician to do. . . . It was not something a scientist would really waste his time doing. They had a distrust of what they called 'black boxes' [i1992, T1 S1].

These attitudes had changed markedly by the mid-1960s.

IN RETROSPECT

Tuve's consistently stated research policies for the DTM clearly informed his judgments of the programs in rock magnetism and isotope dating and were instrumental in the shaping of their direction and of their different fates at the DTM. There are some similarities between the two programs. Both had as their initial, ultimate goals large questions about the Earth and its history. Both soon focused on refining and solving problems associated with the techniques that might answer those questions. Both were initially speculative fields, although rock magnetism was probably more so. Both contributed to their respective fields through publications and interaction with and influence upon other researchers. Few North American researchers pursued rock magnetism but there were many individuals and groups, mostly in universities, who had taken up isotopic dating. What was distinctive about the DTM-GL approach was the simultaneous pursuit of a range of methods, the aim of cross-checking all those methods with respect to a single suite of rocks, and the systematic use of isotope dilution techniques in conjunction with mass spectrometry. Wetherill comments [i1992, T1 S1] that the isotope dilution procedures developed and followed at the DTM indicated errors of 20% or more in ages reported by others, especially those using the Sr-Rb system or using chemical separation techniques instead of ion-exchange columns coupled with mass spectrometry.

The differences between the programs are even more striking, even if we limit ourselves to Tuve's criteria. The rock magnetism program became increasingly isolated from researchers inside and outside the DTM. Although Graham collaborated with A. F. Buddington and J. R. Balsley on magnetostriction, their findings were generally negative; e.g., "that many conclusions that have been offered on the basis of rock-magnetism data, relating to polar wandering, continental drift, secular variation, and reversal of the Earth's magnetic field, are subject to serious doubt" [Graham et al., 1957, p. 465]. The isotope dating program, however, involved extensive collaboration and even synergy among other groups and individuals inside and outside the DTM. The rock magnetism program drifted more and more toward geology and away from physics. Although the same might be said of the isotopic dating program in the 1960s and after, in its first decade the focus was very much physics applied to geophysical problems. No doubt, however, the determinant considerations which led to the termination of the former and the expansion of the latter was that rock magnetism seemed to produce mostly negative results which called into serious question for Graham himself the very worth of the technique, whereas the isotope dating group identified and solved problems and established the technique as significant and reliable.

It is well to remember that what is crucial in fostering research is not so much policies as people. Tuve's emphasis on immediate results—or at least measurable progress—could be misguided and counter-productive. It is striking that just as rock magnetism was being wound down at the DTM it was burgeoning elsewhere and was to be a critical factor in the mid-1960s in the "modern revolution in geology." It might be thought that this illustrates the short-sightedness of Tuve's directorship: if only he had persisted for a few more years, if only he had built up the rock magnetism group through an infusion of new and more enthusiastic young researchers, then could not the DTM have been at the forefront of the "revolution"? Perhaps not, for Graham's personality and his views on paleomagnetism posed severe problems. It would have been necessary to appoint one or even several scientists with sufficient seniority or self-confidence to pursue their own lines of research independently of Graham and to set aside the very objections and problems which Graham was adumbrating. In contrast, the isotopic dating group, by all accounts, though it included some strong personalities, was relatively closely-knit and enjoyed good working relations. Both cases indicate, however, that the successes and failures of the DTM under Tuve's directorship owe at least as much if not more to Tuve's own judgment of people and programs as to any formal or informal policies.

I do not wish to give the impression that Tuve was a paragon of virtue in research policy and management. Tuve's autocratic and sometimes patronizing personal style irritated some staff. No doubt this, as well as scientific issues, caused friction between Tuve and Graham. One former senior scientist recalls that in 1957 Tuve ordered him to go to South America on a seismic expedition in a few days' time. The scientist, who had no special expertise in seismology but whose presence was needed to swell the size of the party which Tuve himself was leading, was worried about trying on such short notice to make arrangements for his family and finances for the trip. When he indicated to Tuve these reservations and expressed reluctance to go.

> . . . [Tuve] said 'Hire a babysitter! Hire a gardener!' I was getting $5000 a year in those days. I was cutting my own hair and was making up powdered milk at night because I couldn't afford to buy regular milk. This [Tuve's response] didn't go over too well [no attribution by request].

Moreover, the hierarchical management structure which he gradually instituted of director, assistant director, and four section heads from a total research staff of 14 was rather cumbersome and could stifle initiatives. Finally, his disdain for sources of financial support outside the CIW increasingly put the DTM scientists at a disadvantage with respect to their colleagues elsewhere who in the late 1950s and 1960s had access to rapidly-growing government funding through the NSF, the ONR, and other bodies. Nonetheless, for his time and his place, the policies which he implemented proved effective.

ARCHIVAL SOURCES

I have made free use of the very rich archives of the DTM, the GL, and the CIW, all of which are held at the CIW, 1600 P Street, Washington, DC. Useful finding aids are available but patient browsing is often rewarded. The Merle A. Tuve papers, deposited in the Manuscripts Division of the Library of Congress, contain a vast quantity of material on Tuve's personal and scientific interests, including correspondence, diaries, memos, and official papers relating to his career at the DTM. These are cited as LC, Tuve. I have made extensive use of interviews with present and former DTM and GL scientists. Those listed in the references, unless otherwise indicated, were conducted by me and recorded on audiotape.

ACKNOWLEDGMENTS

This research was supported in part by Australian Research Council Grant A59031653. I thank especially a number of scientists for granting interviews: L. T. Aldrich, J. R. Balsley, G. L. Davis, A. L. Hales, W. D. Parkinson, G. R. Tilton, F. D. Stacey, and G. W. Wetherill. Reference to audiotaped interviews is made according to the interviewee and date; thus, Aldrich i1987. T1 S2 means Tape 1, Side 2. I also owe much to several individuals and institutions for providing access to relevant materials: C. E. Helsley of the Hawaiian Institute of Geophysics, the American Institute of Physics, the Library of Congress (Manuscript Division), Shaun Hardy of the Department of Terrestrial Magnetism, and Susan Vasquez of the Carnegie Institution of Washington.

REFERENCES

Ahrens, L. H., The Convergent Lead Ages of the Oldest Monazites and Uraninites (Rhodesia, Manitoba, Madagascar, and Transvaal), *Geochem. Cosmochem. Acta*, *7*, 294-300, 1955a.

Ahrens, L. H., Implications of the Rhodesia Age Pattern, *Geochem. Cosmochem. Acta*, *8*, 1-15, 1955b.

Aldrich, L. T., J. B. Doak, and G. L. Davis, The Use of Ion Exchange Columns in Mineral Analysis for Age Determinations, *Am. J. Sci.*, *251*, 377-387, 1953.

Aldrich, L. T. and A. O. Nier, Argon 40 in Potassium Minerals, *Phys. Rev.*, *74*, 876, 1948.

Aldrich, L. T., G. W. Wetherill, G. R. Tilton, and G. L. Davis, Half-Life of Rb^{87}, *Phys. Rev.*, *103*, 1045-1047, 1956.

Aldrich, L. T., interview by H. E. Le Grand, 26 October 1987, i1987.

Aldrich, L. T., interview by H. E. Le Grand, 29 January 1992, i1992.

Dalrymple, G. B., *The Age of the Earth*, 474 pp., Stanford University Press, Stanford, 1991.

Davis, G. L., interview by H. E. Le Grand, 31 January 1992, i1992.

DTM *Annual Report*, Annual Report of the Director of the Department of Terrestrial Magnetism, in *CIW Year Book*.

GL *Annual Report*, Annual Report of the Director of the Geophysical Laboratory, in *CIW Year Book*.

Glen, W., *The Road to Jaramillo: Critical Years of the Revolution in Earth Science*, 459 pp., Stanford University Press, Stanford, 1982.

Graham, J. W., The Stability and Significance of Magnetism in Sedimentary Rocks, *J. Geophys. Res.*, *54*, 131-67, 1949.

Graham, J. W., Evidence of Polar Shift since Triassic Time, *J. Geophy. Res.*, *60*, 329-47, 1955.

Graham, J. W., Paleomagnetism and Magnetostriction, *J. Geophys. Res.*, *61*, 735-39, 1956.

Graham, J. W., A. F. Buddington and J. R. Balsley, Stress-Induced Magnetizations of Some Rocks with Analyzed Magnetic Minerals, *J. Geophys. Res.*, *62*, 465-74, 1957.

Le Grand, H. E., Conflicting Orientations: John Graham, Merle Tuve and Paleomagnetic Research at the DTM 1938-1958, *Earth Sci. Hist.*, *8*, no. 1, 55-65, 1989.

Tilton, G. R., interview by H. E. Le Grand, 5 February 1992, i1992.

Tuve, M. A., interview deposited at American Institute of Physics (Center for History of Physics), i1982.

Wetherill, G. W., An Interpretation of the Rhodesia and Witwatersrand Age Patterns, *Geochem. Cosmochem. Acta*, *9*, 290-292, 1956a.

Wetherill, G. W., Discordant Uranium-Lead Ages I, *Eos Trans. AGU*, *37*, 320-26, 1956b.

Wetherill, G. W., G. J. Wasserburg, L. T. Aldrich, G. R. Tilton, and R. J. Hayden, Decay Constants of K^{40} as Determined by the Radiogenic Argon Content of Potassium Minerals, *Phys. Rev.*, *103*, 987-989, 1956.

Wetherill, G. W., interview by H. E. Le Grand, 30 January 1992, i1992.

Merle A. Tuve's Post-War Geophysics: Early Explosion Seismology

THOMAS D. CORNELL
History Department, P.O. Box 9887
Rochester Institute of Technology, Rochester, N.Y.

INTRODUCTION

During the summer of 1988, while preparing an essay on Merle A. Tuve for the second supplement to the *Dictionary of Scientific Biography* [Cornell, 1990], I noted the piecemeal treatment of his career in the historical literature. The most extensive study was my own dissertation [Cornell, 1986]. But it had been confined to the first half of Tuve's life: his boyhood in South Dakota; his college studies at the University of Minnesota, and then at Princeton and Johns Hopkins; and his work as a staff member at the Department of Terrestrial Magnetism (DTM) of the Carnegie Institution of Washington (CIW) through about 1940, when he shifted his attention to wartime projects.

My interest in Tuve's career had focused on his nuclear physics, especially his work during the 1930s using Van de Graaff generators as particle accelerators [Cornell, 1988]. Similarly, Stewart Gillmor [1981] had considered his ionospheric studies (pursued jointly with Gregory Breit) in the 1920s; Ralph Baldwin [1980] had examined the development of the proximity fuse (led by Tuve) during World War II; and Allan Needell [1987] had contrasted Tuve's post-war approach to radio astronomy with the approach taken by Lloyd Berkner.

Recently, other Tuve studies have appeared. Their diverse origins suggest a growing interest in Tuve's career among historians of science. Nevertheless, these new studies still concentrate on particular aspects of his career. Thus Michael Dennis [1991] has contrasted two large-scale, applied-science programs during World War II (one led by Tuve and one at MIT led by C. S. Draper); Angelo Baracca [1988, 1989] has emphasized Tuve's resistance to the emerging Big-Science style in the mid-20th century; and Homer Le Grand [1989] has examined Tuve's role in supporting the paleomagnetic researches of John W. Graham at the DTM after World War II. Moreover, a list of Tuve's publications has appeared [Poggendorff, 1989].

Although there remains a need for a full-length Tuve biography, my aim here is to discuss the most important facet of his career that has not yet received close examination in the historical literature, namely, his post-war use of conventional explosives to study the Earth's crust.

Seismology emerged as a field of research long before Tuve made his contributions, but its origins lay in the study of earthquakes. For our purposes, one of the earliest events of note was an earthquake near Zagreb in 1909, when Andrija Mohorovicic (1857-1936) noticed that seismometers had recorded two sets of elastic waves, separated in time [Bullen, 1974]. He suggested that these waves had reached the instruments via two different routes, one through the Earth's crust (we would say) and the other through the upper edge of the mantle. That boundary is now known as the Mohorovicic discontinuity.

Also using seismic data from earthquakes, other researchers discovered the boundary between the Earth's mantle and core [Brush, 1980]. Notable here were the contributions of Beno Gutenberg (1889-1960), who left his native Germany in 1930 to join the faculty at Caltech, bringing with him "the European tradition of viewing seismology as a research tool . . . to investigate the physical properties and structure of the earth's interior" [Goodstein, 1984, pp. 201-202]. Gutenberg's move, in turn, made available to him not only the facilities at Caltech but also those of the CIW's Seismological Laboratory (formed in 1926 and turned over to Caltech in 1937).

Explosion seismology was not without precedent prior to 1940 [Byerly, 1956, pp. 125-127]. In the mid-19th century, the pioneering seismologist Robert Mallet (1810-1881) observed that gunpowder explosions in granite and in sand produced seismic waves of different velocities [Dean, 1991, pp. 48-50]. Similarly, in the late 19th century John Milne (1850-1913) carried out explosion studies in Japan [Herbert-Gustar and Nott, 1980, p. 82]. Nevertheless, solid progress came only after World War I, when explosion techniques were developed to locate new oil deposits [Weatherby, 1940]. Refraction methods (using seismic waves that traveled considerable distances along the upper edge of a layer) became common in the mid-1920s, while reflection methods (using seismic waves that "bounced off"

The Earth, the Heavens and the Carnegie Institution of Washington
History of Geophysics Volume 5

the upper edge of a layer) became common in the early 1930s. Scientists whose inclinations were not as commercial also demonstrated their interest. For example, Gutenberg (with Harry O. Wood and John P. Buwalda) [1932] undertook explosion studies in California in the early 1930s (in part, with CIW backing).

But before the end of World War II there had been no systematic programs of experimentation using large explosions to study the Earth's crust. What Tuve did after 1945 was to seize the opportunity for such work made possible by his skill with electronic equipment and by the availability of surplus explosives from the military. Under his leadership, researchers varied the locations and conditions of both the shots and the detectors, thus transforming the landscape into a vast laboratory bench.

Although he began in the West, the logistics of field work quickly led Tuve to shift his attention to the environs of Washington, D.C. Only as he and his co-workers developed suitable seismometers did they extend their efforts into the Appalachian Mountains. In 1948 and 1949 they observed the effects of large blasts in Tennessee and confirmed expectations that the Mohorovicic discontinuity lay deeper under the mountains than under the coastal region. But their seismograms proved more difficult to interpret than they had anticipated. In order to compare their results with the earthquake results that Gutenberg had been collecting, they traveled to California to observe a quarry blast. Moreover, in 1950 they adopted a new field technique. Instead of single, isolated instruments, they began using strings of instruments, evenly spaced in linear arrays. As their confidence in the field results grew, the group began surveying different geological provinces. In 1951 they made field trips to the Mesabi Range in Minnesota and to Puget Sound in Washington State, and in 1954 they surveyed the Colorado Plateau, where they found that the Mohorovicic discontinuity lay far closer to the surface than expected.

In subsequent years, the DTM researchers extended their surveys beyond the contiguous United States. In 1955 they traveled to Alaska [Tuve, 1966]. In 1957, as part of the CIW's contribution to the International Geophysical Year (IGY), they mounted an expedition to South America [Sullivan, 1961, pp.379-382]. The DTM also continued its seismic program in the 1960s. This paper ends with the trip west in 1954.

Several developments after 1954 suggest that the latter half of the decade should be seen as part of a different period in Tuve's geophysical studies. Given his close association with Vannevar Bush (1890-1974), 1955 marked a natural divide because of Bush's resignation after serving as CIW president since 1939 [Owens, 1990]. The 1957 South American trip established another important divide. The IGY altered the general context for all geophysical work. In addition, the CIW trip turned out to be the first in a series of expeditions to South America, thereby introducing a new theme in Tuve's seismic studies [Tuve, 1963]. Also in 1957, Tuve lost his primary co-worker: shortly after returning from the trip, Howard Tatel died from the effects of a brain tumor [CIW, file "DTM—Staff. Tatel, Howard E."].

Finally, not until the latter half of the decade did explosion seismology become directly relevant to the Cold War. From the outset, seismologists had observed the seismic waves produced by nuclear explosions. But only with the rise of widespread support for a test ban did seismology emerge as a possible source of monitoring techniques [Bolt, 1976].

OUT OF THE LAB AND INTO THE FIELD

Although Tuve had been a member of the DTM staff since 1926, his early work was in the laboratory rather than in the field. In his nuclear studies, for example, he strove to create very high voltages to accelerate protons to high energies. Directing a beam of high-energy protons onto atomic nuclei enabled him to study how the repulsive electrical forces between protons break down in the nucleus and are replaced by strong but short-range attractive forces. In the process Tuve designed and constructed complex equipment, thereby rendering the whole enterprise quite intensive technologically. Yet he kept his engineering efforts strictly subordinate to his study of nature's forces.

The DTM's field work was a distinctively different enterprise. Where Tuve's lab studies required large, complex equipment to produce the appropriate conditions, the field work required travel to appropriate sites and measurements under the conditions that nature provided. Large, complex equipment might well be involved, but it tended to take the form of the vehicles (notably, the non-magnetic ship *Carnegie*) that conveyed researchers to the places where they wished to use their instruments.

Given Tuve's greater experience with laboratory work, there was some discussion after World War II regarding how suitable a candidate he was for the position of DTM director. In late 1945 Lloyd V. Berkner (one of Tuve's DTM co-workers) wondered: "Will Tuve have sufficient interest in geophysics to permit appropriate development in that field?" [CIW, file "DTM—Future Program," unattributed TS, Oct. 25, 1945].

Even after his appointment, Tuve continued to emphasize the value of lab work. In mid-1946, for example, he received from his DTM colleague A. G. McNish a memorandum that expressed the field scientist's point of view. "Maybe Goethe was partly right," wrote McNish:

when he said that the way to understand Nature is to study her where she reveals herself, not to cramp her into a laboratory. At least that seems the way as far as geophysics is concerned [LC, Tuve Papers, Box 130, "Comments on Goranson's Memo on Geophys. Lab.," June 21, 1946].

In the margin, however, Tuve fired off his response: "Both [approaches are] needed (she also reveals herself in the lab—if you lead up to it properly!)."

As director, Tuve deliberately sought to reform the DTM. He proclaimed in his first report for the *CIW Year Book*: "The Department now turns to a new emphasis on laboratory and experimental work" [1947, p. 54]. But he also recognized that he could not abandon completely the DTM's traditional emphasis. Thus to the passage just quoted he added: "Much of this [work] will involve field trips, of course, as the earth is too large and varied to be studied in one spot." In short, one reason for Tuve's taking up field work after World War II was that the organization he now headed had long emphasized such studies.

Another question to consider is what experience in geophysics Tuve had had by 1945. Here we may include his ionospheric studies: working at the DTM in the latter half of the 1920s, first with Breit and then with Odd Dahl and Lawrence R. Hafstad, and using the nearby transmission facilities of the Naval Research Laboratory, Tuve had developed the pulse-echo method for measuring the height of the ionized layer of the Earth's atmosphere.

In addition, there was the Sixth Annual Conference on Theoretical Physics, held in Washington, D.C., March 21-23, 1940. Tuve had helped to originate these annual conferences, sponsored jointly by the CIW and George Washington University, and he had continued to play a central role in their planning. The 1940 topic was "The Interior of the Earth," and primary attention was given to the Earth's deep interior [Teller and Tuve, 1940]. Among the questions considered was one in which Tuve had already demonstrated a professional interest, namely: what gives rise to the Earth's magnetic field? He was especially interested in the work that Roy W. Goranson (1900-1957), a staff member at the CIW's Geophysical Laboratory (GL), described at the conference. For several years Tuve had been discussing with Goranson the feasibility of studying how the Curie point for ferromagnetic materials changed at extremely high pressures, and just prior to the conference Tuve had recommended a joint project between the GL and the DTM, with Ellis A. Johnson as the principal DTM collaborator and himself as a consulting collaborator [LC, Tuve Papers, Box 19, file "Extremely High Pressures. Also Bitters"].

But why didn't Tuve resume his nuclear studies after the war? In brief, by 1945 nuclear physics had become too applied to suit Tuve's tastes. "It was a business," he later explained. "Everything was serious engineering work, not just personal experimentation or personal knowledge . . ." [Tuve Interview, 1982]. Although he had opportunities to reenter the field, he chose not to take advantage of them. For example, on May 3, 1946, he declined an invitation to discuss the design of accelerators at an upcoming meeting of the American Physical Society. "The fact is," he wrote, "that I do not anticipate being in a position to make any worthwhile comments on this subject by September, and I therefore extend my regrets" [LC, Tuve Papers, Box 120, Tuve to E. P. Wigner and J. A. Wheeler].

It is true that Tuve followed closely the work of DTM researchers who used radioactive isotopes (produced by the DTM cyclotron) to pursue biophysical studies. It is also true that Norman P. Heydenburg (with Tuve's support) continued using the DTM's Van de Graaff accelerator. But Tuve's personal interest had cooled dramatically. Thus he commented to Paul C. T. Kwei on February 3, 1947:

As you know, the world of nuclear physics changed from a philosophical field to a welter of power politics, and in returning to Carnegie with the determination to go back to "things of the spirit" I have been quite unhappy about nuclear physics [LC, Tuve Papers, Box 123].

To continue examining why Tuve's research interests changed after the war, let us ask yet another question: when did an interest in explosion seismology first emerge as part of his post-war agenda? Apparently, what surfaced first was a general interest in geophysics. Writing on October 31, 1945, to John A.Fleming (then still DTM director) Tuve argued:

Our greatest immediate opportunity may well lie in the direction of a major new emphasis on a broad program in geophysics, building up a great research and training center concerned with the Physics of the Earth by the intensive cooperation of many institutions and groups in the Washington area, with special reference to a comprehensive and balanced coverage, and to the support of new ideas. I would be personally much interested in a joint effort of this kind if it were to be intentionally developed as an example of the fruitfulness of the freedom, scholarship, traditions and youthful atmosphere which are characteristic of a university, when a similar scheme is intelligently applied to specialized research groups and includes selected senior staff members from certain agencies of the Federal government [LC, Tuve Papers, Box 134].

The months that followed were difficult ones for Tuve, marking as they did the period during which the organization he had helped to create for the purpose of developing the proximity fuse ended its wartime existence and took on its peacetime identity (as the Applied Physics Laboratory (APL) of the Johns Hopkins University). Meanwhile, his private, handwritten notes began to reveal the specific features of his geophysical interests. For example, in a memorandum on the subject of a "Revivified Department" [LC, Tuve Papers, Box 129, Apr. 1, 1946], he included under the general heading "Geophysics" the following item:

> "Quantitative Geology"
> Mountain building
> Crustal shifts & history
> Seismology & Volcanology.

But where the other items on the list included the names of possible group leaders, this one included only the marginal expression of the need to "Find one!" [Ibid.]

To my knowledge, the earliest full statement of Tuve's personal commitment to explosion seismology came in a major memorandum that he prepared shortly before assuming his new duties as DTM director. Dated June 22, 1946, the document presented a detailed overview of the DTM's future. One research group proposed was "Section C, Exploratory geophysics of the crust," which (among other things) would concern itself with

> the possibility of learning new facts about the basement levels of the Earth's crust by "deep seismic shots," using for each experiment perhaps 500 tons of TNT detonated in the deep recesses of an abandoned mine [LC, Tuve Papers, Box 121, "Statement Concerning the Scientific Program of the Department of Terrestrial Magnetism for the Immediate Future"].

Moreover, in a personnel chart at the end Tuve listed himself as the group's leader.

Here, then, is the argument so far: just after the end of World War II, with nuclear physics no longer an attractive outlet for his efforts, with geophysics an area in which he had a long-standing interest, and with his becoming the director of an organization that had traditionally emphasized field work, Tuve turned his personal attention to a program of explosion seismology. But why explosion seismology? Part of the answer lies in sheer opportunism. "I had heard that the Navy was going to destroy some three hundred million pounds of TNT," he later explained. "So being a close friend of [Admiral G. F. Hussey, Jr.] the Chief of Naval Ordnance, I . . . asked him to hold some of it for explosion work for Carnegie" [Tuve Interview, 1982].

But where did the idea for the program come from? Unlike the case of nuclear physics, which Tuve had thought about, talked about, and written about long before putting anything into practice, there had been no hint of an earlier interest in explosion seismology. Thus to proceed further we must look elsewhere than the specific details of Tuve's own life and career.

The idea for such a program came to Tuve from Leason H. Adams (1887-1969). Since 1910 Adams had been a staff member at the GL. In 1936 he became acting director (succeeding Arthur L. Day, the first director of the GL), and then in 1938 he became director, until his retirement in 1952. Throughout his long tenure at the GL, Adams's most active area of personal research was the laboratory study of rocks under high pressure. For him, such work was an essential complement to seismic studies. He believed that by combining seismic data with the results of high-pressure studies, geophysicists would be able to determine the character of the materials composing the various layers of the Earth's interior [Yoder, 1990; Adams, 1939; and Adams, 1946, p. 26].

Just after World War I, Adams proposed setting off conventional explosives as a way of studying the Earth's crust at depths below those of interest to petroleum geologists. Writing to Day on March 14, 1923, he had assessed "the situation with regard to sound-ranging the outer crust of the earth" by commenting:

> More is known about the nature of the material at 1000 miles below the surface than at 40 miles, and measurements of the velocity of propagation of an artificial earthquake would remedy this deficiency [CIW, file "GL—Adams, L. H., Seismology," copy appended to Adams, "Memorandum Concerning the Use of Seismic Methods in the Exploration of the Earth's Interior," Apr. 9, 1948].

Although conceding that earthquake studies might eventually yield the desired results, Adams had also emphasized the advantage of well-placed explosions, namely, "that the disturbance is localized, and that it takes place where and when one wants it." He had hoped to use explosives offered by the Army and to set them off "in hard rock, possibly in an abandoned mine-shaft."

In a 1948 memo, Adams explained that the prompt implementation of his earlier idea had been prevented by the lack of suitable instruments and that the Ordnance Department had eventually withdrawn its offer to provide the explosives. Consequently, the memorandum continued, the idea:

> . . . was not again given active attention by the Geophysical Laboratory until October 1945 when Adams at the invitation of the Navy witnessed the explosion of a quarter-million pounds of high explosives at Pocatello, Idaho. At this time he mentioned to several persons, including Dr. E. J. Workman and Dr. S. Karrer, who also witnessed the explosion, the proposal in 1923 for setting off a suitable quantity of high explosives so as to obtain records of vibrations that had penetrated the upper layers of the Earth probably down to a depth of a hundred miles or more, and he was gratified to see an enthusiastic response from persons interested in finding out more about the nature and structure of the Earth's crust [Ibid.].

Adams's companions at the 1945 blast Everly John ("Jack") Workman (1899-1982) was a physicist who had assisted Tuve with the proximity-fuse project. In 1946 he became president of the New Mexico School of Mines (now the New Mexico Institute of Mining and Technology) at Socorro, a position he held until his retirement in 1964 [Baldwin, 1980, pp. 82, 95, and 135-139; Brook and Wilkening, 1983].

In a letter to Adams on June 4, 1946, Workman brought up the subject of explosion seismology:

> Some time has passed since our last conversation relative to the Crust of the Earth Experiment, which we discussed about a year ago on the Idaho trip, and again about a month or six weeks ago. I have been thinking considerably about this experiment and I had an opportunity to talk at some length concerning it on Dr. Karrer's recent visit to Albuquerque [LC, Tuve Papers, Box 121].

Workman also had consulted with Stephen Brunauer at the Navy's Bureau of Ordnance and had located suitable mine shafts for the shots. Hoping that his college's Research and Development Division would be able "to take an active part in this experiment," he concluded:

> . . . I should like to have your opinion as to what the course of action should be to get the thing underway. Do you think it would be good to get together . . . a statement of the problem and initiate a request for materials and undertake discussions with possible co-ordinating agencies? [Ibid.].

The main difficulty from Adams's point of view was the fullness of the GL's post-war research agenda. But in discussing the situation with Tuve, he discovered that his DTM colleague was more than willing to undertake a cooperative effort, with the DTM "to be the active partner, to manage the enterprise, and to supply most of the manpower" [CIW, file "GL—Adams, L. H., Seismology," Adams, "Memorandum Concerning the Use of Seismic Methods in the Exploration of the Earth's Interior," Apr. 9, 1948]. As Adams noted in his 1948 memorandum:

> Tuve entered into this venture with characteristic energy and enthusiasm; and almost immediately obtained the promise of the Navy Department to make available to the Carnegie Institution a large amount of explosive . . . . He also elicited an offer by the Navy to transport, load, and fire the powder . . . [Ibid.].

Accordingly, Adams replied to Workman on June 11, 1946, that the DTM and the GL were planning to "join forces in promoting the large-scale seismic prospecting" [LC, Tuve Papers, Box 121].

What emerged from these discussions between Adams and Tuve was a pair of proposals, each dated June 24, 1946, and each prepared by Tuve. The first involved the study of rocks under high pressure, with Goranson as the principal researcher. "The proposal," Tuve explained:

> arises from the fact that Dr. Goranson used a special equipment at Los Alamos during the war for the production of extremely high instantaneous pressures by the use of shaped charges and by the use of special projectiles in extremely high velocity smooth bore guns. Part of this equipment is a very elaborate electronics set up. . . . Dr. Goranson's war work was concerned with the physical properties of metals and similar materials under extremely high instantaneous pressures. We propose to use this same technique and equipment for studies of olivine, basalt, and other basement rocks of geological interest [LC, Tuve Papers, Box 121, "Statement Concerning Proposed Program, Properties of Rocks and Other Materials under Extremely High Pressures for Short Duration."].

Through informal arrangements with Norris E. Bradbury, J. Robert Oppenheimer's successor as director at Los Alamos [Bradbury, 1980], the equipment that Goranson had used during the war had been set aside for transfer to the CIW. Similarly, Workman had informally agreed to provide a site for Goranson's work at an ordnance proving grounds, near Albuquerque, that the New Mexico School of Mines still operated under a Navy contract. Thus the immediate purpose of Tuve's "Statement" was to obtain formal authorization to proceed.

The second proposal involved explosion seismology. Tuve's aims were deliberately broad and ambitious:

> Summary—During the war members of the Carnegie Institution staff and others who were greatly interested in geophysical problems had many discussions regarding the desirablility of using some of the large surplus stocks of high explosive at the end of the war for two purposes: (a) to attempt experiments which might give new information concerning the deeper layers of the Earth's crust, and (b) to attract conspicuous interest, especially on the part of young scientific investigators, toward the interesting possibilities of scientific geophysics [LC, Tuve Papers, Box 121, "Statement Concerning Proposed Program of Wide Collaboration on Deep Seismic Shots under Carnegie Institution Auspices," June 24, 1946].

Tuve explained that he would personally start the field work during the summer, albeit on a limited basis, in order to define the project more carefully and to develop the equipment needed for other institutions to collaborate. As with Goranson's project, he expected that Workman would play an important role. Tuve planned to use as explosion sites the mine shafts located by Workman.

Bush responded favorably, commenting to Walter M. Gilbert, the CIW's executive officer: "I like the idea of [the] CIW furthering & centering this—with lots of collaboration" [CIW, file "DTM—Seismic Program," n.d.]. Soon Tuve informed Adams that he had obtained the official permission they needed to proceed. In the same letter, he formally requested Goranson's full-time services, as well as the assistance of Joseph W. Greig (1895-1977), a petrologist who would help choose the sites for the shots. Finally, Tuve noted: "In accordance with our conversation, it appears best to centralize the handling of these projects here in the DTM, and I am glad to undertake this responsibility" [LC, Tuve Papers, Box 121, July 12, 1946].

Another preliminary step was to contact Gutenberg. Tuve wanted to acquaint Gutenberg with the Carnegie project, in hopes that the older researcher would be willing to advise them. He also asked for Gutenberg's "sympathetic indulgence regarding our rather brash efforts to enter a complicated field of research which has been so beautifully opened up by your life long efforts" [LC, Tuve Papers, Box 331, Aug. 20, 1946]. With similar deference he stressed the preliminary character of the upcoming work:

> As a means of rapid education in the subject Dr. Goranson and I have undertaken to make up several very simple seismograph units which we are shipping to New Mexico later this week. If all goes well we hope, with Dr. Workman's assistance, to make some very simple measurements of wave amplitude at several distances from a dynamite shot of 2 or 3 tons, which will be well tamped in an abandonned [sic] mine in pre-Cambrian intrusive structures east of Albuquerque [Ibid.].

Those first test shots, using 1-ton charges, came during the period from August 30 to September 6, 1946, at a site near Albuquerque. As Tuve later reported, the tests revealed several problems. Some involved the explosives. Others involved the seismometers. Still others arose due to the field conditions: for example, Tuve noted that "[w]ind makes the mountains noisy with motions and creaks" [LC, Tuve Papers, Box 121, "Topical Report of Tuve's Activities, September 1946," Oct. 9, 1946]. Nevertheless, a beginning had been made. Tuve's program of field studies was underway.

BRINGING THE PROJECT INTO FOCUS

Tuve still hoped to proceed with a pair of 100-ton explosions later in 1946. But additional problems soon led him to postpone these plans. The most significant problem involved a change in how the Navy administered the project. According to the new arrangement, the Bureau of Ordnance continued to handle the explosives. But the funding now came from the Office of Naval Research (ONR), which had just been created on the basis of legislation passed by Congress in August [Schweber, 1988; Sapolsky, 1990, pp. 9-56].

To discuss these new arrangements and to map out plans for the future, Commander George Vaux of the ONR contacted Tuve about scheduling a conference for representatives from the DTM, the ONR, and Ordnance. Held on October 29 (after Goranson's return from New Mexico), the meeting gave Tuve an opportunity to describe the ambitious program he had in mind. As Vaux later reported, Tuve told the assembled group:

> It will . . . be necessary to set off charges whose weight ranges from 100 to 2,000 tons, in order to transfer sufficient energy into the earth's crust to enable detection of the resultant wave. . . . It is desired to set up these simple devices at fifteen or twenty stations in a chain, with the stations located at increasing distances from the source, but with the same azimuth from it. Four or five chains would be employed, each having a different azimuth, so that the total number of stations might approach 100 [LC, Tuve Papers, Box 331, Vaux, "Conference on Deep Seismic Shots, Held 29 October 1946," Oct. 30, 1946].

Also at the meeting, Captain J. L. Phares of Ordnance explained "that it is the duty of the Office of Naval Research to act as liason [sic] agency for the Navy in the organization of the program" [Ibid.].

One complicating factor was the surplus radio equipment that Tuve had requested. Anticipating a large-scale project, he had asked for 100 transmitters. Vaux, in turn, wanted to provide a grant to cover the cost of storing, transporting, and servicing the equipment, in order to avoid the bureaucratically awkward position of having to do so after the Navy had given it away. The problem as Tuve saw it was Bush's strong preference for a post-war policy of avoiding all federal funding at the CIW. Tuve argued the CIW should accept the modest amount the Navy had offered. "I believe," he wrote Gilbert on November 8, 1946:

> that a workable pattern is being evolved which will leave the Institution at all times ready to say that no subsidy of our work by Government funds exists, but also allowing us to say that we are collaborating fully with the Government agencies who are interested in our program [CIW, file "DTM—Seismic Program"].

In a memorandum to the ONR recommending a contract between the CIW and the Navy, Tuve described the full extent of the cooperative efforts he had in mind. Beginning with groups from half a dozen laboratories, he expected upward of three times that number to become involved. In addition to explosion studies on land, there would be work at sea (for which he had in mind W. Maurice Ewing at Columbia University), as well as Goranson's high-pressure work. Tuve hoped that the Navy would arrange separate contracts with each of the major participants. Even so, he counted on the CIW's playing a central role, not only by providing overall coordination but also by acting "as custodian of Navy equipment for the various collaborators" [CIW, file "DTM—Seismic Program," Nov. 8, 1946].

According to Tuve's plan, primary responsibility for the project would be vested in the participating civilian organizations. But the considerable expense of handling the explosives led some of the Navy's officers to resist these arrangements. As a result, Commander Roger Revelle (a geophysicist who had worked for the Ordnance but then recently appointed head of the ONR's Geophysics Section) visited Tuve on December 20, 1946, to discuss the project.

Tuve found the conversation so upsetting that he sent a detailed account to Captain R. D. Conrad of the Naval Ordnance Laboratory (NOL):

> Commander Revelle first expressed his insistence that he would have nothing to do with this project unless agreement was reached that he would have a project engineer in the ONR through whom every contact with the Navy is to be handled, down to the last details of telephone calls and arrangements during the actual shooting, with no other contacts or channels between the participating groups and Navy men, for example, in the Bureau of Ordnance or at the shooting sites.
>
> . . . He strongly indicated his determination that the Navy shall not be "milked" and his attitude that the Navy should participate in, and that he will personally support, only problems out of which the Navy clearly will get specific benefits of technical character, over and above any general benefits arising out of working relationships with active scientific men who are pursuing their own interests. . . .
>
> All of this position I consider exactly opposite to the policy necessary in the armed services if they are to remain actively connected with scientific men (as distinguished from engineers) and if the Service officers are to be continuously in touch with young scientific men and with new developments which will arise during peacetime and whose significance to the Navy will not be evident in advance of the work. In general Commander Revelle's position is that of an engineer, and his concept clearly is that scientific men can and should be persuaded to accept direction in their work from those in charge of (Navy) funds. This position is perhaps appropriate to the other bureaus of the Navy, but it is not a basis on which we are willing to collaborate with the ONR [DTM, GF 1935-1948, file "75. Office of Naval Research," Dec. 23, 1946].

Revelle's remarks angered Tuve because they challenged the deep ties that he had formed with the Navy. One indication of those ties had come toward the end of 1945, when Admiral Hussey conferred on Tuve the Naval Ordnance Development Award in recognition of his wartime service [CIW, file, "DTM—Tuve, Merle A., 1959-1982," Hussey to Tuve, Dec. 10, 1945]. Also, after Tuve resigned his position as director of the APL in 1946, Hussey wrote him an appreciative letter and requested Tuve's continued services as a consultant to the Navy [LC, Tuve Papers, Box 113, Feb. 21, 1946].

A further indication of Tuve's Navy ties came in August 1946 at the laying of the cornerstone for the NOL at White Oak, Maryland. Tuve had been invited to speak at the ceremony, and his remarks emphasized the importance he attached to cooperative efforts between civilian researchers and the military:

> I am convinced that the most important element in the future readiness of the Navy is a continuous and comprehensive set of relationships between Navy men and civilian specialists of every kind throughout the technical activities of our industrial and university

groups [LC, Tuve Papers, Box 362, "In Defense of Freedom," Aug. 15, 1946].

But just as important as the practical need for cooperation was the larger purpose of those efforts. Accordingly, Tuve concluded his speech by listing the beliefs that constituted, for him, "the code of the West":

(1) The right to know,
(2) the right to differ,
(3) the right to be an individual,
(4) the opportunity for well-being, and
(5) the obligation to defend these freedoms for others [Ibid.]

Although he intended the list as a general statement, Tuve might equally have claimed it as his own personal code, which no project in which he participated could violate. Thus he found unacceptable the constraints that Revelle wanted to place on the Carnegie seismic work, and his distress over the matter led him to delay making further arrangements until the ONR had clarified its policy. Meanwhile, however, a trend toward a solution was already underway. Tuve had expected the project to involve nationwide field studies and multi-institutional cooperation. But Goranson's high-pressure work was evolving along lines that suggested a different approach, namely, to downsize the project into a local effort pursued primarily by Carnegie researchers in collaboration with researchers from a handful of other institutions.

During the fall of 1946 Goranson ran into unexpected difficulties. He learned from Bradbury that some of his equipment was still being used at Los Alamos and that duplicating it would be delayed by a shortage of personnel [LC, Tuve Papers, Box 121, Oct. 9, 1946]. Not long afterward, Tuve made a counterproposal: the New Mexico School of Mines would hire one of the Los Alamos technical men, who would remain at Los Alamos long enough to duplicate the equipment. "If this arrangement appears reasonable to you," he wrote Bradbury, ". . . it would enable us to provide the missing items and have the entire equipment assembled at the New Mexico School of Mines laboratory site near Albuquerque before February 1" [LC, Tuve Papers, Box 121, Oct. 30, 1946].

Toward the end of November, however, Goranson suggested yet another possibility. On the one hand, he was responding to a less-than-satisfactory level of cooperation between Los Alamos and the New Mexico School of Mines (apparently due to friction between Bradbury and Workman). On the other hand, he was responding to the interest that the NOL had expressed in housing the equipment from Los Alamos and in hiring the technical personnel needed to support the project. In a memorandum, Goranson described the offer:

> I would be permitted to carry out our program of tests. Our results would be unclassified. For their side of the ledger they stated that they would be more than recompensed by being able to conduct tests in which they are interested and if they could have my advice on techniques of operation and analytical procedures [LC, Tuve Papers, Box 331, "Proposal Concerning Ultra High Pressure Program," Nov. 22, 1946].

Accepting Goranson's recommendation, Tuve moved quickly to formalize the new arrangement, writing letters in early December to Workman, Bradbury, and Captain R. D. Bennett (the Technical Director at the NOL). Implicit in his assessment of how to proceed was his preference for the seismic work. For example, he explained to Bradbury:

> Goranson and I find ourselves more or less submerged in the deep seismic program, with a large and increasing burden before they can be made fruitful. If in addition we attempt to personally inaugurate the program on elastic constants using the special explosive equipment we may find both programs inclined to stall [LC, Tuve Papers, Box 331, Dec. 4, 1946].

But Goranson's proposal had been motivated by another factor as well. "Distance from Washington to New Mexico," he had pointed out, "is such that it would be difficult to keep one's finger on the pulse of operations at New Mexico" [LC, Tuve Papers, Box 331, "Proposal Concerning Ultra High Pressure Program," Nov. 22, 1946]. A similar problem also affected the seismic work. What the Carnegie researchers needed was easier access to the explosion sites. Accordingly, Tuve wrote Vaux on October 23, 1946: "If we can observe a few underwater explosions of depth charges placed in Chesapeake Bay we may be able to make a much more intelligent selection of instruments for the larger shots out west" [LC, Tuve Papers, Box 331].

In a letter to Workman dated January 17, 1947, Tuve summarized the problems associated with the seismic work planned for New Mexico later in the year. To begin with, the larger shots (100 tons each) would be more expensive and would require a larger scale of operations than he had anticipated. "In addition," he noted:

> there have been some differences in point of view with Roger Revelle in the Office of Naval Research who wanted to take a strong hand in respect to the observing program itself, wanting measurements in three

dimensions, absolute calibrations of amplitude, and other items which would not only be very difficult but to my mind are of uncertain interpretation. As a result all of our arrangements with ONR are for the moment held in abeyance. . . . My position was that if the Navy wants certain observations the Navy should ask their own laboratories to make them or hire engineers, but men in pure science will require freedom to do whatever they are interested in and not what they are told [LC, Tuve Papers, Box 331].

Already, however, Tuve had found a temporary alternative much closer to home. After learning that the Navy was, for its own purposes, setting off explosions near Solomons Island in the Chesapeake Bay, he arranged through Brunauer for the Carnegie researchers to participate [LC, Tuve Papers, Box 331, Tuve to Bureau of Ordnance, Dec. 2, 1946]. Moreover, he wrote Workman early in the new year: "They have also sub rosa been shooting several one ton shots exclusively for us" [LC, Tuve Papers, Box 331, Jan. 17, 1947].

By early March 1947 the problem with the ONR had been smoothed over. Writing to Alan T. Waterman, the ONR's Chief Scientist and Deputy Chief, Tuve indicated his satisfaction with the result: "It appears that our discussions with Commander Revelle had a somewhat unfortunate emphasis, but the entire situation is adequately cleared up now" [LC, Tuve Papers, Box 331, Mar. 5, 1947]. At the same time, Tuve outlined to Vaux's replacement, Beauregard Perkins, Jr., a more realistic program, one based on a series of shots in the Chesapeake Bay area rather than in the West [LC, Tuve Papers, Box 331, Tuve to ONR, Mar. 5, 1947].

Meanwhile, Tuve had been addressing another major problem, the development of suitable instrumentation for the Carnegie project. Since the late 19th century, measurements of seismic waves from earthquakes had been made using rather large instruments mounted permanently in observatories and running continuously [Dewey and Byerly, 1969; Benioff, 1948]. By contrast, what Tuve needed were lighter, more rugged instruments that could be easily transported by car or truck for field use at the time of each blast. In meeting this need, Tuve demonstrated what has emerged as one of the distinctive features of his career, namely, "his skill in applying electronics to almost any given job" [Ewing, 1963, p. 287].

During the summer of 1946, one of Tuve's main goals was to test seismometers that employed a specialized vacuum tube invented by Ross Gunn several years before. As a staff member at the Naval Research Laboratory since 1927, Gunn had long been one of Tuve's Washington contacts. Tuve hoped that Gunn's micrometer tube would serve as the basis for the simple yet sensitive instruments he sought for the Carnegie project. Thus he wrote Vaux on August 21, 1946:

> We have made up a simple unit comprising a horizontal and vertical seismometer each using a "micrometer tube" of the movable cathode type developed by Gunn before the war and now in preliminary manufacture by RCA [CIW, file "DTM—Seismic Program"].

But extensive tests revealed that the tube's noise level was too high [LC, Tuve Papers, Box 121, "Projects under Investigation during November 1946," Dec. 11, 1946; Box 124, Goranson, "Memorandum on Deep Seismic Shots," Oct. 24, 1946].

Writing on November 20, 1946, to William F. Sprengnether, whose company manufactured seismic instruments, Tuve described the features desired for the CIW seismometers:

> A simple horizontal component pendulum with a period between one and six seconds, plus an amplifier with two or three channels of different sensitivity, recording on a Brush Development Company (Cleveland) magnetic pen recorder, appears possible as a minimum equipment which would meet the requirements [LC, Tuve Papers, Box 331].

Writing to G. R. Shaw at the Radio Corporation of America (RCA), Tuve described one alternative to the Gunn tube:

> We have been making some tests with the three Type C798B vacuum tubes . . . which are being produced on a preliminary basis by your development group. . . .
>
> If these tubes work out as promising as they look it may be a very direct and convenient answer to the problem of making a great many low cost sets of equipment for observing shock waves through the various layers of the Earth's crust down to 80 kilometers [LC, Tuve Papers, Box 331, Dec. 13, 1946].

But these tubes (which RCA was developing for use in phonographs) also proved too noisy [LC, Tuve Papers, Box 331, Tuve to RCA, Jan. 22, 1947; Tuve to M. Acheson, Feb. 4, 1947].

In late 1946 the lack of suitable seismometers began affecting Tuve's interactions with the ONR. "We are still not too happy regarding our instrumentation," he wrote Vaux on October 23, "and it may prove best to postpone our proposed 100 ton preliminary shots in [New] Mexico until February . . . " [LC, Tuve Papers, Box 331]. A couple of weeks later, he added: "It is hoped that suitable prototypes of instruments can be developed during this winter . . . " [CIW, file "DTM—Seismic Program," Tuve to ONR, Nov. 8, 1946]. Instrumental development also played a role in his December conversation with Revelle. "Commander Revelle," Tuve wrote to Conrad afterward, "evidently wished to discuss with us the possibility of altering our proposed instrumentation (which we have not planned beyond the initial stages of the work) . . . " [DTM, GF 1935-1948, file "75. Office of Naval Research," Dec. 23, 1946]. In short, the lack of suitable seismometers loomed as a potential point of vulnerability.

Fortunately, a breakthrough soon occurred. Writing to Marcus Acheson at Sylvania Electronics Products, Inc. (the company whose tubes he was then using), Tuve described the experience:

> It concerns an accidental magnetic effect on the platecurrent of the 729 rugged pentode which I found at Christmas time . . . In looking around for a simple electrical transducer I brought to the lab a couple of Alnico magnets with toy dogs mounted on them which my little daughter received for Christmas, and tried various rugged tubes for a magnetic effect on the platecurrent when brought adjacent to the pole of one of these magnets [LC, Tuve Papers, Box 331, Feb. 4, 1947].

Following several months of development work, Tuve wrote Perkins: "We are making reasonable progress on our instrumentation for the deep seismic project" [LC, Tuve Papers, Box 331, Tuve to ONR, May 21, 1947]. By then the Carnegie researchers had designed the instruments that they would use for the next several years. "Our first truly successful field seismometer," read a later description:

> . . . was a velocity electromagnetic type. . . . The inertial element was in the form of an inverted pendulum supported by two strips of spring steel which served as the hinge and accounted for a small portion of the restoring force. The major part of the restoring force was supplied by a small bent wire of beryllium copper. . . . The magnet was a radar magnetron magnet. . . . These instruments were magnetically shielded by an iron case which made them very heavy (about 30 lbs) [DTM, Box 1989-46, file "III. Seismic Instruments," unattributed TS essay entitled "Seismometer," n.d. (probably mid-1950s)].

Although the Carnegie program of seismic studies was coming into focus, there remained the problem of whom to put in charge. From the outset Tuve recognized that once the project was underway, his own duties as DTM director would prevent him from providing the full attention it needed. But a growing post-war interest in geophysics on the part of both industry and academia led to difficulties in locating a suitable leader. "With the present shortage of technical men," Tuve wrote Bush on May 19, 1947, "it is simply not possible to find outstanding candidates in other laboratories throughout the country" [LC, Tuve Papers, Box 124]. By then, however, Tuve was willing to offer the position to a staff member at the APL. He recommended Tatel.

Tatel was born in New York City in 1913. After receiving his baccalaureate and master's degrees at MIT, he went to Stanford, where he received his doctorate in 1940 in nuclear physics. Early in World War II, Tatel continued his nuclear studies at the University of Michigan. He then joined the proximity-fuse project, and after the war he remained at the APL. "Throughout all of his activities [during and after the war]," Tuve informed Bush, "he was regarded . . . as one of our very best technical men, as he preserved the analytical and critical viewpoint of a physicist in everything he did" [Ibid.].

Although the offer came in mid-May, Tatel's appointment only took effect in September [LC, Tuve Papers, Box 121, Tuve to Tatel, May 15, 1947]. During the intervening months he continued his work for the APL. In particular, he and James A. Van Allen prepared equipment for use in cosmic-ray studies of the upper atmosphere, equipment that was mounted in the nose cone of a V-2 rocket and launched from the White Sands Proving Ground on July 29, 1947 [DTM, Box 1989-45, file "V-2 Technical," Tatel and Van Allen, "Outline of 24 July 1947 Shoot V-2 No. 30," May 16, 1947; Tatel and Van Allen, 1948; Van Allen and Tatel, 1948; Van Allen, 1983, p. 17].

In early September, after Tatel's arrival in Washington, Tuve circulated a memorandum to senior staff members asking them for suggestions regarding a weekly geophysical seminar at the DTM. "My idea," he told them:

> is that in a period of 2 years we should cover most of the serious areas of geophysics with lecture treatment and outline notes, text material, etc., at the level of an advanced graduate course, with full analytical treatment critically examined. [LC, Tuve Papers, Box 130, Tuve to Tatel, E. A. Johnson, L. V. Berkner, and E. H. Vestine, Sept. 9, 1947].

He also suggested that in preparation for the seminar "Tatel may wish to go to Columbia & Harvard to check on their course content."

Early in October, Tatel made the trip. At Columbia he discussed marine geology with Maurice Ewing. At

Harvard, Francis Birch told him that the "[g]reat problem today is [the] structure of [the] earth—mountains, continents, and oceans" [DTM, Box 1989-44, Tatel, Notebook Entry, Oct. 9, 1947]. There also, Tatel consulted L. Don Leet, who had long been interested in the seismic effects of explosions and who had measured the seismic effects of the 1945 atomic bomb test in New Mexico [Leet, 1946].

Back in Washington, Tatel prepared a series of lectures that drew extensively from Harold Jeffreys's treatise, *The Earth* [1929]. The series began in October and lasted until April 1948 [LC, Tuve Papers, Box 130, "Projects under Investigation during October 1947," Dec. 1, 1947; "Projects under Investigation during April 1948," June 1, 1948]. During the second year, plans were made for various presentations on the general topic of "Historical Geophysics" [LC, Tuve Papers, Box 137, Feb. 1, 1949]. These developed into a seminar entitled "Milestones in the Pre-Cambrian" [Tuve, 1949, pp. 76-77].

Pleased with his colleague's work, Tuve decided to appoint him acting chairman of the DTM's Earth's Crust Section [DTM, GF 1935-1948, file "106. Tatel, Dr. Howard," Tuve to Tatel, Jan. 15, 1948]. Tuve had been the section's first chairman, but in January 1947 Ellis Johnson had taken over. When Johnson began a leave of absence in February 1948, Tatel became acting chairman and on April 30 chairman.

Along with sections on the lower and upper atmosphere, the section on the Earth's crust (which included studies of paleomagnetism, as well as the seismic work) was part of the DTM's new emphasis on experimental geophysics. Under Tuve's leadership, the DTM had been de-emphasizing its long-standing program of observatory studies. Arrangements were made to transfer the DTM's geophysical observatories in Peru and Australia to the governments of those countries. Thus in February 1948 Tuve was able to write the CIW's new executive officer, Paul A. Scherer (an engineer whose father had been president of Throop College, a precursor of Caltech, and whose wife was the daughter of George Ellery Hale [Bush, 1947, p. 14; Embry, 1954]): "The transition from the older program of the Department to fresh experimental projects is proceeding vigorously and with enthusiasm on the part of the staff" [CIW, file "DTM—Director 1941-1949. 1 of 3," Feb. 5, 1948]. Moreover, Tuve commented in the 1948 *CIW Year Book*: "the Department during this report year has begun to function effectively as a physics department with special interests in geophysics and in laboratory physics" [p. 53].

Meanwhile, the seismic studies based on explosions at Solomons Island had matured to the point that the Carnegie researchers were able to present a paper at the 1948 spring meeting of the American Geophysical Union (AGU) in Washington, D.C. They reported that they had largely completed the initial stage of the project, namely, "the development of suitable instrumentation and techniques," and were now ready for an expanded program of field studies [Tuve et al., 1948].

By early May 1948 Tatel and Goranson had prepared a memorandum that outlined where the project was headed. They explained that the Carnegie researchers would continue using explosions at Solomons Island. In addition, they now planned to take advantage of quarry shots to extend their studies inland, "[t]o determine the layering of the crustal structure underlying the Appalachian Highlands" [LC, Tuve Papers, Box 331, "Outline of Earth Crust Program," May 4, 1948].

Also by then, the project had achieved sufficient coherence to render it newsworthy. Accordingly, Ron Ross (a writer for Science Service) reported in *Science News Letter* [1948]:

> Explosions on the bottom of the Patuxent river have been set off at intervals for more than a year. Seconds after a blast, waves through the earth are recorded on the scientists' instruments at points at varying distances and directions from the explosion.
>
> This new information is helping the Carnegie Institution scientists plot a unique map. It will show what the earth is like down to 30 miles under the nation's capital and nearby states.

One more sign that the program had reached maturity came in September 1948 when the American Association for the Advancement of Science held its Centennial Meeting in the nation's capital. During the 1930s Tuve and his DTM co-workers had used similar occasions to show visitors the particle accelerators used for their nuclear studies. Now, in addition to the cyclotron and the Van de Graaff generator, visitors could view the "[e]quipment and techniques used in exploration of [the] earth's crust by seismic methods" [LC, Tuve Papers, Box 136, W. C. Parkinson, "Memorandum for Information of Chairman of Committee on Afternoon Activities during Centennial Meetings of AAAS, September 1948," June 28, 1948].

With the seismic program and the general reorganization of the DTM well underway, Tuve seriously considered accepting a position elsewhere. Since the end of the war, he had been approached by several universities. Each time, however, he had not felt ready to leave the DTM. For example, in early 1946, he had discouraged overtures from the University of Denver, writing "that it is simply not possible for my mind and spirit to face new challenges for many months to come, at the earliest" [LC, Tuve Papers,

Box 118, Tuve to H. Moore, Jr., Jan. 7, 1946]. Similarly, on May 19, 1947, he had written in regard to a position at the University of Minnesota:

> I find we have grown many roots here, and I am falling back on my decision of more than two years ago, during the stress of the war, to return to the Carnegie Institution and live quietly for at least three years, while I re-evaluated the things which I most want to do in my life [LC, Tuve Papers, Box 121, Tuve to H. E.Hartig].

But a year later Tuve responded more favorably when the University of Washington at Seattle approached him about the deanship of the College of Science and Arts. Already he had visited the campus. In May 1947 he had attended the inauguration of the university's new president, Raymond B. Allen, and had participated then in discussions regarding the future of the university's physics department [LC, Tuve Papers, Box 122, "Rough schedule for Tuve's Seattle trip," May 16, 1947; Box 124, "Brief Report of Tuve's Activities, May 1947," June 24, 1947]. Even so, he now delayed a decision, writing Allen that he would be needed at the DTM at least through the summer [LC, Tuve Papers, Box 138, 23 May 1948].

Tuve proceeded with his summer plans by traveling to New Mexico for seismic work in the Manzano Mountains near Albuquerque (where he found that tamping the 1-ton shots with sand was far less satisfactory than tamping with water, the usual practice for the blasts in the East) [CIW, file "DTM—Director 1941-1949. 1 of 3," Tuve to Berkner, July 26, 1948; Tuve, 1948, p. 65]. "We had a good, but somewhat frantic, time arranging some shoots with Tuve when he was out here," Workman wrote to Adams afterward. "We would like to do one or two more, but I am afraid that we have used up most of our good holes" [GL, Box 31, file "Workman, E. J. 1943-48," July 12, 1948].

Tuve also delivered the commencement address at the New Mexico School of Mines, and his remarks revealed something of his general attitude toward Earth studies. "No man," he told his audience:

> can study the record of historical geology and remain the self-centered youth he once was. You do not walk on rocks and dirt like ordinary men, you walk on the pages of the book of time [LC, Tuve Papers, Box 138, "Commencement Address by M. A. Tuve," May 28, 1948].

All the while, Tuve pondered the Seattle offer. Before leaving Washington, D.C., he had privately identified the main criterion he would use in making his decision. At the head of his handwritten list of "Personal *values*" he had placed "Personal participation in research, for the apprehension, appreciation, and growth of truth" [LC, Tuve Papers, Box 138, May 5, 1948]. Accordingly, he wrote Allen from New Mexico:

> Before I can make a decision to leave the Carnegie I will need to know more about the Dean's actual duties and work load. I dare to imagine some of his opportunities (and troubles!). Would I have opportunity for personal participation in research? This is a very real part of me, I know, and I must doubt that I am prepared to become exclusively an administrator. I feel that a scholar must continue to dig for truth on the same basis as any student or he soon goes on the "inactive list," and his judgment and leadership suffers. Direct contact with the materials of his subject, breaking sod that hasn't been ploughed before, is a necessary and humbling experience. I still need to keep on working in physics. [LC, Tuve Papers, May 23, 1948].

Then Tuve traveled briefly to California, where the dedication ceremonies (on June 3, 1948) for the 200-inch telescope at the CIW-supported Mt. Palomar Observatory [Wright, 1952, pp. 179-188] gave him an opportunity to discuss the Seattle offer with Bush and Scherer, who helped him to decide in favor of staying at the DTM. "I wired Ray Allen afterward," Tuve informed Scherer upon returning to New Mexico, "that I could not consider a change from Carnegie" [CIW, file "DTM—Director 1941-1949. 1 of 3," June 12, 1948]. Scherer, in turn, replied: "It was with the greatest pleasure—and relief—that I found your bully letter waiting when we got back from the memorable trip to Palomar" [LC, Tuve Papers, Box 138, June 24, 1948].

To Ernest O. Lawrence, his long-time friend (and, since 1944, a CIW Trustee), Tuve also described the situation:

> I thought mighty hard about Seattle, but finally wired Ray Allen that I could not accept. I like Van Bush and Paul Scherer exceedingly much, and the Institution is an agency of unique character. . . . Anyway, a shift like that represented a change from a highly personal life to a "public" life, in that I would owe so much of my waking hours and my energy to others [University of California at Berkeley, Bancroft Library, E. O. Lawrence Papers, July 2, 1948].

By the time that he got around to writing Lawrence, Tuve was on board the Queen Elizabeth and headed to Europe. One highlight of the trip was a visit to Oslo, Norway, where he addressed a meeting of the International Union of Geodesy and Geophysics on the topic of "Geophysics—Vocation or Avocation?" In his talk he mentioned his Norwegian ancestry, made some general remarks about what he believed should characterize research, and concluded with two examples of the DTM's geophysical projects: paleomagnetism and cosmic-ray studies. The seismic work he described only in passing:

> We measure the thickness and depths of the granitic and basaltic layers down to 50 kilometers or more, and have found very large amplitudes at the critical angles for reflection at various distances. This work has been great fun, but the data are complicated and I will not attempt to present those observations tonight [LC, Tuve Papers, Box 362, Aug. 20, 1948].

Tuve's European travels also included visits to laboratories in several countries, and what he saw tended to confirm his belief that the DTM offered him as good a place as any for the pursuit of research. Writing toward the end of July to Berkner (whom he had recommended to serve as acting director in his absence), Tuve explained that he "had wanted to make this trip with a primary concern [for,] and intensive contacts regarding basic work in geophysics . . . " [CIW, file, "DTM—Director 1941-1949. 1 of 3," July 26, 1948]. But a month later he reported:

> The meetings here have been interesting in spots, but a bit tiresome. The quality of the work and ideas in many places is too low-grade to command respect, and the uncritical acceptance and even lively support of trivial activities is not a great stimulus. . . . The DTM contributions seem to be of a rather high standard, altho[ugh] I am biased, of course [DTM, GF 1935-1948, file "108. Tuve, M. A.," Tuve to Berkner, Aug. 25,1948].

To summarize: Although Tuve had not taken up work in explosion seismology until after the end of World War II, by 1948 he had developed a deep personal stake in the project. His decision against accepting the Seattle offer not only revealed the extent to which research, in general, lay at the heart of his identity as a scientist; it also suggested that the seismic work, in particular, carried much of the weight of his commitment to research. Moreover, his overseas trip revealed how close to the forefront of worldwide geophysical research the project now stood. Thus by the late 1940s it had emerged as just the kind of work that Tuve most wanted to pursue.

THE PROJECT IN FULL SWING

To appreciate more fully the conceptual framework within which the Carnegie seismic project took shape, let us consider an article by Gutenberg that Tuve put at the head of an early list of "Seismic References" [LC, Tuve Papers, Box 113, n.d. (probably 1946)]. The article included an expression of the central assumption that had long guided the work of the seismologists. "To obtain quantitative results," Gutenberg noted:

> we suppose that the earth's crust consists of a number of parallel concentric layers . . . in each of which the velocity v is constant and higher than in the next closer to the surface [Gutenberg, 1943a, p. 474].

During an earthquake (or an explosion), the original disturbance would propagate through the different layers at different velocities. Thus each layer would give rise to a separate seismic wave train. It was this set of wave trains (or "phases") that seismologists hoped to observe in their seismograms.

Another common assumption was that mountains would deform the crustal layers. In the mid-19th century the British astronomer G. B. Airy had hypothesized that the material composing the Earth's crust was less dense than the underlying material. Because the crust floated on this underlying material, projections upward were possible only if compensated by projections downward. Mountains had to have "roots" [Greene, 1982, pp. 240-242]. In his 1943 article, Gutenberg pointed out that the existing data supported Airy's hypothesis, though he added that further studies were needed for full verification [Gutenberg, 1943a, pp. 493-495].

The Carnegie project was initiated within a conceptual framework like the one described by Gutenberg. For example, Adams noted in his 1948 memorandum:

> The present tendency in applying seismic methods to determining the composition and structure of the Earth, particularly the surface layers, is to postulate the existence of certain homogeneous layers, and by simplified geometrical or analytical procedures to determine the velocities of waves moving along paths consisting of series of straight lines. [CIW, file "GL—Adams, L. H., Seismology," "Memorandum Concerning the Use of Seismic Methods in the Exploration of the Earth's Interior," Apr. 9, 1948].

Furthermore, in the 1948 *CIW Year Book* Tuve included Figure 1 [p. 64]: The various "P" waves are pressure (longitudinal) waves (and in the figure their velocities are listed, in kilometers/second). Seismic activity also gives rise to shear (transverse) waves, which are referred to as "S" waves. The boundary between the "granitic" layer and the "basaltic" layers is known as the Conrad discontinuity [Bott, 1971, p. 41]. (Derwood, Maryland, was the location of a field station that the DTM had completed in 1947 [Tuve, 1947, Plate I (facing p. 43) and p. 78].)

Tuve also noted of the Carnegie program: "The first objective [after developing suitable instrumentation] was to determine the layering of the crustal structure underlying the region around Washington, D.C., and the Appalachian Highlands" [Tuve, 1948, p. 60]. But more than just assuming a layered structure and then seeking to measure the depth of the various layers, and more even than measuring how the depth of those layers changed in the vicinity of mountains, the Carnegie researchers hoped that the seismic project would shed light on the question of how mountains originate.

Certainly the question was one with which Adams and Tuve were familiar. In a talk that he gave on April 30, 1947 (as retiring president of the AGU), Adams placed the "[o]rigin of mountain chains" first in his list of "Some Unsolved Problems of Geophysics" [Adams, 1947, p. 675]. Similarly, Tuve noted in his report for the 1947 *CIW Year Book*: "we are still uncertain regarding the processes giving rise to mountain ranges . . . " [p.64]. He also included on a later list of the "Main Questions or Problems" for the Carnegie studies of the Earth's crust, the "[o]rigins and evolution of continental structures," which involved (among other things) the "forces and seat of mountain building" [LC, Tuve Papers, Box 146, Oct. 8, 1951].

Along with the conceptual framework just described, the Carnegie seismic project was initiated within an ideological framework that strongly reflected Tuve's personal preferences. As suggested by his interest in the origin of mountains, one feature of the ideological framework was a style of physics that might best be called "natural philosophy." For example, Tuve told the DTM staff in December 1946:

> We are at the beginning of a new epoch in the Department and all of us are much stimulated by the opportunity to devote all of our energies again toward philosophical goals and scientific studies for their basic values, without regard to practical applications [LC, Tuve Papers, Box 120, "Speech for Friday, December 13," Dec. 13, 1946].

Similarly, he observed in 1948:

> A deliberate aim to work as physicists, in reasonably

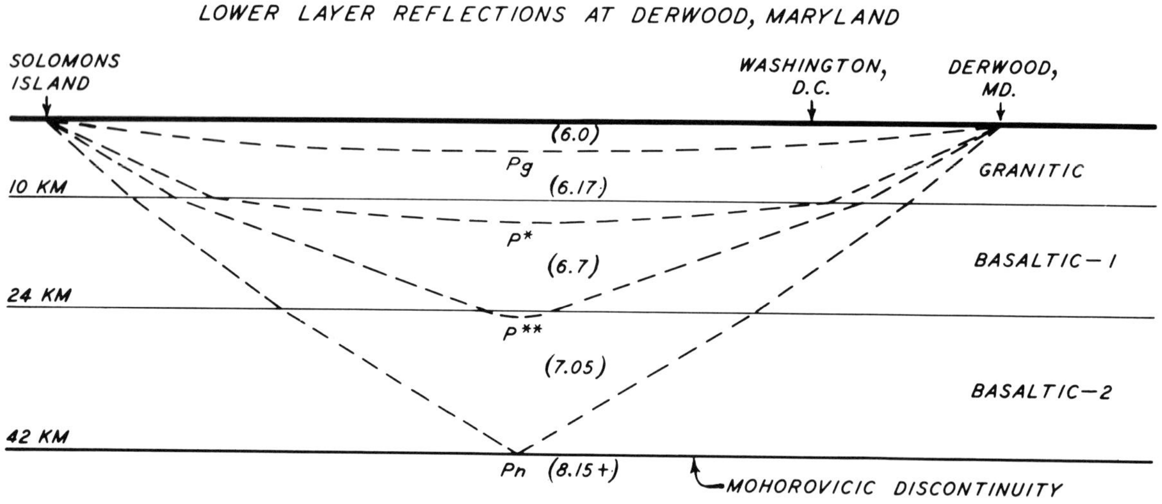

Fig. 1. Tuve's rendering of crustal structure near Washington, D.C.

fresh areas, on significant problems, however difficult, and on problems directed toward philosophical goals without reference to possible applications, is the general policy adopted for guiding the selection and emphasis of the work carried forward [Tuve, 1948, p. 53].

A related feature of the ideological framework arose from the relative autonomy of the CIW. "The Institution," Tuve explained in the 1947 *CIW Year Book*:

> has no external allegiance or obligations, no students, stockholders, voters, or other special groups to serve. This gives us an obligation to accept greater risks than others, and we should accordingly concentrate our efforts primarily on thrusting forward the very frontlines of advancing knowledge [Tuve, 1947, p. 54].

Along similar lines, he later recalled:

> We'd hold a staff meeting once a month, but mostly we'd talk around the lab . . . and I continually held before them: "What are the significant things that a completely free investigator might do . . . where is the proper niche for a nephew of Andrew Carnegie?" [Tuve Interview, 1982].

A third important feature of the ideological framework involved Tuve's estimate of the maximum size for an effective research group. "More and more I am satisfied," he wrote to J. E. Henderson on April 26, 1946, " . . . that real research is almost never done in a group larger than seven" [LC, Tuve Papers, Box 120]. The opinion was one that he offered on other occasions, as well [Tuve, 1947, p. 55; LC, Tuve Papers, Box 362, "Geophysics—Vocation or Avocation?," Aug. 20, 1948].

Finally, the ideological framework emphasized cooperative efforts with other research groups. Unlike the traditional "sealing wax and string" approach (which involved individual researchers and relatively limited means) Tuve's preferred approach involved complex equipment and organizational structures. But unlike the emerging "Big Science" approach (so termed because the massive post-war influx of government funds was permitting the formation of organizations on a "Big Business" scale) Tuve shunned research groups so large that they required a "corporate" structure. Instead, he preferred an approach that involved the coordinated efforts of several small groups, though each group might well be affiliated with a much larger organization (the CIW, the Navy, a university, etc.).

Despite the many explosions at Solomons Island, the first real opportunity to invoke the conceptual and the ideological frameworks came with a series of large explosions set off by the Tennessee Valley Authority (TVA). Dam building was still a primary activity for the TVA, and the one under construction on the South Holston River near Bristol, Tennessee, represented a notable technological achievement. Thus the author of a recent history of the TVA has noted:

> The closest the TVA ever came to totally assimilating its technological constructs into the landscape was with four dams of rolled earth and rockfill [including the South Holston Dam], built during and just after World War II [Crease, 1990, p. 149].

In connection with the construction of the South Holston Dam, the Carnegie researchers learned during the summer of 1948 (while Tuve was traveling in Europe) that 800,000 pounds of explosives were to be set off on July 14. Simultaneous with that blast, however, were quarry blasts elsewhere that disrupted the Carnegie measurements. Fortunately, the TVA engineers expected to set off additional explosions at the same site. Later in July the Carnegie researchers began planning for the next one. After a meeting at the GL (attended by Adams, Goranson, Greig, W. J. Rooney, and J. L. England), Greig began efforts to locate suitable locations for the Carnegie equipment in August and September [DTM, Box 1989-43, file "Greig Reports," Sept. 24, 1948].

Although scheduled for the fall, the second blast (involving nearly 1,500,000 pounds of explosives) was delayed until February 5, 1949. A third blast (involving 500,000 pounds) came on April 16. In each case, the Carnegie researchers worked closely with the TVA engineers. As a result, they were able to determine the exact time of the explosions, a crucial aspect of their work. They also enlisted the support of other researchers, thereby increasing the number of observing stations. Among the participants were colleagues at Columbia University and the New Mexico School of Mines, as well as personnel from the Office of Naval Research and the U. S. Coast and Geodetic Survey [GL, Box 6, file "Department of Terrestrial Magnetism—GL, et al. Experimental Seismic Program. 1949," Rooney, Tatel, Goranson,and Grieg, "Distant Seismic Observations from Large TVA Blasts," 1949; LC, Tuve Papers, Box 332, Tuve, Greig, Rooney, and Tatel, "The Seismic Project of the Carnegie Institution of Washington," June 23, 1949; Tuve, 1949a, pp. 62-63; Leonard, 1950; Roberts and Ulrich, 1951, p. 214; Callahan, 1980, p. 187].

Thus on May 4, 1949, Tuve reported to Bush that the Carnegie researchers were finally achieving a measure of success:

> A question of particular simplicity and interest was whether mountains such as the Appalachians show "roots" of lighter rock projecting down into the heavier rocks below. . . .
> We have made numerous observations near, along, and across the Appalachian chain, and we find that it clearly has a "root" which extends down below the normal depth of the Mohorovicic discontinuity [LC, Tuve Papers, Box 136].

Another indication that the Carnegie seismic project had indeed entered full swing was field work undertaken to settle a discrepancy between results based on explosions and results based on earthquakes. Because the earthquakes of Southern California had been studied so extensively, Gutenberg proposed that the Carnegie researchers bring their portable equipment west to observe a large blast at a quarry of the Minnesota Mining and Metallurgical Company near Corona. Tuve and his co-workers accepted the invitation and made the trip [LC, Tuve Papers, Box 136, Tuve to Bush, May 4, 1949; DTM, Box 1989-44, file "Calif—Seismic," Gutenberg to Tatel, June 13, 1949; LC, Tuve Papers, Box 330, Gutenberg to Tuve, July 5, 1949; Tuve, 1950, pp.63-64].

The California researchers manned their usual stations, while the Carnegie group set up a series of stations from Pasadena east to Yuma, Arizona (which was the site taken by Tuve and his son) [LC, Tuve Papers, Box 330, "Quarry Shot Observers," July 25, 1949]. Making these arrangements was challenging. "We have been remarkably busy," Tuve wrote Scherer just a few days before the blast (which came on August 6, 1949). "Among other things, the scouting for specific location of 31 sites is lots of miles in the mountains," to which he added a single-word commentary: "Fun" [CIW, file "DTM—Director 1941-1949. 1 of 3," Aug. 3, 1949].

Writing afterward to Franklin P. Ulrich of the USCGS, Tuve described another challenge:

> Although we had sufficient equipment, we had very few trained observers; in fact, there were only Messrs. [W.E.] Scott, Rooney, Tatel, and myself, and so part of our problem was to obtain interested personnel. Our most productive contact along this line was Mr. David J. Leeds attached to your Los Angeles office. He quickly learned how to use the equipment and took out a set on all the observation days for which we had explosions [DTM, Box 1989-44, file "Seis. Coop," Sept. 26, 1949].

During the next couple of weeks, the Carnegie researchers remained in California to observe a series of smaller explosions set off by the Navy in the ocean. "[P]revious experience has shown," Tuve explained to the Chief of Naval Research:

> that one explosion with a single line of observations is not sufficient to obtain an unambiguous earth-structure determination. To this end, the Office of Naval Research was asked and provided for the detonation of 13 explosions of 1200 to 2400 pounds of high explosive in Bechers Bay off Santa Rosa Island near Santa Barbara, California. The explosives were placed and set off through the facilities of the Scripps Institution of Oceanography and with the cooperation of the Marine Physical Laboratory of the Naval Electronics Laboratory at San Diego [LC, Tuve Papers, Box 332, Nov. 8, 1949].

Two ships were involved: one provided by the Navy for dropping the depth charges and another provided by Scripps for making the observations [DTM, Box 1989-44, file "Calif—Seismic," J. E. Kahelin to Chief, Bureau of Ordnance, July 22, 1949; LC, Tuve Papers, Box 330, T. C. Thomas to Commander, First Task Fleet, Aug. 9, 1949].

The California trip was not without misfortune: while driving the equipment truck back to Washington, Rooney suffered a heart attack and died [DTM, GF 1935-1948, file "91. Rooney, W. J."; Tuve, 1949b]. Afterward, Tatel informed one of his Navy contacts: "[Rooney] is the man who timed the instant of the explosions and did most of our reduction of data, and the work suffers severely because of his demise" [DTM, Box 1989-44, file "Seis. Corres," Tatel to G. R. Fraser, Oct. 3, 1949]. Nevertheless, Tatel and Tuve quickly decided that the existing earthquake data needed correcting. "[A p]reliminary look," Tuve informed Scherer, "indicates [that the] records are like [those for the] East coast and not like Gutenberg['s] earthquake velocities, etc." [CIW, file "DTM—Director 1941-1949. 1 of 3," Aug. 8, 1949].

Toward the end of April 1950, both groups described their results from the Corona blast at the annual spring meeting of the AGU in Washington, D.C. [Benioff et al, 1950; Tuve and Tatel, 1950a]. In addition, Gutenberg submitted for publication a pair of articles in which he proposed a revision of the earthquake data [Gutenberg, 1951a, b]. But he made it clear that in so doing he was not acting in direct response to the Carnegie fieldwork. For example, he and his colleagues noted in their AGU abstract: "Reinterpretations of travel time curves for Southern California made independently by Gutenberg and Richter are consistent with [the results from the Corona blast]" [Benioff et al., 1950]. Moreover, in his article on

"Revised Travel Times in Southern California" [1951b], Gutenberg presented the change as arising from a new view of the Earth's crustal structure, a view which he had recently published in *Science* [1950].

In his 1958 textbook, *Elementary Seismology*, Charles F. Richter described how his colleague's view had changed. During the war years, Gutenberg had published a series of articles that reflected the traditional approach [1943a, 1944a, 1944b]. (Tuve's early list of "Seismic References" included all three [LC, Tuve Papers, Box 113, n.d. (probably 1946)].) After the war, explosion seismology (including the 1949 Corona blast) helped overturn what Richter called "the 1944 synthesis" [Richter, 1958, p. 285]. But he also pointed out that the explosion data had been accumulating since the 1930s, well before the Carnegie program had begun.

Even more important than the data from the field was a new concept. Specifically, Richter insisted that "Gutenberg cut the knot" [Richter, 1958, p. 286] of the discrepancy between the earthquake and explosion results by introducing the idea of a low-velocity channel. If there existed an intermediate layer through which seismic waves moved more slowly than through the layer immediately above it, then waves moving through the intermediate layer would travel within that layer for a considerable distance before heading toward the surface. Thus the waves that seismologists had taken as coming directly from earthquakes could now be interpreted as having been channeled along the low-velocity layer.

Emphasis on low-velocity channels was a distinctive feature of Gutenberg's geophysics. In a 1959 monograph Gutenberg himself reviewed the subject of "channel waves" [1959, pp. 38-41]. By his death in 1960, his hypothesis of a low-velocity layer in the upper mantle (which he had proposed prior to World War II) had won widespread acceptance, but even Richter had to admit that the status of his hypothesis of a similar channel in the Earth's crust was far less secure [1960,p. 96].

At any rate, Tuve never reconciled himself to the view that the Carnegie results had not directly necessitated their interpretation of the earthquake data. "[Gutenberg] didn't call attention that there had ever been a mistake," he explained many years later. "I don't blame him. We were upstarts. But some of the arrivals that they had identified were not what he had thought they were" [Tuve Interview, 1982]. Similarly, the copy of Richter's book on the shelves of the DTM-GL library bristles with Tuve's marginal comments. "Gutenberg's depths etc," he noted at one point, "were all changed after [the] Tatel & Tuve expedition [of] 1949 showed wrong identification of phases" [DTM-GL library copy (Accession No. 42399) of Richter, 1958, p. 285].

In retrospect, it appears that the difficulty was less a matter of Gutenberg's refusing to recognize the importance of the Carnegie field work than of Tuve's pursuing another line of thinking entirely. Increasingly he and his co-workers were convinced that the crust was not marked by well-defined layers and that much of what they were observing were surface reverberations instead of true returns. To all of this we will return in a moment. But first we need to consider Tuve's response to the emerging Cold War.

By 1950 Tuve found himself working within the context of the Cold War. The full entry of the United States into the Cold War had been marked by the Truman Doctrine (1947), the Berlin airlift (which lasted from the summer of 1948 to the spring of 1949), and the formation of NATO (1949). But in 1950 tensions between the U.S. and the Soviet Union took a turn for the worse. At home (in February) Senator Joseph R. McCarthy initiated a Red Scare, while abroad (in June) the North Koreans invaded South Korea.

One especially worrisome issue for Tuve in all of this was the further development of nuclear weapons. He recognized that any increase in the pace of such work was likely to increase the demands placed on civilian researchers, including those at the DTM. In mid-1946, for example, George Vaux (then at the Navy's Office of Research and Inventions, the agency that Congress was on the verge of transforming into the ONR) requested the services of DTM researcher G. W. Wait in connection with "Operation Crossroads" [DTM, GF 1935-1948, file "75. Office of Naval Research," Vaux to Tuve, July 3, 1946]. This was a program of atomic-bomb tests on Bikini Atoll in the Marshall Islands, and it marked the first detonation of nuclear weapons since their use during the war. One blast came on July 1, 1946, and a second, on July 25. Although a third was planned, it was never carried out [Surcliff, 1947]. "Our Director," Wait replied to Vaux, "has agreed that it would be proper for me to take a limited amount of time to examine and pass upon the records of ionization as requested in your letter" [DTM, GF 1935-1948, file "75. Office of Naval Research," July 26, 1946.]

As the tone of Wait's comments suggests, Tuve basically accepted the need for continued nuclear testing. But he tended to oppose anything that threatened the intellectual independence of private research organizations. He also worried about the effects that the American project might have on the international scientific community. Later in 1946, for example, he declined an invitation to join the board of the Oak Ridge Institute of Nuclear Studies (an association of southern universities that sought to use the facilities at Oak Ridge for educational purposes), explaining to its chairman, Frank P. Graham:

> Our laboratory here in the Carnegie Institution is one of the few laboratories in nuclear physics which can maintain a detached position with respect to the development of atomic energy. . . . Our main concern is to make sure that our laboratory will not be looked on by our own State Department as a place where it might not be too wise to permit foreign guests to work.
>
> . . . It is going to be difficult during the coming years to rebuild scientific relations on an international basis, and if we can contribute toward this end, even by denying ourselves some opportunity for interesting work, we want to do so [CIW, file" DTM—General 1946-1957," Nov. 14, 1946].

During the years that followed, Tuve continued to resist the unrestricted development of nuclear weapons. Thus in a letter to Ernest Lawrence in the spring of 1948 he indicated that while not opposed to using nuclear weapons in cases of military necessity, he objected to the deliberate contamination of explosion sites:

> No political trouble or political system lasts more than a hundred years or so, and if democratic governments ruin major city sites etc for thousands of years it will be a permanent damnation of free institutions.
>
> . . . The use of long-time contaminants would be a crime against every future nation and group, and the character of the freedoms which permitted it to happen would never be forgotten [LC, Tuve Papers, Box 129, Mar. 29, 1948].

By 1948, however, the Cold War was clearly emerging, and with it came the further development of nuclear weapons. As a result, Goranson was invited to resume full-time work at Los Alamos [CIW, file "GL—Director 1936-1949. 1 of 3," N. E. Bradbury to Bush, Nov. 22, 1948]. Initially, he was granted leave of absence for the period from February 15, 1949, to September 15, 1950 [CIW, file "Geophysical Staff. Goranson, Roy W.," Bush to Goranson, Dec. 31, 1948]. But Bradbury continued to request Goranson's services. For example, he wrote Bush on January 12, 1950:

> His contributions to both the experimental and the theoretical aspects of the work of this Laboratory have and continue to be outstanding in a field in which we have all too few real specialists in this country. Because of his presence here, the Laboratory is making not only more rapid progress, but is laying a groundwork of understanding in the whole field of highpressure and shock phenomena [CIW, file "Geophysical Staff. Goranson, Roy W."].

The CIW responded by extending Goranson's leave of absence until the end of 1951, after which his formal CIW ties ended.

Meanwhile Tuve continued to resist the arms race and the fear that fueled it. After President Truman announced on January 31, 1950 that the U.S. would undertake the all-out development of hydrogen bombs, Tuve was one of twelve scientists who asked him to promise that the country would not be the first to use the new type of explosive [New York Times, Feb. 5, 1950, p. 1; York, 1976, pp. 65-74]. Writing on March 15 to Hans Bethe (who had also signed the petition) Tuve noted: "I am grateful to you for your efforts to oppose the hydrogen bomb. I am against it from every standpoint" [CIW, file "DTM—Director 1950-1951"]. But he also expressed concern over the wording of an article that Bethe was about to publish in *Scientific American* [1950]:

> Such concepts as "the destruction of Chicago with all of its suburbs and most of its inhabitants wiped out in a single flash" . . . are needlessly frightening. I do not believe you can frighten people into being good [CIW, file "DTM—Director 1950-1951," Tuve to Bethe, Mar. 15, 1950].

A second issue that the Cold War raised for Tuve was the more general issue of personnel. Developing nuclear weapons was not the only project to require the assistance of scientific researchers. Thus as the Cold War heated up, Tuve had to decide whether or not to mobilize the DTM. His tendency was to resist full mobilization. He encouraged his DTM colleagues to participate in defense-related work, but only to a point. On February 19, 1947, for example, he wrote Bush regarding the Joint Research and Development Board, which Bush chaired and which sought to coordinate post-war military research and development [York and Greb, 1977, pp. 14-16]:

> We have already four or five members of the Department's staff involved in this work, and if the activity for each individual grows beyond a nominal amount of committee work it can represent a considerable inroad on our research team.
>
> . . . I therefore recommend that we allow our staff members to accept consulting fees for government and quasi-government activities only for an average not exceeding two days per month [CIW, file "DTM—Director. 1941-1949. 2 of 3"].

Bush preferred proceeding on a case-by-case basis [CIW, file "DTM—Director. 1941-1949. 2 of 3," Feb. 21, 1947]. Toward 1950, the pressure to participate in defense-related work increased. Thus in early 1948, Ellis Johnson left the DTM for full-time work organizing the Army's Operations Research Office (which he directed until 1961) [Page et al., 1974]. Nevertheless, he retained his formal ties to the CIW until the end of 1956 [Tuve, 1957, p. 146]. As M. B. Smith explained to Tuve: "[Johnson] says it has been a distinct advantage to have CIW as a backstop as he could afford to be quite positive in controversial situations" [LC, Tuve Papers, Box 330, June 19, 1951].

In 1950 Tuve himself turned down the position of Deputy Director and Research Director for the Weapons Systems Evaluation Group (attached to the Joint Chiefs of Staff) [CIW, file "DTM—Director 1952-1953," Tuve to J. E. Hull, Mar. 16, 1950]. He also turned down the position of Chairman of the Ordnance Committee of the Research and Development Board (which in 1947 had taken the place of the Joint Research and Development Board [Rearden, 1984, pp. 96-103]) [CIW, file "DTM—Director 1950-1951," Tuve to W. Webster, Mar. 24, 1950].

Already Tuve, Berkner, and DTM staff member Richard B. Roberts were participating in Project Hartwell (a study of undersea warfare, run by MIT during the summer of 1950 [Schweber, 1988, pp. 28-32]), and in July 1950 he requested that other DTM personnel be cleared for work on secret projects [CIW, file "DTM—Director 1950-1951," Tuve to ONR, July 14, 1950]. But in 1951 he declined to participate directly in Project Vista (a study of the defense of Europe, run by Caltech [Eliot, 1986]). "I am personally not ready . . . to accept responsibility for electronics in your Vista Project," he wrote to William A. Fowler [CIW, file "DTM—Director 1950-1951," May 7, 1951]. But he indicated his willingness to follow the project informally, and he pointed out that he and his DTM colleagues were "more or less continuously called on for opinions as to projects or people, or for help in recruitment, by our convenient location too close to the Pentagon" [Ibid.].

As his remarks to Fowler suggest, the continued demands of the Cold War led Tuve to articulate his reasons for limiting DTM participation in military projects. His personal stance began emerging at the time of his 1948 trip to Europe (while the Berlin airlift was underway). "My chief reaction right now," he wrote Scherer on September 1:

> is that I want to get back to work, as this has been too long a siege, and the international tensions have colored the picture for everyone so much this summer that the scientific problems have not had undivided attention in any laboratory—or by me either [CIW, file "DTM—Director 1941-1949. 1 of 3"].

Tuve's personal response to the increasing tension between the U.S. and the Soviet Union was to renew his commitment to basic research, a stance that he made quite clear when he refused the position with the Weapons System Evaluation Group. Thus he explained to its director, Major General John E. Hull:

> With the intense dedication of physicists to military problems and to atomic energy, and with the large scale subsidy of physics by military funds, I have seen a challenge and a role for me in dedicating my efforts toward peacetime physics [CIW, file "Director 1952-1953," Mar. 16, 1950].

A fuller statement of Tuve's expectations for the DTM came in a letter to Bush dated December 22, 1950. Tuve reported that he had been discussing the situation with his co-workers and that they had agreed unanimously on a number of points. First, they wanted to continue participating in "several specific and limited projects, chiefly items spotlighted by Hartwell and Troy" [LC, Tuve Papers, Box 144]. Second, they wanted to continue their reliance on CIW (i.e., private) funds. Third, they agreed to help each other in their various defense-related activities. Finally, came Tuve's commitment to basic research. "We have also agreed," he told Bush:

> to help each other turn his mind and efforts back to basic research at intervals. In the early stages we may be almost fully occupied with defense, but I would guess that the men think vaguely of spending perhaps one-third of their time on basic research if the emergency settles down, after a year or so, to a long period of steady stress [Ibid.].

Less formal (but equally heartfelt) were remarks that Tuve made at a physics department seminar at the University of Washington in May 1950. He began by expressing his preferred approach to physics. Specifically, he believed that physicists should constantly ask themselves: "What is the nature of Physical Law or the *Laws of Nature?*" [LC, Tuve Papers, Box 139, "Seminar for Un of Wn Physics Dept. 4 PM Monday 28, 1950," n.d.]. In his view, physics was not the same thing as either engineering or what he called the "Politics of Power." Instead, "a real physicist is at heart a philosopher on a quest," whose goal was the solving of "great puzzles" posed by the natural world. "[P]rogress on these [puzzles]," Tuve insisted, "is far more permanent than defense or budgets." Among the contemporary areas of research that he surveyed for his audience were astrophysics, biophysics, and nuclear physics. But the focus of his talk was expressed by its title:

"Why Geophysics?"

In short, Tuve decided to wage war by waging peace. Far from trying to escape the pressures of the Cold War, he attempted to face them in what he saw as the most positive manner possible. Recall that he had opened his "Code of the West" with "[t]he right to know" and had closed it with "the obligation to defend" the freedoms he had listed. True to his code, Tuve now took the pursuit of basic research as his frontlines, and he redoubled his efforts to study the Earth's crust using seismic waves generated by explosions.

As the project matured, Tuve and his co-workers became increasingly confident in the technical performance of their seismometers. Thus Tatel explained:

> In the course of the last two years, instruments have been developed which are usable in the field by moderately skilled operators. It is now the usual practice for as many as seven Carnegie groups to go out in the field and all return with good records. The future will probably bring some minor changes in the equipment which will make for easier operation, but the general outline has been determined [LC, Tuve Papers, Box 136, "Progress and Status of the Carnegie Seismic Project," May 3, 1949].

At about this time the group also modified their equipment. "In the newer instruments," read a later description, "we discarded the heavy steel shields as we found from experience that . . . light brass shields provided all the electrostatic shielding necessary" [DTM, Box 1989-46, file "III. Seismic Instruments,"unattributed TS essay entitled "Seismometer," n.d. (probably mid-1950s)].

Even after clearing up the discrepancy between their data and Gutenberg's, however, the Carnegie researchers continued to have problems interpreting their seismograms. Their response to these problems was complex. To begin with, as the decade of the 1940s gave way to the decade of the 1950s, they abandoned the assumption that the Earth's crust was marked by well-defined layers. Instead, they concluded that if intermediate layers did exist then the boundaries between them were probably irregular in much the same way that the Earth's surface is irregular.

Thus Tuve noted in the draft of a letter to Adams dated May 8, 1950:

> The picture of the earth's crust as a layer cake may be modified to that of a "marble cake" or a"marbelized layer cake". . . . Nature does not always offer the charm of simplicity. A confused structure in the crust may be the actual reality [DTM, file "Archives. Seismic"].

Writing in the 1952 *CIW Year Book*, he conveyed a similar idea in a less figurative fashion:

> The textbook picture of a downward succession of several horizontal granitic and basaltic layers was not confirmed, even in special explosion experiments designed to make such structures prominent features of the records. Evidence was found, instead, for real inhomogeneities in the crustal rocks down to 30 km or more, as though they were irregular in vertical and horizontal dimensions, as the basement rocks appear when exposed at the surface [pp. 67-68].

Originally, the aim of the Carnegie researchers had been to identify in their seismograms the returns from the different layers. But as they lost confidence in the reality of multiple layers, they shifted their attention to identifying strong, unambiguous returns from at least one layer. The idea was to focus on those returns and to see how the depth of the layer varied across the country. After much experimentation, however, they found that the only returns in which they had full confidence were those from the Mohorovicic discontinuity.

More specifically, the Carnegie researchers focused their attention on "critical reflections." These are reflections from the Mohorovicic discontinuity recorded at stations just far enough from the explosion site so that refraction along the Mohorovicic discontinuity sets in. In the course of their experiments, the Carnegie researchers had attempted to observe nearly vertical reflections but without success. Not until the 1970s did reflection studies become possible for such depths [Howell, 1990, p. 79].

Another problem on which Tuve and his co-workers focused was the phenomenon of surface reverberations. They concluded that seismic waves from the interior became transformed at the surface into reverberations that lingered long enough to mask the arrival of subsequent waves from the interior. Moreover, the precise character of these reverberations varied from location to location, giving rise to different seismic results at different recording stations, even if the stations were not far apart.

Another way in which the Carnegie researchers responded to the problem of how to interpret their seismograms was to undertake laboratory studies of vibrating systems. The work with models was largely Tatel's project [Tatel, 1954; Tatel and Tuve, 1954]. Thus Tuve later wrote to Bush:

> Dr. Tatel, recognizing the possibility of using small blocks of homogenous material, such as Armco iron blocks, perhaps 20 or 30 cm on a side, by making use of impulses a few tenths of a microsecond in duration, has seized on the opportunity presented by the high

speed cathode ray oscillograph which is in daily use in our nuclear physics section. Working chiefly at night and over the weekends, when the nuclear physics equipment is not in use, Dr. Tatel has carried out a beautiful set of fundamental measurements on the conversion of compressional waves to surface waves, and the modification of this process by topographic features at the surface [LC, Tuve Papers, Box 152, May 4, 1954].

But the most important response was to borrow a long-standing practice of petroleum geophysicists, namely, the use of a linear array of seismometers [Tuve and Tatel, 1950b; Tuve et al, 1951]. The array (or "string") consisted of a primary line of seismometers (or "jugs") oriented radially from the blast site. In addition, there might be a secondary line placed in either a "T" or "L" pattern at right angles to the radial string. The seismometers were all wired to a multi-channel recorder. With several seismometers registering simultaneously, true returns from the crustal layers could be distinguished from effects that were purely local. Of course, the interest of the petroleum geophysicists (for whom 24-channel systems had by then become standard) lay with sedimentary strata near the Earth's surface. To probe more deeply, the Carnegie researchers required a much larger scale, not only for the array but also for the explosions. Yet the basic idea was the same.

As early as November 1949 Tuve wrote in a private document:

> A large systematic structure [in the Earth's crust] should reflect (or refract) waves systematically so that a large area of surface near the receiver (i.e.an area covered by many receivers) should move coherently and systematically [LC, Tuve Papers, Box 332, notes dated Nov. 20, 1949].

One step toward adopting the new procedure came during January, 1950, when Tuve and Tatel traveled to Louisiana, Texas, and Oklahoma, to confer with geophysicists in the petroleum industry [CIW, file "DTM—Director 1941-1949. 1 of 3," Tuve to F. F.Campbell, Dec. 9, 1949; DTM, Box 1989-44, file "Texas—Notes Jan. '50," "Itinerary for Dr. Merle A. Tuve and Dr. Howard Tatel," Dec. 28, 1949].

Even before the trip, however, the Carnegie researchers had received assistance from the Shell Oil Company of Houston. Through Harold Gershinowitz, Shell's director of exploration and production research, they had borrowed 60 seismometers. Not until the end of 1950 (after they had made for themselves a set of similar seismometers) did they return the borrowed instruments [DTM, Box 1989-44, file "Seis. Coop," Tatel to Gershinowitz, June 8, 1950; Tatel to Gershinowitz, Dec. 19, 1950; Tatel to Gershinowitz, Feb. 28, 1951]. "After what amounts to years of persuasion on your part," Tatel wrote to M. King Hubbert, the company's associate director of exploration and production research:

> we have finally taken your advice, much augmented of course by our own difficulties, and tried a long string of recorders somewhat similar to general oil-field practices [DTM, Box 1989-44, file "Seis. Coop," Oct. 21, 1950].

Contact with petroleum geophysicists had not come out of the blue. In May 1947, for example, Ellis Johnson had visited several commercial geophysical laboratories in Tulsa. "First of all," he reported to Tuve after returning, "the development personnel of the geophysical companies have an extremely lively society [the Society of Exploration Geophysicists] and a very keen interest in the physics of the Earth" [DTM, GF 1935-1948, file "53. Johnson, Ellis A.," May 29, 1947]. Johnson also stressed their interest in new instrumentation, and he pointed out that "all of the top people foresee difficulties or failures in the use of present geophysical methods of prospecting." Although admitting that the companies tended to be secretive about the details of their methods, he also commented that they "were anxious to provide help in any expeditions which we might make out west."

Tuve's contacts included Earle W. Johnson, who had participated in the wartime work on the proximity fuse and who after the war had returned to the General Geophysical Company (which he had founded in 1935) [Moody, 1954]. At about the same time as Ellis Johnson's trip to Oklahoma, Tuve had written Earle Johnson in Houston:

> I am having a lot of fun back in the Carnegie Institution, and I have not the slightest envy for Hafstad and others who are still pushing so hard on defense activities. . . . The purpose of our seismic shots is to study the granite and basaltic layers down to the Mohorovicic discontinuity. The real purpose is of course simply to have a gang of curious physicists become interested in the broad problems of Earth structure. Too bad that you vigorous oil people don't go below the thin skin of the sedimentaries [LC, Tuve Papers, Box 331, May 20, 1947].

In May 1950, Tuve traveled to the West Coast, going first to California and then to Seattle. While in California, he and Tatel lectured at the Institute for Geophysics, which L. B. Slichter headed at the University of California at Los Angeles [CIW, file "DTM—Director 1950-1951," Tuve to Slichter, Mar. 20, 1950; GL, Box 6, file "Department of

Terrestrial Magnetism—GL, et al. Experimental Seismic Program. 1950," Tuve to P. G. Gane, May 26, 1950; Palmer, 1963].

They also approached officials of the United Geophysical Company of Pasadena, a major company in the field of seismic prospecting [Bates et al., 1982, pp. 326-328], and arranged to borrow "two nearly complete sets of your older and obsolescent equipment" for the simultaneous recording of seismic activity by 24 different seismometers [DTM, Box 1989-44, file "Seis. Corres.," Tuve to United Geophysical Company, May 26, 1950]. At the same time, they borrowed from the Consolidated Engineering Corporation "a set of galvanometers to fit the two 24-element cameras which the United Geophysical Company is lending us" [DTM, Box 1989-44, file "Seis. Coop.," Tuve to C. F. Robinson, May 26, 1950].

Tests using the borrowed equipment came during the summer. As Tatel later reported to R. A. Peterson at the United Geophysical Company:

> We spent the major part of June, July, and August setting up the equipment and installing it in one of our big trucks. . . .
>
> Readying 48 amplifiers for a field test was a new experience to us, particularly having them all working at the same place and at the same time. We did not manage to have a 100 per cent score, but throughout the two successful tests we managed to have 44 working on the average . . . [DTM, Box 1989-44, file "Seis. Coop.," Sept. 26, 1950].

As Tatel's reference to the trucks suggests, adopting the new approach confirmed the truck-based character of the Carnegie seismic program. Their field work took full advantage of the vast network of roads that by then crisscrossed the country.

Only after field tests had demonstrated the promise of the new method was Tuve ready to approach Bush regarding additional funds. All along, the seismic work had required funds beyond the usual allotments: $12,000 in 1946; $1,000 in 1947; $8,000 in 1948; and $10,000 in 1949 [LC, Tuve Papers, Box 331, "Statement of Special Appropriations and Expenditures for Seismic Work by Geophysical Laboratory and Department of Terrestrial Magnetism (in Addition to Regular Budget Funds)," Sept. 30, 1950]. At times, the need for special funds was a matter of concern for senior staff members. "Exp[erimen]t[a]l geophysics takes big funds," Ellis Johnson had commented at a meeting with Tuve in early 1948. "If we are to operate on CIW funds [we] can't expect to do big geophys[ics] problems" [LC, Tuve Papers, Box 130, Tuve, notes labeled "Chairmen," Jan. 5, 1948].

Although Tuve had viewed Johnson's concern as a "counsel of despair" [Ibid.], he recognized that continuing the project required being certain about the soundness of their methods. As a result, he delayed presenting his case to Bush until September 28, 1950:

> Complexities in our observations and uncertainties of interpretation . . . have driven us to try out a large-scale adaptation of the oil prospecting technique using a large array of seismic instruments, modified for observations at distances of 30 to 150 km from the explosion instead of the customary 1000 feet used in oil prospecting. . .
>
> Dr. Adams and I feel that the intensive examination permitted by the seismic array procedure will clarify much that we have observed by the spot-by-spot method [LC, Tuve Papers, Box 144].

Tuve indicated that $12,000 (half for the DTM and half for the GL) would allow them to proceed.

As anticipated, however, Bush required assurances that the new method would enable the Carnegie researchers to overcome their earlier difficulties. Thus Tuve wrote the CIW president on October 13, 1950:

> It is true that our intensive coverage of various areas with separate observation points demonstrated the "rough and jumbled" nature of the material under these areas. As a result of the complications, however, we have increased the complexity of our equipment and procedures and now have a technique which enables us to eliminate false indications of structures due to local effects [LC, Tuve Papers, Box 144].

Bush's response was to allot half the requested amount immediately and to promise the other half later [CIW, file "DTM—Seismic Program," Bush to Tuve, Oct. 24, 1950].

Although the Carnegie researchers were moving away from viewing the Earth's crust as composed of uniform layers, their use of the array technique did confirm the existence of the Mohorovicic discontinuity. As a result, they significantly altered their research program. Instead of exploring, in detail, the structure of the Earth's crust in one particular area, they undertook to measure the depth of the Mohorovicic discontinuity in different geological regions.

As early as May 1949, Tuve had requested funds for a trip to study the Canadian shield (the ancient core of the North American continent) [LC, Tuve Papers, Box 136, Tuve to Bush, May 4, 1949]. Adams supported the idea, commenting at a planning session in June that the area around Washington, D.C., was "the world's worst location to study" (due to the thick accumulation of unconsolidated sediments), so that there arose the goal of determining "[w]hat a simple place looks like, e.g., Canada" [LC, Tuve Papers, Box 332, Tuve, "Conference: Seismic Planning," June 29, 1949]. By then the CIW's Executive Committee had agreed to provide additional funding for the seismic project ($10,000 split evenly between the DTM and the Geophysical Lab), thereby enabling Greig to undertake a preliminary survey [CIW, file "GL—Director 1936-1949. 1 of 3," Scherer to Adams, May 20, 1949; Adams, 1950, p. 54]. As Greig himself explained afterward:

> . . . I was looking into the practicability of doing work in the basement complex in Ontario by using shots set off in the course of mining, quarrying, or construction work. . . . I had a set of instruments with me and set them up several times to pick up the disturbance from shots . . . simply to see what sort of background movement we might expect, and what signal strength we could count on at different distances from the shots [DTM, file "Archives, Seismic," Greig to D. R. Derry, Nov. 24, 1950].

From Adams's point of view, Grieg's trip confirmed the importance of the proposed expedition. "The advantages of an investigation of the Pre-Cambrian shield in Canada," he wrote Tuve on November 17, 1949, "now loom up more clearly than ever before . . . " [GL, Box 6, file "Department of Terrestrial Magnetism—GL, et al. Experimental Seismic Program. 1949"]. But the work required to develop the array technique had tempered Tuve's enthusiasm, leaving Adams to note in a private memorandum: "Tuve . . . does not think we are ready yet to tackle the Canadian shield project. He thinks that perhaps we should wait until a year from this summer" [GL, Box 6, file "Department of Terrestrial Magnetism—GL, et al. Experimental Seismic Program. 1950," Jan. 27, 1950].

Not until early 1951 were the Carnegie researchers in a position to plan an expedition to study the Canadian shield. In early March, Tuve wrote to J. Tuzo Wilson at the University of Toronto. Wilson was an authority on the Canadian shield who was also synthesizing what was then known about the Earth's crust [Wilson, 1954, 1985]. Tuve hoped to enlist his support for an expedition to the mining area near Sudbury, but he also explained the contingencies involved:

> Our plans may be disrupted by obligations for national defense, but we hope to fit in a couple of brief trips anyway. Without a radio truck we would despair of obtaining observations, because the high-speed film cannot run for more than half a minute or so. If we obtain a radio license and can make the trip we hope that you or one of your colleagues can join us in observing [LC, Tuve Papers, Box 331, Mar. 2, 1951].

Toward the end of the month an exasperated Tuve wrote that too much red tape stood in the way of arranging "for half a dozen squeaks out of our radio as appropriate warning of the time when mine or quarry charges will be set off" [CIW, file "DTM—Director 1950-1951," Tuve to Wilson, Mar. 26, 1951]. Because advance notice was such a critical aspect of the Carnegie work, he was forced to consider other approaches.

One alternative was to arrange for the Navy to set off a series of depth charges in Lake Superior. Tuve recognized that the blasts might prove harmful to fish. But consultations with the appropriate state officials led to plans for minimizing such injuries. Another hurdle was securing approval from the Canadian government (even though the explosions were to take place in American waters). In this case, the necessary approval was forthcoming. A third problem was obtaining a suitable vessel. Here the Carnegie plans were frustrated by the inability of the Navy to provide one until later in the year [DTM, GF 1948-1952, file "84. Deep Seismic," Chief of Naval Research to Chief of Naval Operations, May 7, 1951; CIW, file "DTM—Director 1950-1951," Tuve to C. O. Alexis, May 21, 1951].

At that point, Tuve and his co-workers decided upon a third approach. Thus Tatel informed Wilson:

> we are setting out on our expedition to explore the pre-Cambrian on June 9, and, as you can see from the enclosed schedule, we propose to use explosions set off in the course of normal mining operations in the Mesabi range [DTM, GF 1948-1952, file "84. Deep Seismic," May 25, 1951].

Following the work in Minnesota, the Carnegie researchers proceeded with a second expedition, to Seattle, where J. E. Henderson helped to arrange for the Navy to set off depth charges in Puget Sound [CIW, file "DTM—Director 1950-1951," Tuve to Geophysics Branch, ONR, May 24, 1951; LC, Tuve Papers, Box 330, Tuve and Tatel, "Tentative Seismic Schedule for Summer of 1951, "May 18, 1951; Tuve, 1951, pp. 70-73; Adams et al, 1952; Tuve, 1952, p. 68].

Both expeditions were successful, thereby contributing to a pair of important presentations, one by Tatel and the other by Tuve. By then Tatel had become a leading spokesman for the Carnegie group. Not only had he shouldered much of the responsibility for the recent field trips, but he had also received Tuve's recommendation for tenure at the DTM. Writing to Bush on June 11, 1951, Tuve asked that Tatel "be considered eligible for indefinite extension or renewal of his appointment," explaining:

> I believe it has been amply clear for a couple of years that Dr. Tatel is well suited by temperament, training, initiative, interests and personality to be a fruitful member of our Department [LC, Tuve Papers, Box 144].

To this, Scherer added: "Tatel did an excellent and difficult job this summer with two departments, the iron mines, the Navy and a large group of staff and automotive equipment" [CIW, file "DTM—Staff. Tatel, Howard E.," Scherer to Bush, n.d.]. Tatel's appointment as "a member of the senior [DTM] staff" took effect later in the year [CIW, file "DTM—Staff. Tatel, Howard E.," Bush to Tuve, Oct. 29, 1951].

At the Brussels meeting of the IUGG in early September, Tatel reported: "Contrary to our earlier views we find no evidence now for any layering or systemization within the crust" [LC, Tuve Papers, Box 330, "Investigation of the Earth's Crust by Explosions," 1951]. In a letter to Tuve afterward, Tatel described a conversation at the meeting between him and Gutenberg regarding the issue of intermediate layers:

> Gutenberg gave out some remarks . . . in which he talked about "layers." Later (in private) I reminded him they no longer existed. He was quite nonplussed and pulled out some explosion data . . . from Corona. Ours was the main bit of data. I observed the data were few and since there were no critical reflections there may be a continuous distribution. He said "O.K." they still are "layers"! . . . So there he is [LC, Tuve Papers, Box 330, Sept. 5, 1951].

In his Brussels paper, Tatel reported the preliminary results from the recent expeditions. The work in Minnesota had yielded a depth for the Mohorovicic discontinuity about the same as the depth found in the Appalachian Mountains, while no discontinuity had been found in Puget Sound. In addition, he emphasized the attention that he and his co-workers had given to the development of suitable field techniques. "The most time consuming part of the work," he concluded (again jabbing at the idea of intermediate layers), "has been to ascertain whether we could determine crustal structure rather than measure a set of parameters based upon an hypothesis" [LC, Tuve Papers, Box 330, Tatel, "Investigation of the Earth's Crust by Explosions," 1951].

The second major Carnegie presentation came on November 9, 1951, at a meeting of the American Philosophical Society in Philadelphia. There Tuve read a paper that he had co-authored with Tatel and Adams. Published two years later, it marked the first major review of the Carnegie program. In addition to describing the methods used and the results obtained, the paper presented the larger philosophical context in which Tuve and his associates were pursuing their work. It opened with their hope that knowledge of the Earth's crustal structure would shed light on "the evolution of these structures during past geological time" [Tatel et al., 1953, p. 658]. Similarly, they speculated: "it seems reasonable to think that the crust throughout its volume may carry the scar tissue, so to speak, remaining from along succession of mountain building epochs" [Ibid., p. 667].

As their field trip for 1952, the Carnegie researchers explored the western edge of the Great Plains. In particular, they arranged with the Magnolia Petroleum Company of Dallas, Texas, to spend a couple of weeks in June working with the company's field crew stationed at Kimball, Nebraska [Tuve, 1952, p. 69]. But the results were disappointing, leading Tuve to report:

> I believe, from our experience, that it will not be feasible to explore the Mohorovicic discontinuity under the high plains using ordinary ten pound reflection shots, but that due to high noise level and moderately high absorption in the sedimentary layers, shots of several hundred pounds will be needed [DTM, GF 1948-1952, file "84. Deep Seismic," Tuve to J. Paige, June 27, 1952].

Also in 1952 Adams retired as director of the GL. From the outset, he had served as the group's senior scientist. Tuve may have been the primary animating spirit, but at every step Adams's judgment had helped to shape and validate the directions in which Tuve wanted to move. By contrast, Adams's temporary successor, George W. Morey, had little interest in the project and formally ended the participation of the GL. "The seismic work has been discontinued here," he wrote Goranson early in 1953, ". . . but we are making steady progress in our basic program" [CIW, file "Geophysical Staff. Goranson, Roy W.," Jan. 9, 1953]. Nevertheless, the project retained sufficient momentum to continue without interruption.

Another development in 1952 involved the DTM fellowship program. All along, Tuve had viewed the seismic work as an opportunity to promote geophysics among the up-and-coming generation of researchers. For example, he had noted in the 1947 *CIW Year Book*:

> One purpose of taking initiative in this direction, perhaps of greater importance than the expected observations themselves, is to interest a wider group of

young students in scientific geophysics . . . [p. 64].

But the goal of bringing recent Ph.D. recipients to the DTM proved difficult to implement. In 1948, for example, Tuve explained to Bush why he was recommending as a DTM Fellow a graduate student from India who was still working on his doctorate at Caltech:

> We have wanted young men who have completed their graduate studies, but the demand for physicists is such that these candidates are accepting well-paid posts of a semi-permanent nature in place of fellowships [LC, Tuve Papers, Box 131, June 26, 1948].

Similarly, Tuve wrote Scherer a year later:

> As you know, with most of the first-class American graduate students in geophysics being diverted to the oil companies, we have for three years been asking Gutenberg, Ewing, Tuzo Wilson (Toronto), and [Perry] Byerly, to help us find first-class men for scientific geophysics as a life career [LC, Tuve Papers, Box 136, June 23, 1949].

Writing on July 7, 1952, Tuve reminded Bush: "Since 1946 many of us here at the Department have been vigorously interested in increasing our contacts with advanced graduate students" [LC, Tuve Papers, Box 149]. Although the fellowship positions had been hard to fill, Tuve indicated that he had now found a promising candidate. Specifically, Francis Birch had recommended one of his students at Harvard as "a man who might profit greatly by doing his thesis work at the DTM" [Ibid.]. As a result, Pembroke J. Hart was offered a DTM Fellowship beginning in September 1952, and in 1955 he received his doctorate on the basis of research that he had conducted at the DTM [LC, Tuve Papers, Box 149, Tuve to Hart, July 30, 1952; Tuve et al, 1954; Hart, 1955].

Meanwhile, the senior researchers at the DTM continued to focus their attention on the problem of reverberations. As Tuve explained to Gordon G. Lill (who since 1947 had headed the Geophysics Branch of the ONR), reverberations affected the signal to noise ratio, so that a better understanding of reverberations was likely to lead to improved field measurements [DTM, GF 1953-1954, file "84. Seismic," Feb. 10, 1953]. Tuve also described the two types of experiments he had in mind. One was to use the horizontal strain seismometer that Tatel was developing to study in more detail how waves from the interior were transformed into surface waves. The other type of experiment was to make simultaneous field observations, on the surface and in deep mines.

Tuve's purpose in describing his plans to Lill was to obtain the special funds needed for additional shots at Solomons Island. The Navy contract under which the Carnegie researchers usually operated had not been funded for Fiscal Year 1953, because "the work load at NOL prohibited them from guaranteeing services ahead of time for the entire year" [DTM, GF 1953-1954, file "75. Navy Department, Office of Naval Research," Chief of Naval Research to Chief of Bureau of Ordnance, n.d.]. Accordingly, a DTM report for early 1953 noted: "The seismic work has come to a halt because neither the Army or Navy are at this time able to provide shots" [LC, Tuve Papers, Box 161, "Projects under Investigation for the Months of February and March 1953," n.d.]. But Tuve continued to lobby for the support the group needed. In May 1953 he asked Bush for another allotment of special funds, and in June he asked Lill for a series of larger explosions at Solomons Island [CIW, file "DTM—Director 1952-1953," Tuve to Bush, May 26, 1953; DTM, GF 1953-1954, file "75. Navy Department, Office of Naval Research," Tuve to Lill, June 25, 1953]. Shortly thereafter, the Navy resumed its contract [DTM, GF 1953-1954, file "75. Navy Department, Office of Naval Research," Tuve to ONR, July 31, 1953].

As the above-mentioned "work load at the NOL" suggests, the Cold War continued affecting the Carnegie program. Although Tuve remained actively engaged in military work, he still preferred to focus on basic science and to avoid fully mobilizing the DTM. Thus Bush explained to MIT president J. R. Killian, Jr.:

> Merle is consulted a great deal on defense matters and he exercises quite an influence in an informal way which is particularly easy for him since he is located right here in Washington. On the other hand, he has strongly resisted any attempt to draw him into the affair on a permanent basis and this he has done through the strong conviction that a few people at least need to be furthering fundamental science without too much distraction [CIW, file "MIT," Jan. 23, 1953].

But the pressure of the Cold War was taking its toll on Tuve. For example, he wrote his older brother during the fall of 1953:

> Some day, perhaps, I can try to be human again, but the atomic bomb situation has had me so depressed and even bitter about the inadequacies and the evasions of our government that I have not felt able to have many warm contacts [even with family members] [LC, Tuve Papers, Box 162, M. A. Tuve to G. L. Tuve, Oct. 15, 1953].

Later that same month Tuve drafted a "Statement Concerning the H-Bomb" regarding the inadequacy of defense measures in the event of an attack involving hydrogen bombs for possible release to the public [LC, Tuve Papers, Box 164, Oct. 29 1953].

During the following year the Cold War forced itself on Tuve in another way: at almost the same time as the Oppenheimer hearings, which lasted from April 12 to May 6, 1954, Tatel's loyalty was called into question. It was not the first time that Tuve had experienced such problems. In 1950 Stephen Brunauer at the Navy's Bureau of Ordnance and his wife Esther Brunauer at the State Department had been among those whose loyalty had been questioned by Senator McCarthy [Fried, 1990, pp. 23-29].

Writing on March 24, 1950, to Senator Millard E. Tydings, who chaired the sub-committee investigating McCarthy's allegations, Tuve expressed his confidence in the Brunauers and voiced his concern regarding the steps that McCarthy had taken:

> Commander Brunauer, known to me for 17 years, has been carefully investigated by all of the resources available to the Government and has been judged fully loyal and dependable. Unless Senator McCarthy has new evidence which throws this judgment into question it is treacherous for him to handle the matter in the newspapers. [CIW, file "DTM—Director 1950-1951"].

In a separate letter regarding Esther Brunauer, he wrote:

> The procedures used by Senator McCarthy open him to the charge that he is but a tool of subversive organs. We cannot unjustly, and with complete intemperance, undermine public confidence in our Government servants, particularly in flagrant cases such as this [CIW, file "DTM—Director 1950-1951," Tuve to Tydings, Mar. 24, 1950].

Nevertheless, in April 1951 Stephen Brunauer was suspended from the Navy as a security risk, an announcement which made front-page news even on the day that Truman relieved MacArthur of command [New York Times, Apr. 11, 1951]. Meanwhile, his wife (precisely because she was his wife) was given leave of absence from the State Department, though "with no charges against her" [LC, Tuve Papers, Box 148, S. Brunauer to Tuve, Apr. 25, 1951].

In Tatel's case, the problem arose when the Screening Division of the Eastern Industrial Personnel Security Board informed him that they were denying "a request for your access to certain classified matter at Applied Physics Laboratory" [LC, Tuve Papers, Box 377, G. Heck to Tatel, Apr. 27, 1954]. In its "Statement of Reasons," the Board cited favorable statements that Tatel had made in the past regarding communism and the Soviet Union, as well as an incident in 1947 when Tatel had objected to the security procedures at the White Sands Proving Ground [LC, Tuve Papers, Box 377, "Statement of Reasons," enclosed with G. Heck to Tatel, Apr. 27, 1954].

The first "Reason" dated back to Tatel's undergraduate years. In 1934 he had attended an anti-Nazi demonstration in Boston, after which he was stopped by policemen, who beat him and put him in jail. Subsequently, he was tried for "inciting to riot" and given a suspended sentence. The White Sands incident can likewise be linked to his anti-Nazi feelings. Specifically, he was angered because the technicians who had been brought from Germany after the war were allowed to move about more freely than he and the other visiting scientists [LC, Tuve Papers, Box 377, Tatel, "Affidavit of Howard E. Tatel" (draft), May 27, 1954]. In a supportive affidavit, Van Allen recalled the incident:

> Upon arriving at White Sands, learning that a large group of German engineers and technicians worked there in a classified area, and then—as a U. S.citizen with a long record of classified work—being required to sign a very restrictive security agreement (involving among other things an agreement to submit to search at any time), he refused to sign the agreement in an outbreak of personal indignation. Later, after discussing the matter with me and with other members of our field party he agreed to sign the agreement and did so [LC, Tuve Papers, Box 377, May 12, 1954].

In a supportive affidavit of his own, Tuve vigorously defended his colleague. "There is not an ounce of opposition to our Government in Howard Tatel," he wrote, adding: "He is a man of great democratic strength and convictions, a loyal American citizen and an excellent security risk" [LC, Tuve Papers, Box 377, May 21, 1954]. But after realizing that the DTM was no longer party to any classified contracts, Tatel decided to allow his clearance with the Navy to be withdrawn, which occurred in July 1954 [LC, Tuve Papers, Box 377, Tatel to R. C. Sullivan, June 2, 1954; CIW, file "DTM—Staff General 1946-1963," Tuve to R. F. Lynch, Apr. 5, 1955].

Meanwhile, the Carnegie researchers proceeded with a pair of field trips. Even before embarking, Tuve was quite optimistic. "On our seismic program," he wrote Bush on May 4, 1954, "we are beginning to see the landscape more clearly after years of traveling in dark and cloudy weather" [LC, Tuve Papers, Box 152].

In late June, Tuve and his co-workers cooperated with the

Hanley and Bird Company of Bradford, Pennsylvania, to carry out additional reverberation studies. They traveled to northwestern Pennsylvania, where they observed the effects of explosions in deep oil and gas wells. "We have examined the records," Tuve wrote J. M. Bird on July 8:

> and they verify our expectations from the model experiments, namely that most of the reverberation observed at the surface is in the form of surface waves, not volume scattering from buried structures [DTM, GF 1953-1954, file "84. Seismic"].

Not only had the trip gone well technically, but the cooperation had been genuinely cordial, leading Tuve to add:

> As for our own feelings, hunting good seismic traces beats trout fishing all hollow. It was a pleasure to be with you, and I look forward to coming up again [Ibid.].

The pleasure Tuve derived from the success of the trip clearly revealed the extent to which he had mastered the cooperative approach to large-scale research. By 1954 he and his Carnegie co-workers had developed considerable skill in making arrangements with a wide variety of institutions, not only commercial outfits such as the Hanley and Bird Company but also universities and government organizations. Summer after summer, they had successfully assembled the "coalitions" that their fieldwork required.

After the trip to Pennsylvania, the Carnegie researchers headed west, where they made the New Mexico School of Mines their main base. Their goal was to observe the effects of explosions set off by the Phelps Dodge Mining Company at Morenci, Arizona, where open-pit copper mining was pursued on a massive scale [Cleland, 1952, pp. 244-260]. As Tuve explained to the company's general manager:

> By a careful search in the various compass directions from Morenci, with stations at various points 80 to 120 miles distant, we hope to locate reflections from the so-called "Mohorovicic discontinuity" underneath the Rocky Mountains in your vicinity. This interesting question of the "roots of the mountains" is the primary reason for our expedition to your territory [DTM, GF 1953-1954, file "84. Seismic," Tuve to C. R. Kuzell, June 16, 1954].

Arrangements were also made to observe the effects of explosions at a mine operated by the Kennecott Mining Company near Silver City, New Mexico.

At first, the news of their results only trickled out. On July 27, 1954, Tuve telegraphed one correspondent: "Have been three weeks inaccessible in mountains on strenuous field trip" [LC, Tuve Papers, Box 167, Tuve to C. Billman]. But he gave no hint of what had transpired. Not until mid-August (after a trip to Bingham Canyon, Utah, for additional observations at the Phelps Dodge mine there) did he begin offering brief accounts of the discovery that he and his co-workers had made.

Thus he wrote Scherer on August 18:

> I am not sure that our enthusiastic snatches of conversation about our results reached you from our DTM phone calls. I shall not attempt a full outline here either, but the gist is simple. We expected, if we were lucky, to find reflections from the Moho under the Colorado plateau at a depth of 60 to 80 kilometers. . .
>
> Even our first shot, with stations NE of Silver City, N. Mex., showed no reflections up near Albuquerque, where they should stand out (for 60 km depth) but a possible reflection, with arrival time close to the ground wave . . . near Socorro! This turned out to be the case. . . . [The] Moho [is] at somewhere between 29 and 33 km . . . Wow! The whole 6000 ft of Colorado Plateau . . . is light . . . [CIW, file "DTM—Director 1954-1959. 3 of 3"].

Toward the end of the letter Tuve added: "Howard and I have the initial impression that this is a basic and unexpected discovery" [Ibid.]. Their early assessment has proven sound, and the discovery remains one of the most important achievements of the early Carnegie seismic program. For example, in his 1985 essay on the "Seismic Exploration of the Crust and Upper Mantle of the Basin and Range Province," L. C. Pakiser, Jr. (a senior geophysicist at the United States Geological Survey), observed:

> Tuve and his coworkers at the Carnegie Institution of Washington . . . made the first determination of crustal thickness in the transition zone between the Basin and Range province and the Colorado Plateau in 1954. . . . In their initial investigations in Arizona, New Mexico, and Utah, Tuve and his coworkers discovered one of the characteristic features of the Basin and Range province—the unexpectedly thin (28-34 km thick) crust [Pakiser, 1985, pp. 453-454].

During the fall of 1954 Tuve reported the Carnegie discovery at a symposium on "The Crust of the Earth" held at Columbia University, October 14-16, 1954, an event which we may take as marking the natural culmination for his early seismic work. The paper, co-authored with Tatel and entitled "Seismic Exploration of a Continental Crust," summarized the results from field trips to various parts of the country (including the recent western trip). Based on these surveys, they argued against the existence of intermediate layers above the Mohorovicic discontinuity. In addition, they argued that the depth of the Mohorovicic discontinuity showed "striking regional differences" [Tatel and Tuve, 1955, p. 50].

Before presenting their field results, however, Tuve and Tatel described their experimental procedure. The explosions, of course, had always been the responsibility of cooperating organizations (the Navy, the oil companies, or the mining companies). What the Carnegie researchers characteristically provided, in addition to overall planning and coordination, had been the equipment for recording the results of the blasts (which on their recent trip west had involved a pair of seismometers at each site [Tuve, 1955, p. 51]).

Summarizing their method with the confident brevity of practiced hands, Tuve and Tatel wrote:

> The choosing of an observing position is sometimes an arduous task. The site must be in a location as far from man or natural sources of ground noise as can be found. Usually, several attempts must be made to find one of sufficiently low noise level that the weak impulses, sometimes only a few Angstrom units in displacement amplitude, may be detected.
>
> Once the site is selected, and the equipment setup, the chronometer is checked by recording its second impulses simultaneously with the time signals from the National Bureau of Standards Radio Station WWV. The seismogram is then recorded along with the chronometer second impulses, and another comparison is made between the chronometer and WWV. Times are thus reliable to 0.03 second or better.
>
> Distances are obtained from the best maps available. Errors rarely exceed 0.5 km and are usually far less [Ibid., pp. 39-40].

Implied here is a striking intellectual achievement, namely, the full transformation of the terrain for miles around into one vast experimental setup. That such was his aim, Tuve had made clear earlier in the year in an outline for a projected monograph on the Carnegie seismic project. After expressing his hope that DTM would serve as "a living example of an old-fashioned highly *personal* research laboratory" in which researchers would "[l]earn to love a subject by investing [them]selves in it," he contrasted experimentation to observation and then indicated that his goal was to "[a]sk questions of the earth by doing experiments" [LC, Tuve Papers, Box 163, "Seismic Monograph Outline—B," Feb. 18, 1954].

CONCLUSION

In 1955 the National Academy of Sciences conferred on Tuve its Bernard Medal, awarded every five years "for Meritorious Service to Science" [CIW, file "Bursars Office—Tuve Files," G. Kirk to Tuve, Apr. 12, 1955]. Although Tuve could place the new award alongside the many others that he had already received, the Bernard Medal stands out because it was the first to recognize explicitly his contributions to geophysics. As such, it highlights one of the main themes of this paper, namely, the biographical theme of how Tuve personally entered the field during the decade following the end of World War II.

Another biographical point involves Tuve's deep commitment to the seismic project. We have seen how Tuve refused several post-war offers of academic positions and how he resisted the full mobilization of the DTM as the Cold War heated up. In large measure, it was the seismic work that kept him at the DTM. Although the project did not dominate his post-war career to the same extent that nuclear physics had before the war, it still occupied a central position. As a result, to understand the man during this period in his life requires examining closely the Carnegie seismic project.

Apart from the detailed results of the field work, what Tuve found important about the seismic project was the style of the work. By the time that I met him, very late in his life, he had developed a distinctive and clearly articulated approach to science. He viewed research as a sport that emphasized personal participation and the progressive development of technique. "[D]o it yourself," he remembered telling his DTM colleagues. "Get in trouble. Go out and make some measurements. Make some mistakes. That's the way you learn to love a subject" [Tuve Interview, Jan. 13, 1982].

Although Tuve's seismic project fit that pattern, I am still not prepared to say which came first, the specific project or the general style. But perhaps it is enough to conclude that much of the significance of Tuve's seismic work lies in its close relationship to his general view of research. Each reflected the other to such an extent that whenever Tuve described to others his preferred style of research, no one could justly accuse him of not practicing what he preached.

Tuve and his Carnegie co-workers may not have succeeded in addressing directly the question of how mountains originate, but they had proceeded according to Tuve's "philosophical" desires. Through the development of suitable equipment and through reliance upon extensive cooperation, they had learned how to experiment with the Earth. Using large explosions to create seismic waves and using sensitive instruments to record the passage of those waves over long distances, they had probed the deep structure of the Earth's crust. Amidst the Cold War, they had surveyed the depth of the Mohorovicic discontinuity across an entire continent.

ARCHIVAL SOURCES AND ACKNOWLEDGMENTS

For facilitating the archival research on which this paper rests, I gratefully acknowledge the assistance of the staff at the Manuscripts Division of the Library of Congress (LC), where the Merle A. Tuve Papers are held; Ray Bowers and Susan Y.Vasquez at the Carnegie Institution (CIW); and Shaun J. Hardy at the Department of Terrestrial Magnetism (DTM), who also helped me with the records of the Geophysical Laboratory (GL). In the citations from the DTM archives, GF stands for "General Files."

For the financial assistance that enabled me to travel to Washington, D.C., I thank the College of Liberal Arts, Rochester Institute of Technology. For his comments on a preliminary draft of this paper, I am indebted to Ronald E. Doel.

The Tuve Interview, 1982 cited in the text was conducted by T. D. Cornell, Jan. 13, 1982. The tape and transcript are available at the Center for History of Physics, American Institute of Physics).

REFERENCES

Adams, L. H., Elastic properties of materials of the Earth's crust, in *Internal Constitution of the Earth*, edited by B. Gutenberg, pp. 71-89, McGraw-Hill, New York, 1939.

Adams, L. H., Report of the GL director, *CIW Year Book [for] 1945-1946*, No. 45, pp. 23-35, CIW, Washington, D.C., 1946.

Adams, L. H., Some unsolved problems of geophysics, *Trans. AGU*, *28*, 673-679, 1947.

Adams, L. H., Report of the GL director, *CIW Year Book [for] 1949-1950*, No. 49, pp. 27-59, CIW, Washington, D.C., 1950.

Adams, L. H., J. W. Greig, P. A. Scherer, H. E. Tatel, and M. A. Tuve, Studies of crustal structure by explosion seismic waves near the Mesabi Range and near Puget Sound (abstract), *Trans. AGU*, *33*, 315, 1952.

Baldwin, R. B. *The Deadly Fuze: The Secret Weapon of World War II*, 332 pp., Presidio Press, San Rafael, Calif., 1980.

Baracca, A., "A differentiation between "big science" vs. "little science": Lawrence and Tuve, first experiments with deutons," *Organon*, *24*, 237-243, 1988.

Baracca, A., "Big science" vs. "little science" in post war physics, in *Proceedings of the International Conference on The Restructuring of Physical Science in Europe and the United States, 1945-1960*, [held at] Universita "La Sapienza," Rome, Italy, 19-23 September 1988, edited by M. De Maria, M. Grill, and F. Sebastiani, pp. 150-160, World Scientific, Singapore, 1989.

Bates, C. C., T. F. Gaskell, and R. B. Rice, *Geophysics in the Affairs of Man: A Personalized History of Exploration Geophysics and Its Allied Sciences of Seismology and Oceanography*, 492 pp., Pergamon Press, Oxford, 1982.

Benioff, V. H., Seismological instruments developed at C.I.T, *Engineering and Science Monthly*, *11*, 24-25 & 31, Feb. 1948.

Benioff, V. H., B. Gutenberg, and C. F. Richter, Progress report, Seismological Laboratory, California Institute of Technology, 1949 (abstract), *Trans. AGU*, 31, 324, 1950.

Bethe, H. A., The hydrogen bomb: II, *Sci. Am.*, *182*, 18-23, Apr. 1950.

Bolt, B. A., *Nuclear Explosions and Earthquakes: The Parted Veil*, 309 pp., W. H. Freeman, San Francisco, 1976.

Bott, M. H. P., *The Interior of the Earth*, 316 pp., St. Martin's Press, New York, 1971.

Bradbury, N. E., Los Alamos—the first 25 years, in *Reminiscences of Los Alamos, 1943-1945*, edited by L. Badash, J. O. Hirschfelder, and H. P. Broida, pp. 161-175, D. Reidel Publishing Co., Dordrecht, 1980.

Brook, M., and M. Wilkening, E. J. Workman, *Phys. Today*, *36*, 72, Apr. 1983.

Brush, S. G., Discovery of the Earth's core, *Am. J. Phys.*, *48*, 705-724, 1980.

Bullen, K. E., Andrija Mohorovicic, in *Dict. Sci. Biogr.*, *9*, edited by C. C. Gillispie, pp. 443-445, Scribners, New York, 1974.

Bush, V., Report of the CIW president, *CIW Year Book [for] 1946-1947*, No. 46, pp. 3-15, CIW, Washington, D.C., 1947.

Byerly, P., Subcontinental structure in the light of seismological evidence," in *Advances in Geophysics*, edited by H. E. Landsberg, pp. 105-152, Academic Press, New York, 1956.

Callahan, N., *TVA: Bridge over Trouble Waters*, 420 pp., A. S. Barnes and Co., South Brunswick, N.J., 1980.

Cleland, R. G., *A History of Phelps Dodge: 1834-1950*, 307 pp., Knopf, New York, 1952.

Cornell, T. D., Merle A. Tuve and his program of nuclear studies at the Department of Terrestrial Magnetism: the early career of a modern American physicist, Ph.D. diss., 552 pp., Johns Hopkins Univ., 1986.

Cornell, T. D., Merle Antony Tuve: pioneer nuclear physicist, *Phys. Today*, *41*, 57-64, Jan. 1988.

Cornell, T. D., Merle Antony Tuve, in *Dict. Sci. Biogr.*, *18*, edited by F. L. Holmes, pp. 936-941, Scribners, New York, 1990.

Creese, W. L., *TVA's Public Planning: The Vision, the Reality*, 388 pp., Univ. of Tennessee Press, Knoxville, Tenn., 1990.

Dean, D. R., Robert Mallet and the founding of seismology, *Ann. Sci.*, *48*, 39-67, 1991.

Dennis, M. A., A change of state: the political cultures of technical practice at the MIT Instrumentation Laboratory and the Johns Hopkins University Applied Physics Laboratory, 1930-1945, Ph.D. diss., 465 pp., Johns Hopkins Univ., 1991.

Dewey, J., and P. Byerly, The early history of seismometry (to 1900), *Bull. Seismol. Soc. Am.*, *59*, 183-277, 1969.

Elliot, D. C., Project Vista and nuclear weapons in Europe, *Int. Secur.*, *11*, 163-183, 1986.

Embry, L. A., Paul A. Scherer, *Science*, *119*, 231-233, 1954.

Ewing, W. M., Twenty-fifth award of the William Bowie Medal [to Tuve]: citation, *Trans. AGU*, *44*, 287-288, 1963.

Fried, R. M., *Nightmare in Red: The McCarthy Era in Perspective*, 243 pp., Oxford Univ. Press, New York, 1990.

Gillmor, C. S., "Threshold to space: early studies of the ionosphere," in *Space Science Comes of Age: Perspectives in the History of the Space Sciences*, edited by P. A. Hanle and V. D. Chamberlain, pp. 101-114, National Air and Space Museum, Washington, D.C., 1981.

Goodstein, J. R., Waves in the Earth: seismology comes to southern California, *Hist. Stud. Phys. Sci.*, *14*, 201-230, 1984.

Greene, M. T., *Geology in the Nineteenth Century: Changing Views of a Changing World*, 324 pp., Cornell Univ. Press, Ithaca, N.Y., 1982.

Gutenberg, B., Seismological evidence for roots of mountains, *Geol. Soc. Am. Bull.*, *54*, 473-498, 1943a.

Gutenberg, B., Earthquakes and structure in southern California, *Geol. Soc. Am. Bull.*, *54*, 499-526, 1943b.

Gutenberg, B., Travel times of principal P and S waves over small distances in southern California, *Bull. Seismol. Soc. Am.*, *34*, 13-32, 1944a.

Gutenberg, B., Reflected and minor phases in records of near-by earthquakes in southern California, *Bull. Seismol. Soc. Am.*, *34*, 137-159, 1944b.

Gutenberg, B., Structure of the Earth's crust in the continents, *Science*, *111*, 29-30, 1950.

Gutenberg, B., Travel times from blasts in southern California, *Bull. Seismol. Soc. Am.*, *41*, 5-12, 1951a.

Gutenberg, B., Revised travel times in southern California, *Bull. Seismol. Soc. Am.*, *41*, 143-163, 1951b.

Gutenberg, B., *Physics of the Earth's Interior*, 240 pp., Academic Press,

New York, 1959.

Gutenberg, B., H. O. Wood, and J. P. Buwalda, Experiments testing seismographic methods for determining crustal structure, *Bull. Seismol. Soc. Am.*, *22*, 185-246, 1932.

Hart, P. J., Variation of velocity near the Mohorovicic discontinuity under Maryland and northeastern Virginia, Ph.D. diss., 96 pp., Harvard Univ., 1955.

Herbert-Gustar, A. L., and P. A. Nott, *John Milne: Father of Modern Seismology*, 196 pp., Paul Norbury Publications, Ltd., Tenterden, England, 1980.

Howell, B. F., Jr., *An Introduction to Seismological Research: History and Development*, 166 pp., Cambridge Univ. Press, Cambridge, 1990.

Jeffreys, H., *The Earth: Its Origin, History and Physical Constitution*, 2nd ed., 346 pp., Macmillan, New York, 1929.

Le Grand, H. E., Conflicting orientations: John Graham, Merle Tuve and paleomagnetic research at the DTM 1938-1958, *Earth Sci. Hist.*, *8*, 55-65, 1989.

Leet, L. D., Earth motion from the atomic bomb test, *Am. Sci.*, *34*, 198-211, 1946.

Leonard, G. K., Record blast provides 1.8 million cu yd of rock for South Holston Dam, *Civ. Eng.*, *20*, 43-47, Mar. 1950.

Moody, C. L., Earle Winston Johnson (1902-1953), *Am. Assoc. Pet. Geol. Bull.*, *38*, 184-186, 1954.

Needell, A. A., "Lloyd Berkner, Merle Tuve, and the federal role in radio astronomy," *Osiris*, 2nd Ser., *3*, 261-288, 1987.

Owens, L., Vannevar Bush, in *Dict. Sci. Biogr.*, *17*, edited by F. L. Holmes, pp. 134-139, Scribners, New York, 1990.

Page, T., G. S. Pettee, and W. A. Wallace, Ellis A. Johnson, 1906-1973, *Oper. Res.*, *22*, 1141-1155, 1974.

Pakiser, L. C., Jr., Seismic exploration of the crust and upper mantle of the Basin and Range province, in *Geologists and Ideas: A History of North American Geology*, edited by E. T. Drake and W. M. Jordan, pp. 453-469, Geological Society of America, Boulder, Col., 1985.

Palmer, C. E., Louis Byrne Slichter: builder of the Institute of Geophysics and Planetary Physics, *J. Geophys. Res.*, *68*, 2867-2870, 1963.

Poggendorff, J. C., Merle Antony Tuve, in *Biographisch-Literarisches Handwörterbuch der Exakten Naturwissenschaften*, *7b*, Pt. 8, pp. 5584-5586, Akademie-Verlag, Berlin, 1989.

Rearden, S. L., *The Formative Years, 1947-1950*, 667 pp., Vol. 1 of History of the Office of The Secretary of Defense, edited by A. Goldberg, Office of the Secretary of Defense, Washington, D.C., 1984.

Richter, C. F., *Elementary Seismology*, 768 pp., W. H. Freeman, San Francisco, 1958.

Richter, C. F., Memorial to Beno Gutenberg (1889-1960), *Proc. Vol. Geol. Soc. Am.*, 93-104, 1960.

Roberts, E. B., and F. P. Ulrich, Seismological activities of the U.S. Coast and Geodetic Survey in 1949, *Bull. Seismol. Soc. Am.*, *41*, 205-220, 1951.

[Ross, R.], Man-made quake waves studied after explosions, *Sci. News Lett.*, *53*, 281, 1948.

Sapolsky, H. M., *Science and the Navy: The History of the Office of Naval Research*, 142 pp., Princeton Univ. Press, Princeton, 1990.

Schweber, S. S., The mutual embrace of science and the military: ONR and the growth of physics in the United States after World War II," in *Science, Technology, and the Military*, edited by E. Mendelsohn, M. R. Smith, and P. Weingart, pp.3-45, Kluwer Academic Publishers, Dordrecht, 1988.

Shurcliff, W. A., *Bombs at Bikini: The Official Report of Operation Crossroads*, 212 pp., William H. Wise and Co., New York, 1947.

Sullivan, W., *Assault on the Unknown: The International Geophysical Year*, 460 pp., McGraw-Hill, New York, 1961.

Tatel, H. E., Note on the nature of a seismogram—II, *J. Geophys. Res.*, *59*, 289-294, 1954.

Tatel, H. E., and M. A. Tuve, Note on the nature of a seismogram—I, *J. Geophys. Res.*, *59*, 287-288, 1954.

Tatel, H. E., and M. A. Tuve, Seismic exploration of a continental crust, in *Crust of the Earth* (A Symposium), Special Paper No. 62, edited by A. Poldervaart, pp. 35-50, Geological Society of America, New York, 1955.

Tatel, H. E., and J. A. Van Allen, Cosmic-ray bursts in the upper atmosphere (letter to the editor), *Phys. Rev.*, *73*, 87-88, 1948.

Tatel, H. E., L. H. Adams, and M. A. Tuve, "Studies of the Earth's crust using waves from explosions," *Proc. Am. Phil. Soc.*, *97*, 658-669, 1953.

Teller, E., and M. A. Tuve, The sixth Washington conference on theoretical physics, *Science*, *91*, 621-623, 1940.

Tuve, M. A., Report of the DTM director, *CIW Year Book [for] 1946-1947*, No. 46, pp. 43-81, CIW, Washington, D.C., 1947.

Tuve, M. A., Report of the DTM director, *CIW Year Book [for] 1947-1948*, No. 47, pp. 53-85, CIW, Washington, D.C., 1948.

Tuve, M. A., Report of the DTM director, *CIW Year Book [for] 1948-1949*, No. 48, pp. 57-80, CIW, Washington, D.C., 1949a.

Tuve, M. A., William Joseph Rooney, 1890-1949, *J. Geophy. Res.*, *54*, 397, 1949b.

Tuve, M. A., Report of the DTM director, *CIW Year Book [for] 1949-1950*, No. 49, pp. 61-81, CIW, Washington, D.C., 1950.

Tuve, M. A., Report of the DTM director, *CIW Year Book [for] 1950-1951*, No. 50, pp. 65-94, CIW, Washington, D.C., 1951.

Tuve, M. A., Report of the DTM director, *CIW Year Book [for] 1951-1952*, No. 51, pp. 65-97, CIW, Washington, D.C., 1952.

Tuve, M. A., Report of the DTM director, *CIW Year Book [for] 1954-1955*, No. 54, pp. 41-94, CIW, Washington, D.C., 1955.

Tuve, M. A., Report of the DTM director, *CIW Year Book [for] 1956-1957*, No. 56, pp. 83-147, CIW, Washington, D.C., 1957.

Tuve, M. A., Cooperative geophysics in the Andes, *Trans. AGU*, *44*, 290-300, 1963.

Tuve, M. A., Preface by Merle A. Tuve [to A. L. Hales and T. Asada, Crustal structure in coastal Alaska, pp. 420-432], in *The Earth Beneath the Continents: A Volume of Geophysical Studies in Honor of Merle A. Tuve*, Geophysical Monograph No. 10, edited by J. S. Steinhart and T. J. Smith, pp. 420-422, AGU, Washington, D.C., 1966.

Tuve, M. A., and H. E. Tatel, Seismic observations, Corona (California) blast, 1949 (abstract), *Trans. AGU*, *31*, 324, 1950a.

Tuve, M. A., and H. E. Tatel, Coherent seismic wave patterns (abstract), *Science*, *112*, 452-453, 1950b.

Tuve, M. A., R. W. Goranson, J. W. Greig, W. J. Rooney, J. B. Doak, and J. L. England, Studies of deep crustal layers by explosive shots, *Trans. AGU*, *29*, 772, 1948.

Tuve, M. A., H. E. Tatel, and L. H. Adams, Coherent explosion wave patterns at 100 kilometers (abstract), *Trans. AGU*, *32*, 316, 1951.

Tuve, M. A., H. E. Tatel, and P. J. Hart, Crustal structure from seismic exploration, *J. Geophys. Res.*, *59*, 415-422, 1954.

Van Allen, J. A., *Origins of Magnetospheric Physics*, 144 pp., Smithsonian Institution Press, Washington, D.C., 1983.

Van Allen, J. A., and H. E. Tatel, The cosmic-ray counting rate of a single geiger counter from ground level to 161 kilometers altitude, *Phys. Rev.*, *73*, 245-251, 1948.

Weatherby, B. B., The history and development of seismic prospecting, *Geophysics*, *5*, 215-230, 1940.

Wilson, J. T., The development and structure of the crust, in *The Earth as a Planet*, edited by G. P. Kuiper, pp. 138-214, Univ. of Chicago Press, Chicago, 1954.

Wilson, J. T., Development of ideas about the Canadian shield: a personal account," in *Geologists and Ideas: A History of North American Geology*, edited by E. T. Drake and W. M. Jordan, pp. 143-150, Geological Society of America, Boulder, Col., 1985.

Wright, H., *Palomar: The World's Largest Telescope*, 188 pp., Macmillan, New York, 1952.

Yoder, H. S., Jr., Leason Heberling Adams, in *Dict. Sci. Biogr.*, *17*, edited by F. L. Holmes, pp.1-3, Scribners, New York, 1990.

York, H. F., *The Advisors: Oppenheimer, Teller, and the Superbomb*, 175 pp., W. H. Freeman, San Francisco, 1976.

York, H. F., and G. A. Greb, Military research and development: a postwar history, *Bull. At. Sci.*, *33*, 13-26, Jan. 1977.

Isotope Geology at Carnegie 1950-1970: Dating Earth Processes

L. T. ALDRICH
DTM-CIW, 5241 Broad Branch Rd. NW., Washington, DC 20015

Merle Tuve (Fig. 1) became director of the Department of Terrestrial Magnetism (DTM) in 1946. His primary fields of study had been in his highly successful probing of the nucleus and of the ionosphere in which he had the lively support of many colleagues. He and a small group of his pre-war colleagues had completed the construction of a small cyclotron with which they hoped to continue these pioneering efforts. This group included R. B. Roberts, P. H. Abelson, D. B. Cowie, and N. P. Heydenburg. They agreed that the post-war future of the cyclotron was severely constrained by the comparative budgets of the CIW and the Atomic Energy Commission, but they were physicists and were encouraged by Tuve to consider other scientific pursuits for which their training prepared them. The first three opted for studies in biophysics, but they continued, with Tuve, to look for other scientific ventures in which training in physics, the limited resources of the CIW, and the unlimited freedom given Tuve by Vannevar Bush could be channeled. Tuve arranged a series of seminars in 1948-1949 under the umbrella title "Milestones in the precambrian." He was already supporting the construction of a Nier-type mass spectrometer (Fig. 3a) to be used for Urey-type studies of the geological and biological effects on the isotope ratios $^{13}C/^{12}C$ and $^{18}O/^{16}O$, and this was in use by Keith and Davis, of the Geophysical Lab (GL) staff.

Tuve had persuaded H. E. Tatel, a war-time colleague at the Applied Physics Laboratory, to join the DTM staff, and they had embarked on joint efforts with several GL staff in attempts to study the Earth's structure using depth-charge explosions and simple electronic equipment. Tuve and Tatel agreed that an expansion of the geophysical studies at DTM to include isotope dating was both feasible and desirable. It was my good fortune that I had been able to do my thesis work with A. O. Nier at the University of Minnesota which included a study [Aldrich and Nier, 1948] of radiogenic argon in potassium minerals. As a further measure of my good fortune, Tuve chose me to be the first part of that expansion. A mass spectrometric study [Aldrich, 1953] of the physics of the evaporation of alkaline earth elements, Ba and Sr, at the University of Missouri was preparation for the Rb-Sr work. The opportunity was provided for freedom to pursue new methods and problems in the determination of ages of rocks in an atmosphere of highly personal and successful research by other physicists in diverse fields of science. Until Tatel's untimely death in late 1957, he was Tuve's right arm in the pioneering explosion studies of the Earth's crust, the investigation of the distribution of hydrogen in our galaxy, and the beginnings of Tuve's South American adventures. The astronomical studies in this period included both radio and optical instrumentation efforts. In partnership with the U.S. Bureau of Standards and the RCA laboratories, the earliest electronic image intensifiers were developed, tested, and provided to observatories all over the world. The radio noise in Jupiter was identified. The nuclear physics research gained a new impetus in the Coulomb excitation studies for which the DTM Van de Graaff generator was especially suited. Indeed, it was only a normal week when some new scientific study in which the DTM was at least a partner was discussed at the Lunch Club.

The early studies of C. N. Fenner and C. S. Piggot [1929] at the GL were the first isotope dating efforts in the CIW, using the isotope measures of Aston in England and their determinations of the ratios of lead to uranium. Dating of ocean cores using the short-lived isotopes of uranium daughter products was carried out at the GL by Piggot and Urry [1940] in the 1930s. The need for the cooperative nature of the program which came to pass in the 1950s was predicted by L. H. Adams [1946] when he wrote

> —[in this] group we place problems that, although of real interest in geophysics, are less closely related to [our] main project— . . . Examples of this class are radioactivity of rocks, . . .

Piggot and Urry did not return to the GL after World War II, but Davis [1950] continued to analyze these relatively short-lived radioisotopes using the equipment they had had built. Fig. 2 shows the participants, their status, laboratory association, and the time of their participation in studies outlined below. The appendix lists the publications of the group in this period. These do not include the annual reports of the CIW which duplicate the work reported.

The Earth, the Heavens and the Carnegie Institution of Washington
History of Geophysics Volume 5

Fig. 1. Merle Anthony Tuve (1901-1982). Photo taken in the early 1960s. Presented here to identify the primary supporter of these and the several efforts discussed below.

(NAME) - BL STAFF (NAME) - PRE-DOCTORAL

YEAR	DIRECTOR	PROF	SUPPORT	FELLOWS	VISITORS
	*ADAMS		STEINER		
<1950	TUVE	*DAVIS, *KEITH	DOAK		
1950		+ ALDRICH			
1951		- KEITH + TILTON	+ HAASE	*NICOLAYSEN*(PRINCETON)	HERZOG*(MIT)
1952				NICOLAYSEN*	HERZOG
	- ADAMS				
1953	+ *ABELSON	+ WETHERILL		NICOLAYSEN*,	GAST*(COLUMBIA)
1954					NICOLAYSEN
1955				JEFFERY	GAST(U.MINN.)
1956		*TILTON		*FAUL(MIT,USGS)	
1957				FAUL	
1958				BASS(PRINCETON)	COMPSTON(U.W.AUST.)
					PATTERSON, WASSERBURG,
1959		+ *HOERING		BASS, HART(MIT)	SILVER(CIT), *JAGER(U.BERN)
					DEUTCH, LEDENT(U.BRUSSELS)
1960		- WETHERILL(TO UCLA)		HART, *DOE	NICOLAYSEN(BPI)
1961		+ HART		DOE	HAYASE(JAPAN)
			SAKKO(FINLAND)		
1962			- HAASE	KUOVO(FINLAND), DOE	BAADSGAARD(ALTA.)
			+ RINEHART		
1963				RAMA(INDIA), *STEIGER(SWITZ)	GHERKIN(BRASIL)
					RICHARDS(ANU)
1964				KROGH(MIT), YAMAGUCHI(JAPAN)	GAST(U.OF MINN)
				STEIGER	
1965		- TILTON	BANNISTER(CHILE)	KROGH, MUNIZAGA*(CHILE)	*MORSE(F.&M.)
		(TO UCSB)			OHMOTO*(PRINCETON)
1966	- TUVE	+ *KROGH		STUEBER(WASH.U.), BROOKS(ANU)	MORSE
	+ BOLTON				
1967			+ BURRHUS	BROOKS, ISHIZAKA(JAPAN)	*PUCHELT(TUBINGEN)
			- DOAK		
1968				ERLANK(CAPETOWN)	BROOKS(U.MONT.)
1969				*HOFMANN(BROWN), SINHA(UCSB)	
1970		- ALDRICH		*MATTINSON(UCSB), SINHA	*SALOMONS(GRONINGEN)
1971		+ HOFMANN			

Fig. 2. A diagram showing the participants in the isotope geology work at the Carnegie Institution for the period 1950-70.

Fig. 3. (a) The instrument built for stable isotope studies adapted from the design of Nier. This machine was altered to allow one to analyze Sr and Rb. (b) An instrument designed and built in the late 1950s specifically to analyze Pb. Instruments of this design were built and shipped to Finland, Chile, and Japan. Shown in the figure is G.L. Davis.

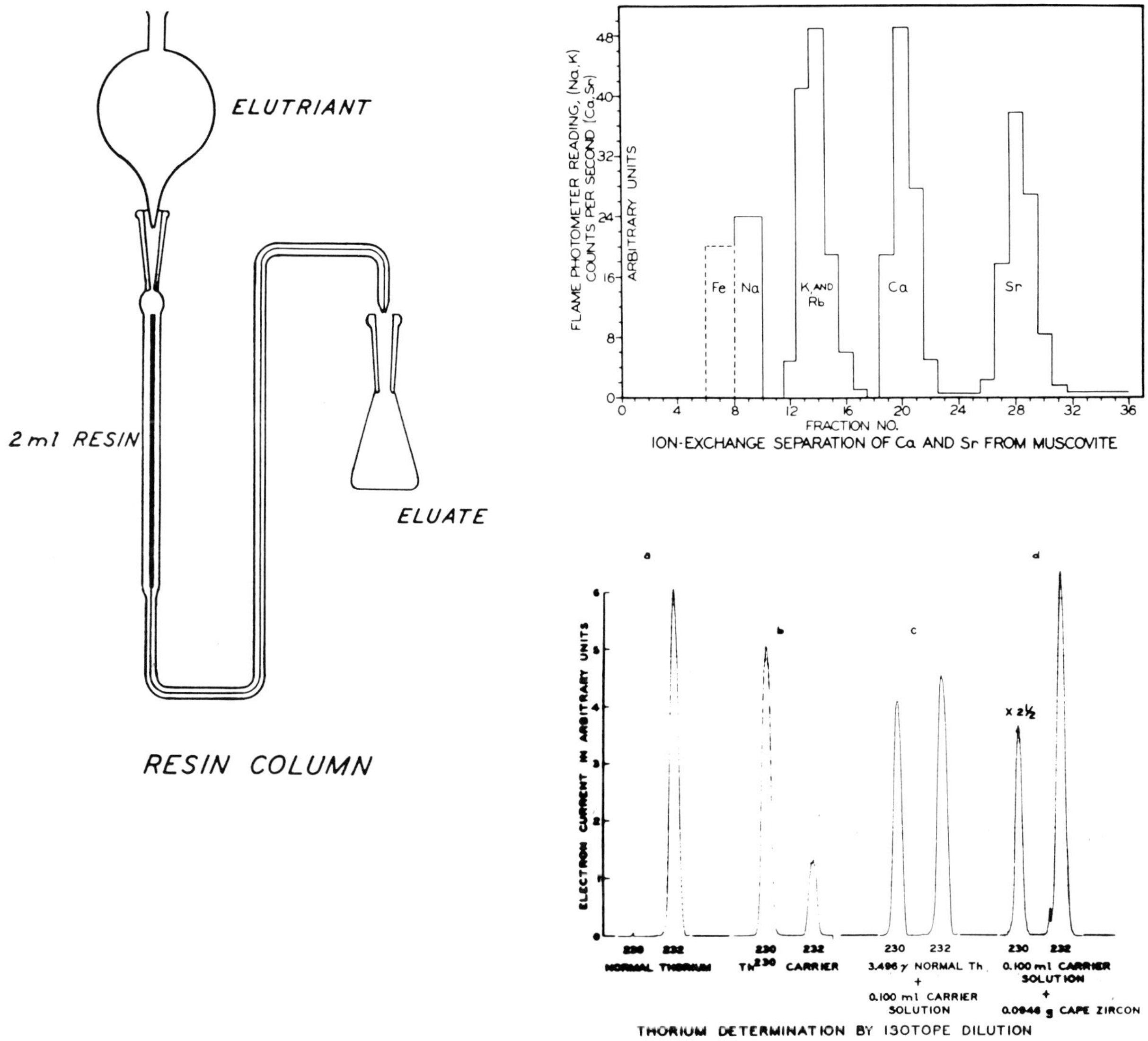

Fig. 4. Elements of the "physicists'" chemistry developed for separating microgram quantities of Sr from milligram quantities of Rb. *Above left:* Schematic of a resin ion-exchange column. *Above right:* Figure shows the separation of Sr from Rb using such a column. The sample was added in a small volume of acid to the top of the column. When the sample had been adsorbed, the amount of acid required to elute everything up to Sr was added. At this time a new, clean beaker was used to collect the Sr for analysis. *Bottom:* An example of the method of stable isotope dilution. ^{230}Th has a long enough half-life to be considered stable for these purposes. Normal Th has a single isotope, ^{232}Th, and by mixing it with a known weight (here denoted by a volume) of the ^{230}Th tracer, one may determine quantitatively the amount of ^{232}Th in the sample from the isotopic ratios of the tracer and the mixture.

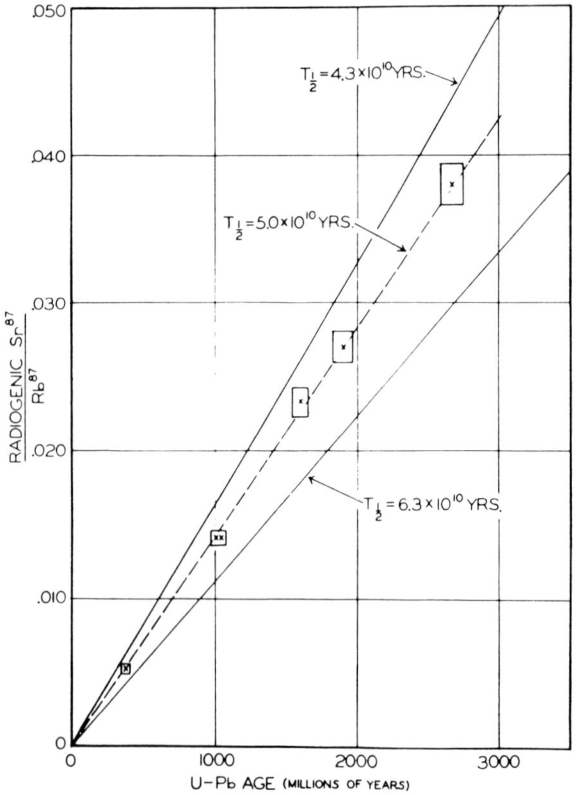

Fig. 5. Data from which the half-life of ^{87}Rb was determined using geologically related Uranium and Rubidium minerals. The half-life determined agrees within experimental error with that determined by careful direct measures of the value indicated.

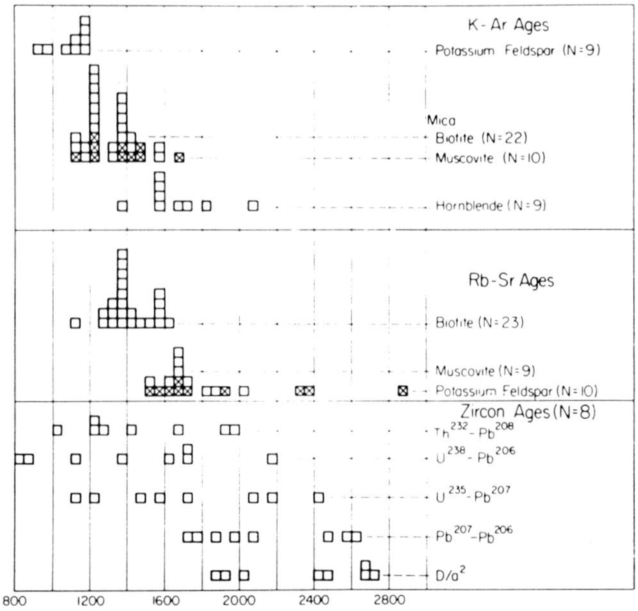

Fig. 6. Histograms of measures of the ages from samples collected in metamorphic region in Dickinson Co., Michigan. Note the spread in all the ages showing the effect of the regional metamorphism in this area.

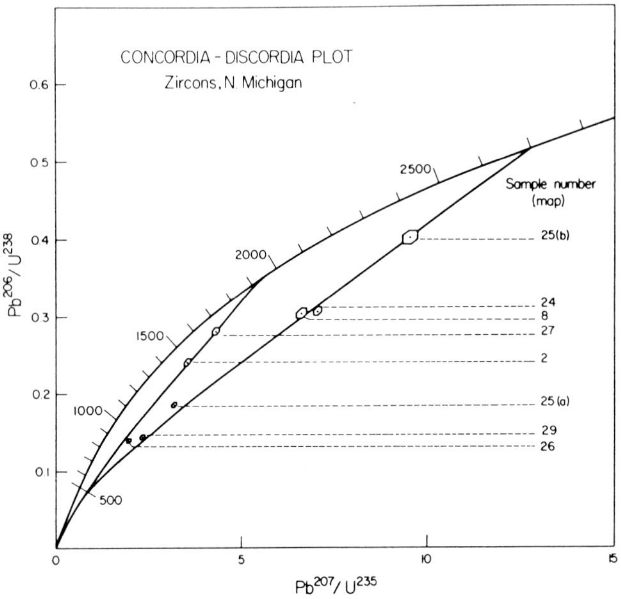

Fig. 7. The data of Fig. 6 for the zircons is plotted on the Wetherill-Tilton concordia plot. Two upper intercepts, at 2700 and 1950 my are believed to be related to orogenic events.

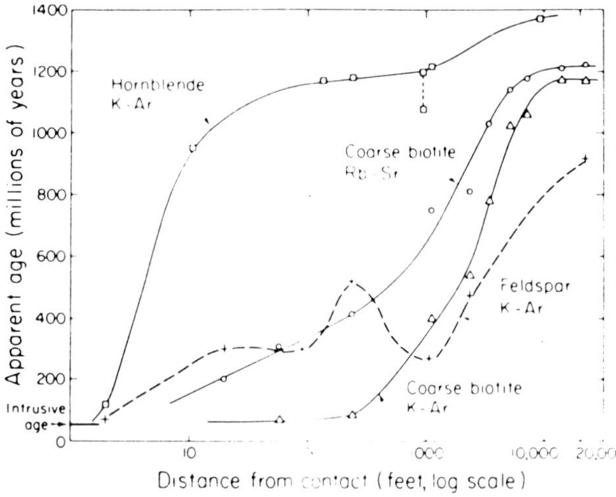

Fig. 8. Distribution of ages by mineral and decay system in a region near Boulder, Colorado. Again the pattern of ages follows the inferred temperature variation due to the intrusion. After Hart et al. [1968]:

These are as inclusive as GEOREF and my files permit.

An encouraging presence in our early work was that of W. W. Rubey, then president of the National Research Council (NRC) of the National Academy of Sciences and chief geologist of the U.S. Geological Survey. The evening seminars in his home on the current state of geology led to the creation of the subcommittee on Nuclear Geophysics of the Committee on Nuclear Science of the NRC. This subcommittee sponsored a number of small conferences concerning isotopic studies. The proceedings of these conferences were published [Aldrich, 1953] since the attendance was small and limited to those invited. A strong influence and voice in these meetings was that of H. C. Urey, then at the University of Chicago. The trio of Urey, Harrison Brown, and M. G. Inghram attracted many students including George Tilton and George Wetherill. Thus it was that Tuve, acting on the advice of Urey and Nier, collected the group which joined G.L. Davis in the work discussed here. It was soon apparent that small meetings could not satisfy the reporting needs of the community of isotope practitioners. We were fortunate to have the help of J. F. Shairer and Waldo Smith in the broadening of the American Geophysical Union's (AGU) section of Volcanology to include wider aspects of geochemistry that provided the principal avenue for both publication and presentation of papers at the AGU's meetings.

In the discussion of early influences on the work, another perhaps surprising part of Tuve's philosophy of geophysics affected all of us. He, a laboratory nuclear physicist, insisted that sample collection was an essential part of our laboratory efforts and that these efforts should not be wasted on non-pedigreed rocks. The part played by J. F. Shairer of the GL in our training in field collecting was both one of great assistance and of great enlightenment.

As a neophyte in geophysics, an early trip to the intellectual "garden," then located in Cambridge, Mass., provided us with stimulus and needed colleagues to undertake the first Rb-Sr work. In an early paper, Ahrens [1949] had demonstrated that optical spectrographic techniques could provide ages of lepidolites, which, as shown by Mattauch [1937] contain nearly pure ^{87}Sr, with a precision of 5-10%. Other micas and alkali feldspars with ratios less than 50 required isotopic measurements of the Sr to provide the determination of the radiogenic ^{87}Sr needed for the age. We agreed to look at some of the Sr concentrates which he had prepared to provide the needed data. The first Sr concentrate of biotite was loaded into the CO^2 machine, revised for solid sample use. We learned after three days of evaporating Rb that a) the ^{87}Sr was about 10% radiogenic, and b) that a more sophisticated chemical procedure was required. Our colleagues in biophysics were aware of this problem due to another Tuvian invention, the DTM lunch club, where we shared simple food, fresh science, and respect for our differing capacities. The biophysicists had been using cation-exchange chromatography for separating the components of the E. Coli stews which they were studying and were aware of the power of this tool in our relatively simple chemical systems. This resulted in the study [Aldrich et al., 1953], shown schematically in Fig. 4a.

L. F. Herzog was a pre-doc at the DTM from the MIT lab of P. M. Hurley, and in his thesis [Herzog et al., 1953] determined the radiogenic ^{87}Sr in a series of minerals at the DTM using these chemical and mass-spectrometric techniques. These samples had had their ratios Sr/Rb determined using the optical spectroscopy methods of Ahrens in the theses of two of his students, Whiting [1951] and Holyk [1952]. The Sr concentrates processed at DTM by Herzog were provided by these MIT colleagues. This work showed a) the feasibility of biotite and other micas with high (>100) ratios of Rb/Sr and b) the difficulty presented by alkali feldspars and celestites for mineral age determination.

The availability of stable isotopes with "unnatural" abundances from the Oak Ridge Isotope Separator, and the available advice of chemists, Tilton on the DTM staff, and Davis and Zies at the GL led us to the next step in developing the Rb-Sr method. The isotope dilution method using ^{230}Th and described by Tilton et al. [1956] is used to demonstrate this procedure since it has the simplicity of having, essentially, a single parent isotope and a tracer with a high ratio of ^{230}Th/^{232}Th, as shown in Fig. 4b. The complete process, collecting samples, separating suitable minerals, dissolving them, spiking the dissolved sample with known amounts of both ^{87}Rb and ^{84}Sr tracer solutions, drying the sample, separating the Rb and Sr fractions on the ion-exchange columns, and making the two mass spectrometer runs needed for the determination of the ratio ^{87}Sr*/^{87}Rb was developed. L. O. Nicolaysen, a Princeton and Carnegie pre-doctoral fellow at the time, had collected a series of samples from South Africa to which these techniques were first applied, the results of which were listed in Tuve's annual report of 1953 and are shown on the next page in Table 1.

Meanwhile Tilton had been setting up the U-Pb, Th-Pb chemistry and Wetherill the ^{40}Ar purification line and mass spectrometers had been constructed with the resolution required for Pb analyses, Fig. 3b, and the cleanliness needed for Ar analyses. These capacities enabled us to make a "geological" determination of the ^{87}Rb half-life [Aldrich et al., 1956], Fig. 5, and to contribute to a similar determination of the ^{40}K decay constant for electron capture [Wetherill et al., 1956]. Studies of many areas of

TABLE 1: Indicated Age of South African Pegmatites (Nicolaysen)

Location	Age(million years)
Popes Claim, S.Rhodesia	3740
Letaba, Transvaal	3850
Hombolo, Tanganyika	3250
M'bale (I), Uganda	2460
M'bale (2), Uganda	2370
Lunya, Uganda	2340
Kinderzitt, Namaqualand	1210
Jakkalswater, Namaqualand	1100
Muika (I), Belgian Congo	1220
Muika (2), Belgian Congo	1210
Karibib-Usakos II, SW.Africa	600
Sahatany, Madagascar	690
Alto Ligonha, Mozambique	570
Dogon Daji, Nigeria	625

precambrian rocks in the U.S., Canada, South America, Finland, Australia, and South Africa using the radioactive isotopes ^{40}K, ^{87}Rb, ^{232}Th, ^{235}U, and ^{238}U were undertaken by fellows and staff. A predoctoral study of Gast [1955] determined that the ratio Rb/Sr in sources of oceanic sediments was about 0.1, a value an order of magnitude less than that which had been estimated.

Early work and more extensive studies at CIW had shown the common discordance of the two U-Pb ages and the derived $^{207}Pb/^{206}Pb$ age. The regularity of the pattern led Wetherill [1956] to look at models which would predict the pattern. He demonstrated that episodic loss of Pb or gain of U would produce the patterns of data from many areas. Tilton [1960] using a model of continuous diffusion showed that the assumptions of his model predicted both the pattern shown, and explained the low-age intercept found for many terranes for which no such age was obvious.

The last phase of the work to be discussed here involves the effects of metamorphic events on the pattern of ages derived from different minerals. The first such terrane to be studied was that in the Iron Mountain, Michigan region which had been intensively mapped by H. L. James [1955] and his colleagues [James et al., 1959]. Samples were collected from the various identified metamorphic zones with the help of James and K. L. Wier. The results [Aldrich et al., 1965] were ambiguous. The mica ages increased as one progressively sampled toward zones of higher grade metamorphism, the inverse of that expected. Further studies showed that the complexity of the region had been underestimated and that, in fact, at least two periods of heating had affected the region. The data for the different systems and minerals are shown in Fig. 6. Fig. 7 shows the application of Tilton's "Concordia" formulation to the zircon data from this suite.

A second study was undertaken by Hart et al. [1968] in the Eldora region near Boulder, Colorado. Fig. 8 is from their paper. The same pattern as that found in north Michigan seems to be followed in this region, but with a smaller linear scale and with higher temperature gradients.

This is a sampler of the isotope activity at the CIW in the 1950s and 1960s, but does not include the stable isotope studies undertaken by Hoering at the GL. This extension of the work was encouraged by Tuve to the extent that he authorized the revision in the DTM shop of the instrument used by Keith and Davis to be ready for Hoering's arrival at the GL. Hoering made excellent use of the machine.

One aspect of the data shown in Fig. 2 should be emphasized. Among the fellows are included scientists from South Africa, Australia, Switzerland, Belgium, Finland, Japan, India, Brazil, and Chile, as well as Canada and the U.S. We were encouraged by Tuve to help in any possible manner to initiate the type of studies in which we were engaged in these countries. This assistance took the form of constructing whole machines for use in Finland, Chile, and Japan (later sent to India), and the University of California, Santa Barbara. One of the more successful of these transplanting operations was that with Finland and this success is demonstrated by the following annual report from Dr. Olavi Kouvo [personal comm.] of the Finnish Geological Survey

> All runs on your machine Pb-U on Source 1, Sm-Nd on source 2 Today run no. 13153 on the original first 9" machine. Thanks for past years' fruitful works here in Otaniemi. Good times!.

We were also encouraged to participate in the meetings and committees planning the meetings of international organizations as well as those of U.S. national societies.

Another view of these times at the DTM is given in a recent letter from L. O. Nicolaysen, present director of the Bernard Price Institute of the University of Witwatersrand in Johannesburg

> . . . I could write a great deal about the heavy atmosphere in the labs and about the powerful influence loud discussions and arguments between Tuve, Tatel, Roberts, and Aldrich had on me,- and every other impressionable young person during that period. The rapid switches between science, and the great social and political themes of the early 1950s exercised a strong influence on everyone participating in those lunch hour sessions. —Vigorous debate about science and vigorous debate about social and political issues changed my entire way of thinking.

Needless to say, this was the most rewarding time of my personal scientific life.

REFERENCES

Adams, L. H., Annual report of the Director, Geophysical Laboratory, *CIW Year Book*, No. 45, 34, 1946.

Ahrens, L. H., Measuring geologic time by the strontium method, *Bull. Geol. Soc. Am.*, *60*, 217-267, 1949.

Aldrich, L. T., The evaporation products of barium oxide from various base metals and of strontium oxide from platinum, *J. Appl. Phys.*, *22*, 1168-1174, 1953.

Aldrich, L. T., ed., Proceedings of the Conference on Nuclear Processes in *Geologic Settings, A report of the Committee on Nuclear Science*, Nat. Res. Council, Washington, D.C., 1953.

Aldrich, L. T., G. L. Davis, H. L. James, Ages of minerals from metamorphic and igneous rocks near Iron Mountain, Michigan, *J. Petrology*, *6*, 445-471, 1965.

Aldrich, L. T., J. B. Doak, G. L. Davis, The use of ion-exchange columns in mineral analyses for age determinations, *Am. J. Sci.*, *251*, 377-387, 1953.

Aldrich, L. T., A. O. Nier, Argon-40 in potassium minerals, *Phys. Rev.*, *74*, 876-77, 1948.

Aldrich, L. T., G. W. Wetherill, G. R. Tilton, G. L. Davis, Half-life of 87Rb, *Phys. Rev.*, *103*, 1045-1047, 1956.

Davis, G. L., Meteorites, Pt.3, Radium content of ultramafic igneous rocks, *Am. J. Sci.*, *248*, 107-111, 1950.

Fenner, C. N., C. S. Piggot, The mass spectrum of lead from broggerite, *Nature*, *123*, 793-94, 1929.

Gast, P. W., Abundance of ^{87}Sr during geologic time, *Bull. Geol. Soc. Amer.*, *66*, 1449-1454, 1955.

Hart, S. R., G. L. Davis, R. H. Steiger, G. R. Tilton, A comparison of mineral age variations and petrologic changes induced by contact metamorphism, in *Radiometric Dating for Geologists*, edited by E. I. Hamilton and R. M. Farquhar, pp. 73-110, Interscience Publishers, London and New York, 1968.

Herzog, L. F., L. T. Aldrich, W. K. Holyk, F. B. Whiting, L. H.Ahrens, Variations in strontium isotope abundances in minerals, part 2: Radiogenic ^{87}Sr in biotite, feldspars, and celestite, *Trans. AGU*, *34*, 461-470, 1953.

Holyk, W. K., Geologic ages by the biotite Sr method, Ph.D. Thesis, Mass. Inst. Tech., Cambridge, Mass., 1952.

James, H. L., Zones of regional metamorphism in the precambrian of northern Michigan, *Bull. Geol. Soc. Amer.*, *66*, 1455-1458, 1955.

James, H. L., C. E. Dutton, F. J. Pettijohn, K. L. Wier, Geologic map of the Iron River-Chrystal Falls district, Iron County, Michigan, U.S.Geol. Survey Map MF225, 1959.

Mattauch, J., Das paar, ^{87}Rb-^{87}Sr und die Isobarenregel, *Naturwissenschaften*, *25*, 189-191, 1937.

Piggot, C. S., W. D. Urry, The radium content of an ocean bottom core, *J. Wash. Acad. Sci.*, *29*, 405-410,1940.

Tilton, G. R., Volume diffusion as a mechanism for discordant lead ages, *J. Geophy. Res.*, *65*, 2933-2945, 1960.

Tilton, G. R., L. T. Aldrich, M. G. Inghram, Mass spectrometric determination of Thorium, *Analytical Chemistry*, *26*, 894-898, 1956.

Wetherill, G. W., Discordant U-Pb ages.1, *Trans. AGU*, *37*, 320-326, 1956.

Wetherill, G. W., G. J. Wasserburg, L. T. Aldrich, G. R. Tilton, R. J. Hayden, Decay constants of 40K as determined by the radiogenic argon content of potassium minerals, *Phys. Rev.*, *103*, 987-989, 1956.

Whiting, F. B., The biotite method of geologic age determination, Ph.D. Thesis, MIT, Cambridge, MA, 1951.

APPENDIX: PUBLICATIONS LIST: ISOTOPE GEOLOGY, CARNEGIE, 1950-1970.

Aldrich, L. T., J. B. Doak, G. L. Davis, The use of ion exchange columns in mineral analysis for age determination, *Am. J. Sci.*, *251*, No. 5, 377-387, 1953.

Aldrich, L. T., G. L. Davis, J. B. Doak, L. F. Herzog, Mass spectrometric analysis of mineral sources of strontium, Pt. 1 of Variations in strontium isotope abundances in minerals, *AGU Trans.*, *34*, No. 3, 457-460, 1953.

Aldrich, L. T., G. L. Davis, L. O. Nicholaysen, C. C. Patterson, G. R. Tilton, Comparison of U-Pb, Pb-Pb, and Rb-Sr ages of Precambrian minerals, in Symposium on Precambrian correlation and dating, *Geol. Assoc. Canada Proc.*, *7*, 7-13, 1955.

Aldrich, L. T., G. L. Davis, L. F. Herzog, W. H. Pinson, Jr. Radiogenic ^{87}Sr in micas from granites, *AGU Trans.*, *36*, No. 5, 875-876, 1955.

Aldrich, L. T., Measurement of radioactive ages of rocks, *Science*, *123*, No. 3203, 871-875, 1956.

Aldrich, L. T., G. W. Wetherill, G. R. Tilton, G. L. Davis, Half-life of ^{87}Rb, *Phys. Rev.*, *103*, 1045-1047, 1956.

Aldrich, L. T., G. W. Wetherill, G. L. Davis, Determinations of radiogenic ^{87}Sr and ^{87}Rb of an interlaboratory series of lepidolites, *Geochimica et Cosmochimica Acta*, *10*, No. 4, 238-240, 1956.

Aldrich, L. T., G. W. Wetherill, G. L. Davis, Occurrence of 1350 million-year-old granitic rocks in western United States, *Geol. Soc. Am. Bull.*, *68*, No. 5, 655-656, 1957.

Aldrich, L. T., G. W. Wetherill, Geochronology by radioactive decay, *Ann. Rev. Nuclear Science*, edited by E. G. Segre, *8*, 257-298, in cooperation with Natl. Research Council, Natl. Acad. Sci., Stanford, CA, 1958.

Aldrich, L. T., G. R. Tilton, G. W. Wetherill, Radioactive ages of micas from granitic rocks by Rb-Sr and K-A methods, *AGU Trans.*, *39*, No.6, 1124-1134, 1958.

Aldrich, L. T., G. F. Brown, Distribution of ages in the Arabian segment of the African shield, *Geol. Soc. Am.*, *71*, No. 12, pt.2, 1813, 1960.

Aldrich, L. T., G. W. Wetherill, Rb-Sr and K-A ages of rocks in Ontario and northern Minnesota, *J. Geophys. Res.*, *65*, No. 1, 337-340, 1960.

Aldrich, L. T., Geochronology, *AGU Trans.*, *41*, No. 2, 287-288, 1960.

Aldrich, L. T., G. L. Davis, H. L. James, Ages of minerals from metamorphic and igneous rocks near Iron Mountain, Michigan, *J. Petrology*, *6*, Pt. 3, 445-472, 1965.

Brooks, C., Relationship between feldspar alteration and the precise post-crystallization movement of rubidium and strontium isotopes in a granite, *J. Geophys. Res.*, *73*, No. 14, 4751-4757, 1968.

Brooks, C., I. Wendt, W. Harre, A two-error regression treatment and its application to Rb-Sr and initial ^{87}Sr/^{86}Sr ratios of younger Variscan granitic rocks from the Schwarzwald massif, southwest Germany, *J. Geophys. Res.*, *73*, No. 18, 6071-6084, 1968.

Brooks, C., S. R. Hart, T. E. Krogh, G. L. Davis, Carbonate contents and ^{87}Sr/^{86}Sr ratios of calcites from Archaean meta-volcanics. *Earth and Planetary Sci. Letters*, *6*, No. 1, 35-38, 1969.

Brooks, C., S. R. Hart, T. E. Krogh, G. L. Davis, The initial ^{87}Sr/^{86}Sr of Michipicoten greenstones and its bearing on the development of the mantle, *Canadian Mineralogist*, *10*, Pt. 1, 127-128, 1969.

Brooks, C., S. R. Hart, T. Krogh, Implications about the mantle from K, Rb, Sr concentrations and ^{87}Sr/^{86}Sr ratios of Archean volcanic rocks, *Geol. Soc. Am. Spec. Paper 121*, 39, 1969.

Davis, G. L., Meteorites, [Pt.] 3 of Radium content of ultramafic igneous rocks, *Am. J. Sci.*, *248*, No. 2, 107-111, 1950.

Davis, G. L., L. T. Aldrich, Determination of the age of lepidolites by the method of isotope dilution, *Geol. Soc. Am. Bull.*, *64*, No. 3, 379-380, 1953.

Davis, G. L., C. A. Hopson, G. R. Tilton, G. W. Wetherill, Age of the Baltimore gneiss, *Geol. Soc. Am. Bull.*, *69*, No.12, Pt.2, 1550-1551, 1958.

Davis, G. L., G. R. Tilton, G. W. Wetherill, Mineral ages from the Appalachian province in North Carolina and Tennessee., *J. Geophys. Res.*, *67*, No.5, 1987-1996, 1962.

Davis, G. L., S. R. Hart, G. R. Tilton, The effect of contact metamorphism on zircon ages, *AGU Trans.*, *45*, No. 1, 115-116, 1964.

Davis, G. L., S. R. Hart, G. R. Tilton, Some effects of contact metamorphism on zircon ages, *Earth and Planetary Sci. Letters*, *5*, No. 1, 27-34, 1968.

Doe, B. R., S. R. Hart, Effect of contact metamorphism on the isotopic composition of lead in K feldspars, *Geol. Soc. Am. Spec. Paper 68*, 163-164, 1962.

Doe, B. R., Relationships of lead isotopes among granites, pegmatites, and sulfide ores near Balmat, New York, *J. Geophys. Res.*, *67*, No. 7, 2895-2906, 1962.

Doe, B. R., Distribution and composition of sulfide minerals at Balmat, New York, *Geol. Soc. Am. Bull.*, *73*, No. 7, 833-854, 1962.

Doe, B. R., G. R. Tilton, C. A. Hopson, Lead isotopes in feldspars from selected granitic rocks associated with regional metamorphism, *J. Geophys. Res.*, *70*, No. 8, 1947-1968, 1965.

Doe, B. R., S. R. Hart, Vliyaniye kontaktnogo metamorfizma na izotopnyy sostav svintsa kaliyevykh polevykh shpatov v rayone shtoka el'dora, kolorado (The effect of contact metamorphism on lead in potassium feldspars near the Eldora Stock, Colorado). Moscow, *Atomizdat*, 229-239, 1969.

Doe, B. R., G. R. Tilton, C. A. Hopson, Izotopy svintsa v polevykh shpatakh iz nekotorykh granitoidnykh porod, svyazannykh s regional'nym metamorfizmom (Lead isotopes in feldspars from selected granitic rocks associated with regional metamorphism), In Izotopy Svintsa v Rudnykh Mestorozhdeniyakh. Moscow, *Atomizdat*, 240-259, 1969.

Gast, P. W., Abundance of ^{87}Sr during geologic time, *Geol. Soc. Am. Bull.*, *66*, No.11, 1449-1454, 1955.

Gast, P. W., Isotope abundance of ^{87}Sr in igneous rocks. *Geol. Soc. Am. Bull.*, 67, No. 12, Pt. 2, 1698, 1956.

Gast, P. W., Rubidium-strontium age method and the age of ancient granitic rocks, *Geol. Soc. Am. Bull.*, *67*, No. 12, Pt. 2, 1814, 1956.

Hart, S. R., R. T. Dodd, Jr., Excess radiogenic argon in pyroxenes, *J. Geophys. Res.*, *67*, No. 7, 2998-2999, 1962.

Hart, S. R., The petrology and isotope-mineral age relations of a contact zone in the Front Range, Colorado, *J. Geol.*, *72*, No. 5, 493-525, 1964.

Hart, S. R., Potassium, rubidium, and strontium in the ultramafic rocks of St. Paul's islands (Atlantic Ocean), *Geol. Soc. Am., Spec. Paper No. 82*, 86-87, 1965.

Hart, S. R., A test for excess radiogenic argon in micas, *J. Geophys. Res.* *71*, No. 6, 1769-1770, 1966.

Hart, S. R., Current status of radioactive age determination methods, *Eos Trans. AGU*, 47, No. 1, 280-286, 1966.

Hart, S. R., L. T. Aldrich, Fractionation of Potassium/Rubidium by amphiboles: Implications regarding mantle composition, *Science*, *155*, 325-327, 1967.

Hart, S. R., G. L. Davis, R. H. Steiger, G. R. Tilton, A comparison of the isotopic mineral age variations and petrologic changes induced by contact metamorphism, in *Radiometric dating for geologists*, edited by E. I. Hamilton and R. M. Farquhar, London and New York, Interscience Publishers, 73-110, 1968.

Hart, S. R., G. L. Davis, Zircon U-Pb and whole-rock Rb-Sr ages and early crustal development near Rainy Lake, Ontario, *Geol. Soc. Am. Bull.*, 80, No. 4, 595-616, 1969.

Hart, S. R., K, Rb, Cs contents and K/Rb, K/Cs ratios of fresh and altered submarine basalts, *Earth and Planetary Sci. Letters*, *6*, No. 4, 295-303, 1969.

Hart, S. R., A. J. Nalwalk, K, Rb, Cs and Sr relationships in submarine basalts from the Puerto Rico trench, *Geochim. Cosmochim. Acta*, *34*, No. 2, 145-154, 1970.

Hart, S. R., C. Brooks, T. E. Krogh, G. L. Davis, D. Nava, Ancient and modern volcanic rocks - a trace of element model, *Earth and Planetary Sci. Letters*, *10*, No.1, 17-28, 1970.

Hart, S. R., K, Rb, Cs, Sr and Ba contents and Sr isotope ratios of ocean floor basalts, in A discussion on the petrology of igneous and metamorphic rocks from the ocean floor. R. Soc. Lond., *Phil. Trans.*, Ser. A, *268*, No. 1192, 573-587, 1971.

Krogh, T. E., Whole rock rubidium-strontium studies in the northwest

Grenville area of Ontario, *Eos Trans. AGU,*, 47, No. 1, 206-207, 1966.
Krogh, T. E., G. L. Davis, Age determinations of the pre-Grenville basement by the Rb-Sr whole rock method (abstract), *Geol. Ass. Can.-Mineral. Ass. Can., Int. Mtg.*, *46*, 1967.
Krogh, T. E., G. L. Davis, Old isotopic ages in the northwestern Grenville province, Ontario. In Age relations in high-grade metamorphic terrains, *Geol. Assoc. Canada Spec. Paper 5*, 189-192, 1969.
Tilton, G. R., C. C. Patterson, G. L. Davis, Isotopic composition of lead in olivine bombs, *Geol. Soc. Am. Bull.,*, *65*, No. 12, Pt. 2, 1314-1315, 1954.
Tilton, G. R., L. T. Aldrich, M. C. Inghram, Mass spectrometric determinations of Thorium, *Anal. Chem.*, *26*, 894-898, 1954.
Tilton, G. R., Volume diffusion as a mechanism for discordant lead ages, *J. Geophys. Res.*, *65*, No. 9, 2933-2945, 1960.
Tilton, G. R., M. N. Bass, G. L. Davis, G. W. Wetherill, 1000-million-year-old minerals from the eastern United States and Canada, *J. Geophys. Res.*, *65*, No. 12, 4173-4179, 1960.
Tilton, G. R., Summary of discussion of discordant zircon ages: In geochronology of rock systems, *New York Acad. Sci. Annals*, *91*, Art. 2, 276-278, 1961.
Tilton, G. R., G. W. Wetherill, G. L. Davis, Mineral ages from the Wichita and Arbuckle Mountains, Oklahoma, and the St. Francis Mountains, Missouri, *J. Geophys. Res.*, *67*, No. 10, 4011-4019, 1962.
Tilton, G. R., G. W. Reed, Radioactive heat production in eclogite and some ultramafic rocks, in *Earth Science and Meteoritics*, North-Holland Publishing Co., Amsterdam, pp.31-43, 1963.
Tilton, G. R., S. R. Hart, Geochronology, *Science*, *140*, No. 3565, 357-366, 1963.
Tilton, G. R., B. R. Doe, Lead isotopes and the age of the Earth, *Geol. Soc. Am. Spec. Paper 82*, 207, 1965.
Tilton, G. R., R. H. Steiger, Lead isotopes and the age of the Earth, *Science*, *150*, No. 3705, 1805-1808, 1965.
Tilton, G. R., Compilation of Phanerozoic geochronological data for Eastern North America, in Geochronology of North America, Natl. Acad. Sci. -Natl. Research Council Nuclear Sci. Ser. Rept. 41 (NAS-NRC Pub.1276), 181-220, 1965.
Wetherill, G. W., $^{40}A/^{40}K$ ratios of feldspars and micas from the same rock, *Geochim et Cosmochim Acta*, *8*, 171, 1955.
Wetherill, G. W., Discordant U-Pb ages. I., *Eos Trans. AGU*, *37*, 320-326, 1956.
Wetherill, G. W., An interpretation of the Rhodesia and Witwatersrand age patterns, *Geochim. et Cosmochim Acta*, *9*, No.5-6, 290-292, 1956.
Wetherill, G. W., Potassium-argon method, *Geol. Soc. Am. Bull.*, *67*, No. 12, Pt. 2, 1828-1829, 1956.
Wetherill, G. W., G. J. Wasserburg, L. T. Aldrich, G. R. Tilton, R. J. Hayden, Decay constants of ^{40}K as determined by the radiogenic argon content of potassium minerals, *Phys. Rev.*, *103*, 987-989, 1956.
Wetherill, G. W., Radioactivity of potassium and geologic time, *Science*, *126*, No.3273, 545-549, 1957.
Wetherill, G. W., Age of the base of the Cambrian, *Nature* (London), *187*, No. 4731, 34-36, 1960.
Wetherill, G. W., Age measurements on the Cutler batholith, Ontario, Canada, in Geochronology of rock systems, *New York Acad. Sci. Annals*, *91*, Art. 2, 423-429, 1961.
Wetherill, G. W., L. T. Aldrich, G. L. Davis, G. R. Tilton, Comparison of radioactive age measurements on pegmatites, *Eos Trans. AGU,*, *36*, No. 3, 533, 1955.
Wetherill, G. W., L. T. Aldrich, G. L. Davis, G. R. Tilton, New determinations of the age of the Bob Ingersoll pegmatite, Keystone, S. Dakota, *Geochimica et Cosmochimica Acta*, *9*, No. 5-6, 292-297, 1956.
Wetherill, G. W., G. L. Davis, C. A. Hopson, G. R. Tilton, Mineral ages in the Baltimore-Washington area (Md.-D.C.), *J. Geophys. Res.*, *64*, No. 8, 1129-1130, 1959.
Wetherill, G. W., P. W. Gast, G. R. Tilton, Precambrian age measurements in Finland, *Geol. Soc. Am.*, *B. 71*, No. 12, Pt. 2, 2000-2001, 1960.
Wetherill, G. W., G. L. Davis, G. R. Tilton, Age measurements on minerals from the Cutler batholith, Cutler, Ontario, *J. Geophys. Res.*, *65*, No. 8, 2461-2466, 1960.
Wetherill, G. W., G. R. Tilton, Age measurements on rocks from the Finnish Precambrian, *J. Geol.*, *70*, No. 1, 74-88, 1962.
Wetherill, G. W., G. R. Tilton, G. L. Davis, S. R. Hart, C. A. Hopson, Age measurements in the Washington-Baltimore area, *Eos Trans. AGU*, *46*, No. 1, 177, 1965.

Archival Sources for the History of Geosciences

DEBORAH DAY
Scripps Institution of Oceanography Archives
Library C-075C, La Jolla, CA 92093

Gregory Good has noted that geophysics did not exist as an independent science in the middle of the last century.

> It had no clear methods or goals and no specialists. Activities now recognized as part of geophysics, e.g. seismology and geomagnetics, were then associated with other sciences. They were fringe sciences . . . [Good, 1985].

Yet today, geophysics is a powerful, creative and productive science. It is composed of an estimated 26,000 professional geophysicists [American Geological Institute, 1987]. It has spawned a vigorous international network of research and educational institutions. Geophysics commands tremendous resources in both the public and private sector, although these resources fluctuate in response to trends in the general economy. Its scientific results in the last century have so fundamentally changed our view of the Earth that they have been characterized as a revolution. It is no wonder that geophysics has captured the imagination of scientists and, indeed, the public at large.

Historians of science have been attracted to the study of geophysics. They have studied it as an example of a science undergoing a paradigm shift, as a Kuhnian scientific revolution in progress, as a science with tremendous potential for intellectual advances as we enter a new century. The history of geophysics has attracted stellar academic talent—Stephen Brush, Charles Bates, Henry Frankel, Robert Friedman, Anthony Hallam, Homer LeGrand, Rachel and Larry Laudan. Many histories, autobiographies, and first person accounts of the Earth science revolution have been written by geophysicists themselves. These two factors—a vigorous period of scientific achievement that transforms a major scientific discipline, coupled with great interest expressed by historians of science—has captured the attention of science archivists. In the last ten years, archivists and geophysicists have expressed great concern about documenting the field of geophysics. Two things have became clear—important manuscript material is being destroyed, and the institutional history of geophysics in the private sector is not being collected. In 1982, I was surprised to learn from Tuzo Wilson that he had not donated his correspondence and other personal papers to an archives. At least he had kept his papers, and in 1986 he donated them to the University of Toronto Archives [Lagrimas Ulanday, personal communication, 1992]. Jacob Bjerknes' papers were lost at UCLA after his death. William von Arx has destroyed "virtually all working and preparatory material." Horace Byers thinks that his have been discarded as well. K. O. Emery wrote me that he has systematically discarded his correspondence after three years. "Seldom needed after that, it occupies more space than it is worth," he said. Benjamin Howell wrote the American Institute of Physics, "I have little worth preserving."

Institutional records documenting geophysics seem even more perishable than the papers of distinguished geophysicists. The greatest employer of American geophysicists has traditionally been the petroleum and mining industries, but almost none of the large American oil or mining companies have archives, although the records of some firms have been collected by the University of Wyoming. Lamont Doherty Geological Observatory does not have an archives. The Archives of the University of California, Los Angeles has few records of the UCLA's Institute of Geophysics. The administrative records of the Colorado School of Mines were discarded in the 1950s. Tremendously valuable archival records of the Bergen School of Meteorology, including papers of that giant Vilhelm Bjerknes, were found in an out of doors shed by Ralph Jewell in May 1976 and preserved by his heroic efforts [Jewell, 1981].

The American Geophysical Union (AGU) established a Committee on the History of Geophysics, and in 1988 this committee began working with the American Institute of Physics Center for the History of Physics to locate, study, and preserve documentation. In 1988, AIP prepared a survey questionnaire which was distributed to all living fellows of the AGU born before 1918. Ninety-seven geophysicists received the survey, 31 replied [Joan Warnow-Blewett, personal communication, 1992]. With the intercession of the AGU and AIP, the papers of many important geophysicists have found an archival home. Harold Jeffreys, Robert Dietz, Athelstan Spilhaus, and

The Earth, the Heavens and the Carnegie Institution of Washington
History of Geophysics Volume 5

many other important geophysicists have been persuaded to save their papers and deposit them in archives.

There is still, however, much persuading to do. When asked about the effort to preserve the correspondence and other papers of scientists, one geophysicist replied:

> I can see this is a viable WPA or other make-work project. I do not have the feeling that it is sufficiently promising to warrant support in a period of full employment. The people involved who are scientists should be doing science. The ones who are "historians of science" should find some more useful way to market their writing and public relations skills [Joan Warnow-Blewett, personal communication, 1992].

It is not necessary to collect the papers of every geophysicist or preserve all the administrative records of every organization that does geophysical research in order to document the history of the field. Archivists have developed theories and procedures of appraising manuscripts to determine what should be saved and what can be discarded. A great deal of thought has been given to the appraisal of contemporary scientific records beginning in 1960 when a group of American historians and other scholars attended a Conference on Scientific Manuscripts sponsored by the History of Science Society [Anonymous, 1960]. More research is needed on how historians and others use archives. Archivists need advice from historians, sociologists and other scholars on what records and papers are most critical to a study of geophysics. How did geophysics emerge from other sciences? What institutions and individuals played key roles in the development of geophysics? What are the current trends in geophysical research? How important are expeditions and experiments to geophysics?

Archivists traditionally appraise records within their institutional context to decide what ought or ought not to be kept. Traditional archival methodology focuses archival attention on saving historical documentation of important geophysicists and geophysical institutions. This has limitations in some sciences, including geophysics, which are very international and progress through vast experiments and expeditions that depend on the work of scientists from many institutions working cooperatively. The IGY is a perfect example. We need to develop archival strategies that will make it possible to target the documentation of large international cooperative projects or expeditions for retention. The AIP has a grant to study multi-institutional collaborations in science and is using several multi-institutional geophysics collaborations as case studies. This study will help archivists decide how past and future great geophysical experiments can best be documented.

In 1983, another landmark study on the appraisal of scientific manuscripts was undertaken by a group of historians and archivists responsible for science archives. This group was known as JCAST, the Joint Committee on the Archives of Science and Technology [1983]. Archivists Helen Samuels [1986], Larry Hackman and Joan Warnow-Blewett [1987], and others have developed a new method to improve identification, selection and retention of archival records called a documentation strategy. Documentation strategy has been used to appraise scientific records with good results. Its particular strength is that it gathers together scientists, historians, sociologists of science and archivists to analyze documentation strengths and weaknesses and to develop viable action plans.

Geophysics generates enormous data bases with very long term scientific value, which are very costly to maintain. There is also the problem of older geophysical data found among the papers of individuals and institutions. Scientists and archivists would profit from a discussion of how to create new databases, and how to improve access to older databases in a cost effective way. Scientists designing new databases could construct compact and archivally sound ones with advice of archivists trained in the preservation of machine readable records. Archivists need the advice of scientists to appraise older databases and make them available to geophysicists.

The AGU and other professional geoscience organizations can do a great deal to identify the papers of geophysicists worth saving for historical purposes. The AIP can contribute to the preservation of the records of geophysics by helping us to understand how geophysicists create and use records, how geophysical experiments and expeditions are initiated and organized, and how geophysical institutions cooperate. But we have a very basic problem which was perhaps best described by Mary Janzen [1980]:

> Efforts to preserve scientific records cannot succeed unless there is an increase in the number of repositories willing to acquire such records.

Efforts to preserve the history of geophysics will fail unless new archives can be created and established archives persuaded to collect and preserve the records of the institutions which are the cornerstones of geophysics, the papers of scientists who have expanded the field, and the records of the great geophysical experiments and expeditions that have characterized geophysics in this century. Archives collecting documentation on the history of geophysics need the aid and encouragement of the scholarly community, and they need resources—a tall order especially in these recessionary times.

The Carnegie Institution of Washington celebrates its ninetieth anniversary not only with this distinguished symposium, but also in another very meaningful way—by its effort to identify, collect, and preserve the historical records of the Institution, its Geophysical Laboratory and its Department of Terrestrial Magnetism. The CIW is a cornerstone of geophysics. Surely the history of geophysics in this century could not be written without reference to the important work done with CIW support.

Archival work is labor intensive and expensive. A study done by William Maher [1982] found that processing costs for a cubic foot of personal papers at the University of Illinois was $46. He found that the processing of institutional records costs one-third that of personal papers. However, these processing costs do not include figures for suitable archival storage.

Preserving the historical records of geophysics and the papers of prominent geophysicists will require the cooperative efforts of the scientific, historical, and archival communities. All three must work together to identify the really valuable materials. Scientists and historians have a particularly important role in identifying valuable collections. Scientists know which institutions, expeditions, and scientists have done critical work in the field. Historians often know of historically valuable collections that are still in the hands of scientists and their families. Archivists are not always familiar with the cutting edge of scientific disciplines. Scientists and historians can advise archivists on collections in peril that deserve preservation.

The three groups must work together, too, to identify material that need *not* be kept. The cost of keeping everything would be ruinous. Archivists can help scientists when expeditions are planned and large geophysical projects undertaken to build efficient record keeping into the operation that will minimize the bulk of records created and collected. Archivists, historians, and scientists need to cooperate more closely, especially in the design of large geophysical databases to minimize the costs of preserving and accessing these records.

The three groups must work together to locate the financial resources necessary to preserve a historical record of geophysics. We cannot look to the National Science Foundation or the National Endowment for the Humanities to provide the entire funds necessary for the work. We must attract new resources: donations from individual donors and foundations to help identify, process, and preserve the historical records.

I learned today that Lucy Bauer Pirtle donated more Bauer papers, once presumed lost, to the Carnegie Institution this year. This gift is surely a sign of great promise, for if these papers have survived, then the papers of other important scientists and the papers of other important scientific institutions presumed lost, may still be found. We have only to look for them and find the space and resources to care for them and make them available for research.

REFERENCES

American Geological Institute, *Summary, North American Survey of Geoscientists, U.S. Section, Survey Results and Forecast of Employment Trends*, 35 pp., 1987.

Anonymous, The Conference on Science Manuscripts, *Isis*, *53*, 157 pp., 1960.

Good, G. A., Geomagnetics and Scientific Institutions in 19th Century America, *EOS Trans. AGU*, *66*, 521-524, 1985.

Hackman, L. and J. Warnow-Blewett, The Documentation Strategy Process: A Model and a Case Study, *American Archivist*, *50*, 12-47, 1987.

Janzen, M. E., Scientific Records in a General Repository, *The Midwestern Archivist*, *5*, 29-37, 1980.

Jewell, R., University of Bergen, unpublished report, 1981.

Joint Committee on Archives of Science and Technology, *Understanding Progress as Process: Documenting the History of Post-War Science and Technology in the United States*, 64 pp., Society of American Archivists, 1983.

Maher, W. J., Measurement and Analysis of Processing Costs in Academic Archives, *College and Research Libraries*, *43*, 59-67, 1982.

Samuels, H., Who Controls the Past? *American Archivist*, *49*, 109-124, 1986.

Sources for the History of the Carnegie Institution of Washington at the Office of Administration

JOHN STROM
Carnegie Institution of Washington
1530 P St. NW, Washington DC 20005-1910

The library and archives of the Carnegie Institution of Washington, housed in the Administration Building at 1530 P Street, NW, Washington, DC 20005 contain the following:

LIBRARY

Publications of the Institution, including the Year Books, which annually present reviews of the current work of the Institution's scientists, and the monograph series, including over 600 titles in fields of scholarship in which the Institution has been engaged since 1902.

ARCHIVES

Minutes of the meetings of the Board of Trustees and of the Executive Committee of the Board, dating from the meeting of the Incorporators on January 4, 1902.

Collections of correspondence, reports, and other material, dating from the founding of the Institution primarily derived from the office of the Institution's president. These are divided into two broad categories:

- The general files, which contain material related to the overall operation of the Institution as an entity and relationships with individual grantees and others affiliated with the Institution but not with any particular department.
- The departmental files, which contain material related to the operations of particular departments and their relationships with the central administration of the Institution. These include:

Department of Plant Biology: Desert Laboratory, opened in 1903, became headquarters of the Department of Botanical Research in 1905. Name changed to the Laboratory for Plant Physiology in 1923. Reorganized in 1928 as the Division of Plant Biology, including ecology. Name changed to Department of Plant Biology in 1951.

Observatories of the Carnegie Institution of Washington: Mt. Wilson Observatory organized in 1904. Unified operation with the Palomar Observatory of the California Institute of Technology began in 1948 under the name Mt. Wilson and Palomar Observatories; in 1970, name changed to Hale Observatories. Upon dissolution of the arrangement with Cal Tech in 1980, Carnegie Observatories organized as Mt. Wilson and Las Campanas Observatories, to include Carnegie's Southern Hemisphere observatory. In 1989, name changed to Observatories of the Carnegie Institution.

Department of Terrestrial Magnetism: Originally organized as Department of International Research in Terrestrial Magnetism on April 1, 1904. Name changed to Department of Terrestrial Magnetism in 1905.

Department of Genetics: Station for Experimental Evolution opened in 1904. Name changed to Department of Experimental Evolution in 1906. Combined with Eugenics Record Office in 1921 to form Department of Genetics. In 1963, department reorganized and renamed the Genetics Research Unit, which was closed in 1971.

Geophysical Laboratory: Laboratory organized in 1906 and opened in 1907.

Department of Embryology: Department organized in 1914.

Department of Historical Research: Department originally organized as a "bureau" in 1903. Became a "department" in 1905, but terminated as a department and incorporated as the Section of United States History in a new Division of Historical Research in 1930.

Division of Historical Research: Established in 1930, included Section of United States History, Section of Aboriginal American History, which continued the archaeological work begun by Sylvanus G. Morley in Central America and by E. H. Morris in the southwestern United States, and Section of the History of Science. Division became the Department of Archaeology in 1951, closed in 1958.

Department of Economics and Sociology: Department organized in 1904 and terminated in 1916.

Department of Marine Biology: Department established in 1904. Name changed to Tortugas Laboratory in 1923. Activities terminated in 1939.

Department of Meridian Astrometry: organized in 1906, became Committee on Meridian Astrometry in 1936,

The Earth, the Heavens and the Carnegie Institution of Washington
History of Geophysics Volume 5

terminated 1938.

Nutrition Laboratory: Organized in 1907 and opened in 1908. Activities terminated January 1, 1946.

Over the past five summers, a project to index the files at the Administration Building in a computerized data base was completed. It is now possible to access archival file information by a computerized search according to file names, key words, and some topics.

It should also be noted that each research department is responsible for maintaining or preserving its own archives, and several have chosen to deposit their collections with other organizations.

The Carnegie Institution of Washington's Contributions to the Physical Sciences: Archival Sources at the Huntington Library

R. S. BRASHEAR
Huntington Library, 1151 Oxford Rd.
San Marino, CA 91108

The Huntington Library is mainly known as a research institution for the study of American and British history and literature. However, the Manuscripts Department of the Library has much material relating to the history of the physical sciences at the Carnegie Institution of Washington (CIW). The main collections on this subject are those of Edwin Powell Hubble, George Ellery Hale, Francis Gladheim Pease, the CIW Observatories, Frederick Eugene Wright, Bailey Willis, and the Pumpelly family.

EDWIN POWELL HUBBLE COLLECTION

The Hubble Collection was given to the Library by his widow, Grace Burke Hubble. It consists of nearly 4000 pieces including manuscripts, correspondence, photographs, astronomical working papers, logbooks, a scrapbook, reprints, and miscellaneous printed matter. Hubble's papers reflect both his scientific as well as his public interests. The inclusion of Grace Hubble's correspondence, journals, and memoirs also illuminate Hubble's private life. The first eight boxes of the collection are devoted to manuscripts by Hubble and others. The majority of Hubble's own manuscripts deal with scientific subjects. Many of the significant unpublished works have been recently published in *The Edwin Hubble Papers: Previously Unpublished Manuscripts on the Extragalactic Nature of Spiral Nebulae*, edited, annotated, and with an historical introduction by Norriss S. Hetherington (Tucson: Pachart Publishing House, 1990). Edwin and Grace's correspondence takes up the next twelve boxes in the collection. Unfortunately, only a few letters survive from the 1920s and 30s, but there are significant ones to and from Harlow Shapley, Vesto Melvin Slipher, Willem de Sitter, Henry Norris Russell, Arthur Stanley Eddington, Joel Stebbins, and Otto Struve. There is also an important run of correspondence with Nicholas Ulrich Mayall from the 1930s regarding the survey of galaxies they made. There are also five boxes of Hubble's working papers from the period 1925 to 1939. In this group is almost complete documentation for his three papers on cepheid variables (1925-1929), for his 1930 paper "The Distribution of Luminosity in Elliptical Nebulae," and for his 1934 paper "The Distribution of Extra-Galactic Nebulae." A copy of the summary report can be obtained from the Manuscripts Department for the cost of reproduction and mailing.

GEORGE ELLERY HALE COLLECTION

The Hale Collection is a group of 152 items (primarily letters from the period 1887 to 1937) acquired from Richard Hale Goodwin in 1958. These letters include Hale's correspondence with his close friend Harry Manley Goodwin, physicist and graduate dean of MIT; astronomical and scientific concerns; the administration of scientific research, organizations, and publications; and Hale's private and family life.

FRANCIS GLADHEIM PEASE COLLECTION

The Pease Collection, purchased in 1991, includes 558 letters (incoming and outgoing) and a few miscellaneous manuscripts and photographs. It covers 1905 to 1937, when Pease was at Mt. Wilson Observatory. They mainly concern Pease's scientific work and his contacts with leading engineering and optical companies. Some of the correspondents represented by at least ten pieces are: Corning Glass Works, Andrew E. Douglass, General Electric, J. B. McDowell Co., Spencer Lens Co., Frederick E. Wright, and William H. Wright.

THE CIW OBSERVATORIES COLLECTION

The Carnegie Observatories' Collection has been placed on permanent deposit at the Huntington Library with the assistance of a grant from the John Randolph Haynes and Dora Haynes Foundation of Los Angeles, California. Not all of the collection is now available to scholars, but the following record groups can be accessed:

The Earth, the Heavens and the Carnegie Institution of Washington
History of Geophysics Volume 5

(1) George Ellery Hale Papers: These consist mainly of 26 boxes of Hale's correspondence as director of Mt. Wilson Observatory from 1904 to 1923. Also included are ten volumes of Hale's letterpress copy books from the Kenwood Observatory (1891-1896) and as secretary of the CIW Advisory Committee on Astronomy (1902-1903). These volumes are not reproduced in the microfilm edition of Hale's papers.

(2) Walter Sydney Adams Papers: Consists of 109 boxes (50,000+ items) of correspondence and manuscripts, the bulk of it from Adams's tenure as director of Mt. Wilson Observatory (1923-1945). Twenty-one of the boxes deal with the work and research carried out by Observatory staff under government contracts during World War II.

(3) Ira Sprague Bowen Papers: Consists of 88 boxes (30,000+ items) of correspondence and manuscripts from Bowen's years as director of Mt. Wilson and Palomar Observatories (1946-1964) and his retirement, when he was still active in designing telescope optics and instruments (1964-1973).

(4) Frederick Hanley Seares Papers: Consists of 21 boxes (10,000+ items) of correspondence, manuscripts, and working papers from Seares's period as astronomer and assistant director at Mt. Wilson Observatory (1909-1945). Much of the material deals with Seares's work in photographic photometry, including extensive correspondence with Jacobus Cornelius Kapteyn, Pieter Johannes van Rhijn, Frank Elmore Ross, Edward Charles Pickering, and members of the International Astronomical Union. There are also many letters from women applying to be computers at the Observatory.

(5) Alfred Harrison Joy Papers: Consists of 10 boxes (2000+ items) of correspondence and manuscripts from Joy's years as astronomer and secretary at the Mt. Wilson Observatory (1915-1948) and his retirement years when he worked a great deal with the Astronomical Society of the Pacific (1948-1972). Of particular interest are two volumes of notes taken by Joy from his classes with Henry Norris Russell in 1910.

(6) Milton Lasell Humason Papers: Consists of 5 boxes (2000+ items) of correspondence, mostly from Humason's first years as secretary of Mt. Wilson Observatory (1948-1952). Unfortunately, there are only a few items of scientific interest in this collection.

(7) Wilhelm Heinrich Walter Baade Papers: Consists of 22 boxes of uncataloged correspondence, manuscripts, notes, and working papers from Baade's years in Göttingen and Hamburg (1915-1931) to his time as astronomer at Mt. Wilson (and Palomar) Observatory (1931-1958). Also included are lectures that Baade gave in the 1950s including the period after his retirement from the Observatories.

(8) Frederick Gladheim Pease Papers: Consists of 14 boxes of uncataloged correspondence, notebooks, and working papers from Pease's years at Yerkes Observatory (1901-1904) and Mt. Wilson Observatory (1904-1938). There is much material on his collaborations with A. A. Michelson and his notes on the construction of the telescopes at Mt. Wilson.

(9) Edison Pettit Papers: Consists of 6 boxes of correspondence (and a few manuscripts) from when Pettit was an astronomer at Mt. Wilson Observatory, but ending before his retirement (1920-1946). There is a large amount of correspondence relating to Pettit's ultraviolet work, especially with the Carnegie Desert Laboratory and the Desert Sanatorium of Southern Arizona, and his solar prominence work with Robert Raynolds McMath.

(10) Paul Willard Merrill Papers: Consists of 58 boxes, 15 of which contain his correspondence (1919-1961) as an astronomer at Mt. Wilson (and Palomar). These papers are uncataloged, but the correspondence is available to scholars in a very rough organization.

(11) Observatories' Photograph Collection: Consists of 3271 historical photographs of the Mt. Wilson and Palomar Observatories, the staff, and visiting astronomers, and a 51-minute videotape of footage of the Mt. Wilson Observatory from the 1930s, taken from several reels of 35 mm nitrate film.

(12) Observatories' Reprint Collection: To supplement the manuscript holdings, there is an extensive collection of reprints of scientific articles (including the Director's Annual Reports) by Observatory staff and visiting astronomers.

FREDERICK EUGENE WRIGHT COLLECTION

Another holding at the Huntington of interest to historians of modern astronomy are the Frederick Eugene Wright-Carnegie Moon Committee Papers given to the library by Frederick Hamilton Wright in 1990. This collection in seven boxes consists of correspondence, reports, observations, and notebooks relating to the Moon Committee's work. The Committee was organized in the 1920s to begin a systematic study at Mt. Wilson of the lunar surface. Wright, of the CIW's Geophysical Laboratory, chaired the committee, which operated until the mid-1930s.

BAILEY WILLIS COLLECTION

The Willis Collection consists of 11,799 pieces (excluding printed material and photographs) of correspondence, manuscript articles, and reports. It covers the period 1856-1957, concentrated on 1880-1949, when Willis was with the Northern Pacific Railway (1880-1882), the U.S. Geological Survey (1882-1915), and Stanford University (1915-1922).

During his career, Willis had many contacts with the Carnegie Institution. From 1903 to 1904, Willis obtained a Carnegie grant for the geological exploration of eastern China. He was a Carnegie Research Associate from 1927 to 1938, for comparative seismology projects around the world. Willis also served on the CIW's Advisory Committee on Seismology from its inception in 1921 until 1937, when the Seismological Laboratory was taken over by the California Institute of Technology. Significant correspondents include Thomas C. Chamberlin, Arthur L. Day, Henry F. Osborn, Raphael Pumpelly, Charles D. Walcott, and Robert S. Woodward. There is also a good deal of correspondence with members of his family when he was on his expeditions. The collection also contains several thousand photographs, most of which were taken on CIW expeditions.

PUMPELLY COLLECTION

The Pumpelly collection consists of 15 boxes of journals and correspondence. It is not completely organized but is available to scholars. The main figure is the geologist/engineer Raphael Pumpelly. His connection to the Carnegie Institution is his organization and leading of the CIW Trans-Caspian Expedition. Only a small portion of the collection is about Raphael Pumpelly himself; the bulk of the correspondence concerns family matters, most of it generated by Raphael's wife, Eliza Shepard Pumpelly. Materials relating to the Trans-Caspian Expedition are few, mainly nine of Pumpelly's notebooks on the Expedition and two diaries of Eliza's, who accompanied the Expedition. Raphael Pumpelly's correspondence consists of only 14 professional letters (1866-1916) and several letters to family members.

Sources for History of Geophysics at the Center for History of Physics, American Institute of Physics

RONALD E. DOEL
Center for History of Physics, American Institute of Physics
American Center of Physics, One Physics Ellipse
College Park MD 20740

For several years the AIP's Center for History of Physics and Niels Bohr Library have included the history of geophysics within their sphere of interests. The Center has tried to locate the papers of geophysicists and records of relevant institutions, and to place these at appropriate archives. The Center has also obtained information on existing collections of interest to historians of geophysics; this information is available to researchers at the Center and will become accessible on-line through the Research Libraries Information Network (RLIN) in 1993. Finally, the Center has sponsored historians who have conducted oral history interviews with geophysicists. More detailed information on these topics appears below.

• *Professional papers*: Although fewer collections of correspondence and related materials are available for geophysicists than for physicists and astronomers—lamentably, given the large size of the American geophysical community—more collections are becoming available. A database at the Center holds information on collections of papers of physicists and allied scientists, including geophysicists, based on surveys of several hundred archives worldwide. The Center also has finding aids detailing some of these collections. At this writing information on these collections is being transferred to RLIN, where it may be accessed by scholars through their home institutions or through neighboring research facilities. The Niels Bohr Library does not usually collect archival materials itself, but certain items, including manuscript biographies, are available for some geophysicists. A full guide to the collections held by the Center for History of Physics is in preparation; researchers interested in particular topics are invited to contact the staff of the Niels Bohr Library for further information. An overview of the state of collections in the history of modern science since World War II is in Warnow-Blewett, 1992.

• *Oral History Interviews*: In recent years oral history interviews with geophysicists have been deposited at the Niels Bohr Library; these interviews are available to qualified scholars within the limits of access set by the interviewee. Substantial interviews are normally transcribed and reviewed by the subject and interviewer. The history of climatology (global warming) and of academic geophysics in the US is well represented in these interviews, although other topics are covered as well. Interviewees include Jule Charney, M. King Hubbert, Syukuro Manabe, Athelstan Spilhaus, Merle Tuve, and John Verhoogen. Other interviews with geophysicists are being made as part of the Center's current Study of Multi-Institutional Collaborations in Space Science and Geophysics, which is examining major projects initiated in the 1970s and 1980s. These interviews, more focused than other AIP interviews, are transcribed but generally unedited; they should become available to researchers in 1994. Collaborations examined include the Greenland Ice Sheet Project (1970s and 1980s, atmospheric sciences) and the Incorporated Research Institutes for Seismology (1980s, solid earth geophysics). A list of oral history interviews accessible at the Center is available.

At present topical and name indexes are being produced. Researchers interested in finding the status of particular collections in the history of geophysics, or who wish to obtain a current list of oral history interviews, are invited to visit or contact the Niels Bohr Library for further information [335 E. 45th Street, New York, NY 10017; telephone 212/ 661-9404; fax 212/ 986-5334; email bhs@aip.org]. Prospective patrons are advised that the AIP is scheduled to relocate by late summer 1993; calling ahead to determine the status of the library and availability of materials is strongly recommended. The new address will be AIP, American Center of Physics, One Physics Ellipse, College Park MD 20740.

The Earth, the Heavens and the Carnegie Institution of Washington
History of Geophysics Volume 5

REFERENCE

Warnow-Blewett, Joan, Documenting Recent Science: Progress and Needs, *Osiris* [second series], 7, 267-298, 1992.

The Earth and Space Sciences at Carnegie: A Pictorial Sampler from the First Six Decades

Compiled by SHAUN J. HARDY
Department of Terrestrial Magnetism
5241 Broad Branch Rd., Washington, D.C. 20015-1305

The historic images shown here represent a cross-section of research activities in the Earth and space sciences during the first sixty years of the Carnegie Institution of Washington. The photographs and interpretive information for Mt. Wilson Observatory were provided by Ronald Brashear from the collections of the Carnegie Observatories at the Huntington Library. All other materials were selected from the archives of the Department of Terrestrial Magnetism, Geophysical Laboratory, and CIW Administration.

Staff and associated researchers of Mt. Wilson Observatory, 1905. From left to right: H. L. Miller (construction superintendent), C. G. Abbot (Smithsonian Observatory astronomer), G. E. Hale (Mt. Wilson director), L. R. Ingersoll (Abbot's assistant), F. Ellerman (Mt. Wilson solar observer), W. Adams (Mt. Wilson astronomer, later director), E. E. Barnard (Yerkes Observatory astronomer), and C. S. Backus (Mt. Wilson observing assistant). [See articles by Brashear, DeVorkin, Hetherington, Lankford, and Gingerich, this volume.]

Physical Laboratory adjacent to the Observatory offices in Pasadena, 1908. The use of laboratory apparatus to simulate conditions in space was a key feature of the new "astrophysical" observatories. At right-center is a DuBois electromagnet with spark terminals between its poles, used to study the Zeeman effect in sunspots. To its left is a 30-ft. spectrograph, and behind it, a Moissan-type inclined arc electric furnace, used to study the spectra of elements under sunspot-like conditions.

Snow Telescope building, ca. 1912. The first permanent telescope at Mt. Wilson was erected in 1905 inside a canvas-covered steel framework. In 1911, the canvas was replaced by sheet metal, as shown here. During use, the telescope mirrors (inside the shelter at left) would be uncovered, to direct sunlight or starlight down the long building to the spectrograph. In the background are the 150-ft. and 60-ft. tower telescopes.

The 150-ft. tower telescope, ca. 1925. Mirrors located in the dome, high above ground-related atmospheric turbulence, directed sunlight to a diffraction grating in a pit below the tower. The reflected solar spectrum could then be analyzed with instruments at the base of the tower. Using this telescope, G. E. Hale conducted his search for the general magnetic field of the Sun.

Visit of Albert Einstein to Mt. Wilson, January 29, 1931. Shown with Einstein on the porch of the "Monastery" are C. St. John (Mt. Wilson solar astronomer), W. Adams (director), E. Hubble, W. Mayer (Einstein's colleague), and A. Joy (Mt. Wilson stellar astronomer).

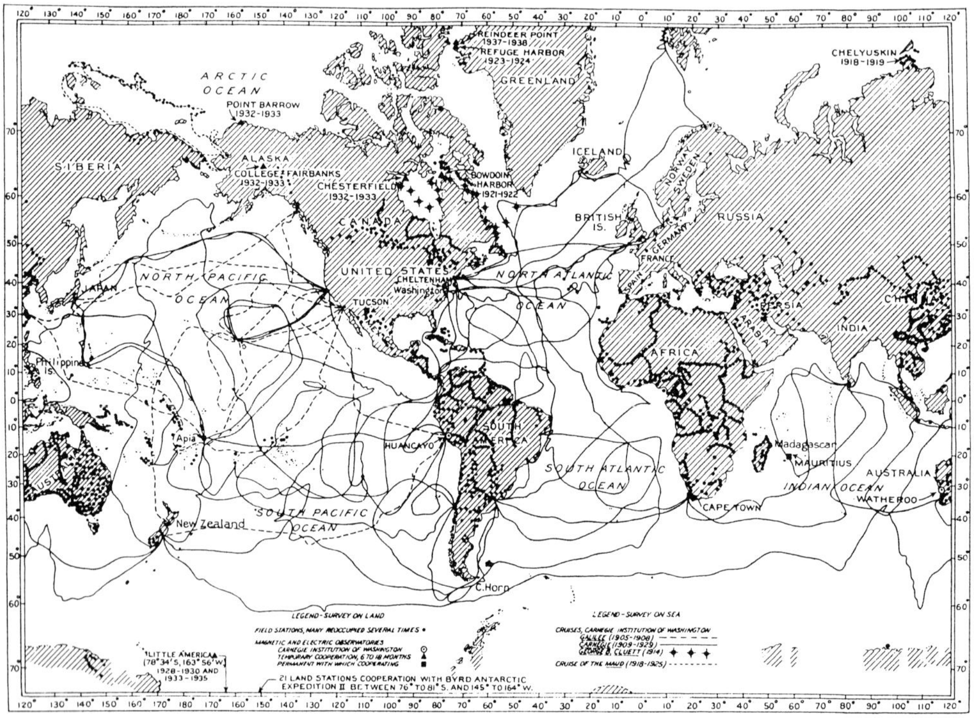

World magnetic and electric survey of the Department of Terrestrial Magnetism, 1905-1938. [See Good, this volume.]

J. P. Ault making magnetic observations on Cruise I of the *Galilee*, 1905. DTM's first marine geomagnetic work was conducted on this chartered brigantine, during three Pacific cruises between 1905 and 1908. Instruments mounted on an open platform on deck and the presence of magnetic construction materials made observing difficult, motivating construction of a specially-designed research vessel (the *Carnegie*).

Magnetic instruments enroute overland in Hainan, China, 1906. In the first four decades of this century, DTM personnel made observations at over six thousand stations on land, often traversing remote and difficult terrain.

Crossing Australia by camel train with E. Kidson's expedition, 1914. Kidson's magnetic survey followed less than a decade after survey teams blazed the "Canning Stock Route" through often hostile parts of Western Australia. His exploratory work is reflected today in the names of several natural features in the region, including Kidson Bluff and Kidson Basin.

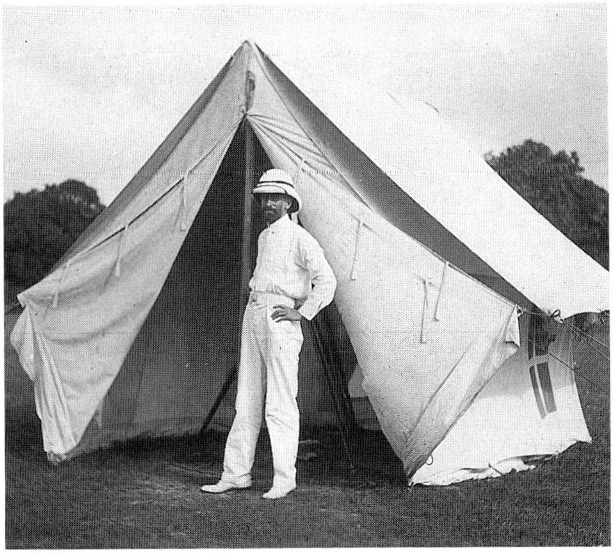

Louis Agricola Bauer, first director of DTM, doing field work at Colombo, Ceylon, 1911.

Launching of the research vessel *Carnegie*, June 12, 1909. Dorothea Louise Bauer, the 17-year old daughter of DTM's director, christened the ship. Over 3,000 guests attended the launching ceremony.

Carnegie under sail on her trial cruise, Long Island Sound, 1909. To reduce errors in geomagnetic measurements, the ship was specially constructed of non-magnetic materials, including a 150 h.p. bronze engine.

Scientific party on the bridge, showing after observatory, August 22, 1909. Included in the group are Capt. W. J. Peters and L. A. Bauer (top row, left and center), and J. P. Ault (front row, second from left), who would succeed Peters as commander of the ship.

Shore party from *Carnegie* comparing land- and ship-borne instruments at Sweetwater Inlet, Colon, Panama, 1915. Observer I. A. Luke is shown operating the earth inductor; H. E. Sawyer is at the galvanometer.

Top: Observing on board *Carnegie* : E. Kidson and H. F. Johnston taking atmospheric electricity measurements on deck, 1911. *Bottom:* L. L. Tanguy observing magnetic declination, 1918.

S. E. Forbush with Vening-Meinesz gravity apparatus, 1929. The device's use on board *Carnegie* was its first application on a surface vessel; previous investigations had been done on submarines [see Oreskes, this volume]. Forbush would later achieve eminence for his cosmic ray studies, particularly for his recognition of the "Forbush effect" in 1937——the decrease in the intensity of cosmic rays observed on Earth, due to magnetic effects of solar flares.

Chemist-biologist H. R. Seiwell collecting biological specimens with a dip net, from *Carnegie* 's "boom-walk," 1928. An extensive refitting for Cruise VII (1928-29) and an enlarged crew equipped *Carnegie* to add chemical and physical oceanography and marine biology to its observing program.

Members of the MacMillan Baffin Land Expedition, with the schooner *Bowdoin*, 1922. MacMillan was one of several arctic and antarctic expeditions in which DTM collaborated, through the establishment of temporary observatories. In addition to collecting geophysical and meteorological data, rock samples were brought back for analysis at the Geophysical Laboratory. CIW magnetic observer R. H. Goddard appears at far right. [See Doel, this volume.]

The *Carnegie* ablaze in Apia Harbor, Samoa, November 29, 1929. A gasoline explosion occurred while refueling, destroying the ship and taking the lives of Capt. J. P. Ault and the ship's cabin boy. In her 20 years, *Carnegie* had cruised 342,681 miles of the world's oceans.

Ionospheric research facilities at Huancayo Magnetic Observatory, Peru, 1935. Huancayo was one of two permanent geophysical observatories established by DTM in the southern hemisphere after World War I. Like its counterpart at Watheroo, Australia, Huancayo was equipped for a broad range of investigation: terrestrial magnetism, atmospheric electricity, meteorology, ionospherics, earth currents, cosmic rays. Ownership was transferred to the Peruvian government in 1947. [See articles by Home and Parkinson, this volume.]

L. V. Berkner with multifrequency ionospheric equipment, at DTM's Kensington Experimental Station, 1937. The Department's investigations of the upper atmosphere had begun in the mid-1920s, with the confirmation of radio echoes from the "Kennelly-Heaviside layer" (ionosphere) by G. Breit and M. A. Tuve. [See Gillmor, this volume.]

Two periods of paleomagnetism studies at DTM: *Top:* A. E. Johnson collecting clay varves at E. Windsor Hill, Connecticut, 1939. ***Bottom:*** J. W. Graham, with specially equipped "Rock Magnetism Mobile Laboratory," 1951. [See Le Grand, this volume.]

Left: J. B. Doak with DTM seismic truck, showing vertical and horizontal velocity-type seismometers, 1948. *Right:* Close-up of equipment for early explosion seismology work: seismometer, amplifiers, chronometer, recorder, and batteries. DTM's study of crustal structure using artificial earthquakes was begun in 1947; experiments were conducted throughout the Washington, D.C. region and the Appalachian Highlands. [See Cornell, this volume.]

Wurzburg radar dish, erected at DTM in 1952. This 26-ft. stationary parabola was used for 21 cm hydrogen-line sky surveys, and was the first in a series of radio telescopes operated by the Department over the next quarter-century. Especially noteworthy was the discovery of Jupiter as a radio source in 1955, using a "Mills Cross" dipole array on a site near Seneca, Maryland.

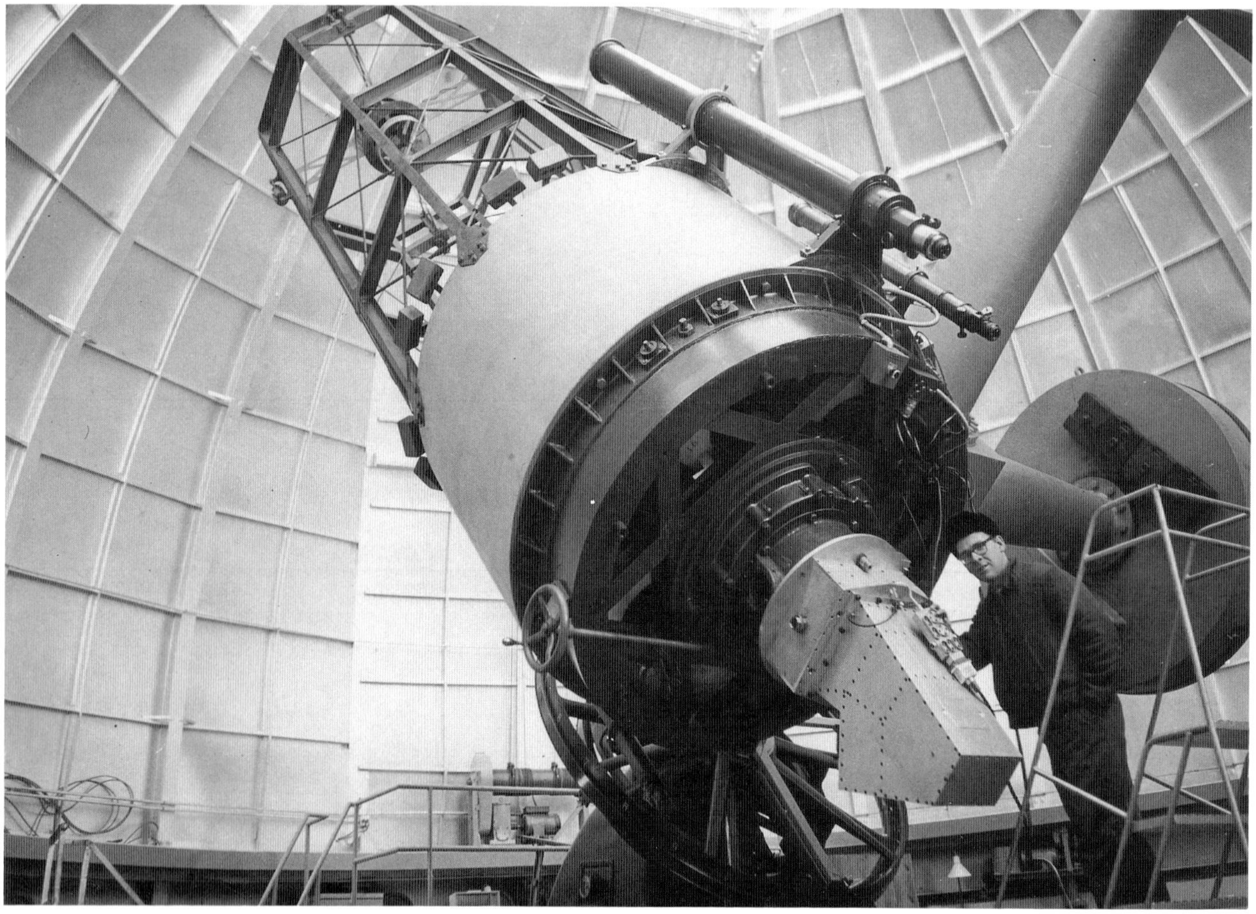

Astronomer W. K. Ford inspecting a Carnegie image intensifier and spectrograph on the Perkins Telescope, Flagstaff, Arizona, ca. 1965. In the 1950s and 1960s, DTM spearheaded a multi-institutional project to develop and test photoelectric image tubes to extend the capabilities of optical telescopes. By 1969, over three dozen systems were delivered and installed at observatories worldwide through a CIW/National Science Foundation distribution program.

First staff members of the Geophysical Laboratory, photographed in 1906 at the Lab's temporary quarters at the U. S. Geological Survey in downtown Washington. Front row: G. R. Hoffman, E. T. Allen, A. L. Day (director), W. P. White. Back row: W. Beck , C. W. H. Ellis, F. E. Wright, B. D. Chamberlain, J. K. Clement, A. F. Susan. The photograph was found in an album compiled by E. S. Shepherd, entitled "The Human Side of Geo-Physics at Washington, D.C., 1906" ——complete with Shepherd's humorous captions for his colleagues at work and play. [See Yoder, this volume.]

Arthur L. Day, first director of the Geophysical Laboratory, and an unidentified colleague (probably E. S. Shepherd), with a carbon arc and resistance furnace, ca. 1906. Day reported that with this apparatus, "30 horsepower can be concentrated into a square inch."

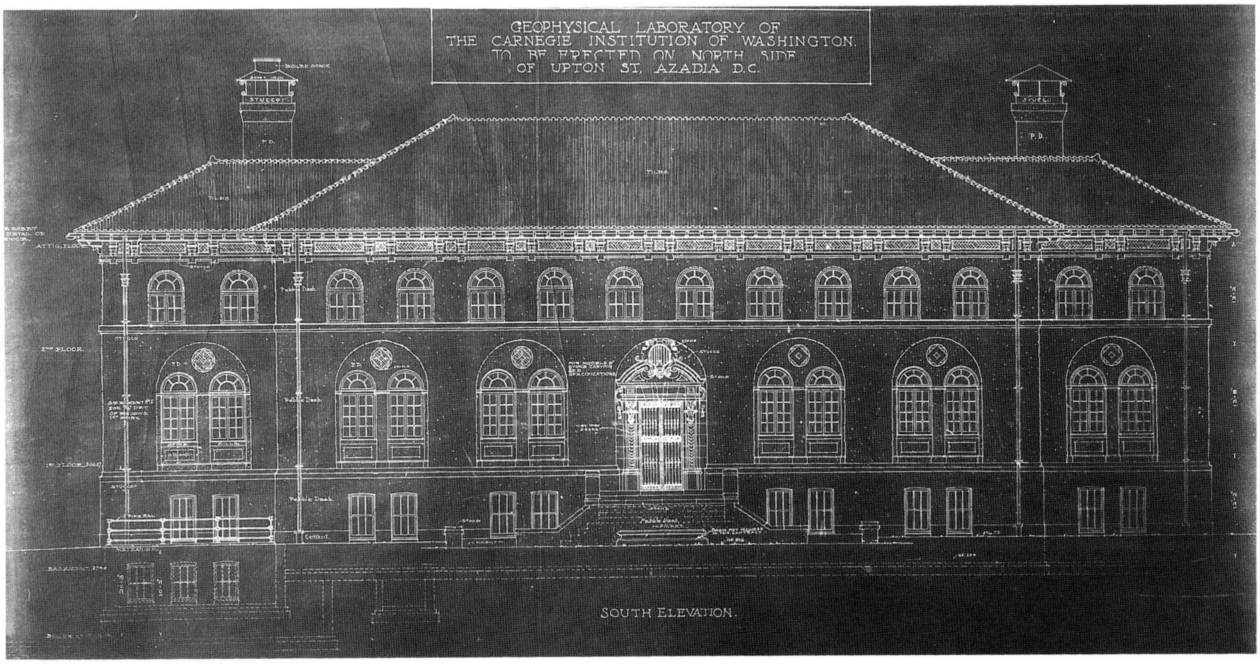

Plans for the Geophysical Laboratory's first permanent home, at 2801 Upton Street, Washington, D.C. Built and equipped for $154,223 in 1906-07, the building would be synonymous with the Lab for the next 83 years.

Undated portrait of petrologist H. S. Washington in Greek costume (1920s?). Washington's interest in volcanic rocks evolved through his archaeological "tours" of the Mediterranean Region. His monumental compilation of 30 years of data, *Chemical Analyses of Igneous Rocks* (1917), was based on an original system of classification ("CIPW"), developed by Washington in conjunction with Cross, Iddings, and Pirsson.

G. W. Morey, N. L. Bowen, and E. G. Zies, enjoying a lighter moment in the lab, ca. 1922. This same year saw the publication of Bowen's "The reaction principle in petrogenesis" ——later called "the most important contribution to petrology of the [20th] century."

Members of an expedition from the Geophysical Laboratory and the Volcanological Survey of the Netherlands East Indies, collecting volcanic gases at Papandayan volcano, Java, 1928. GL staff member E. G. Zies is seated at the right. Kilauea, Lassen Peak, the Valley of Ten Thousand Smokes and Yellowstone were among the sites of volcanism and hydrothermal activity studied actively in the pre-World War II era.

First undisturbed vertical cores of the ocean bottom, obtained by C. S. Piggot, 1936. A gun-fired sample tube and hoisting gear developed at the Geophysical Laboratory were used to collect the samples from North Atlantic depths as great as three miles.

J. F. Schairer with high temperature quenching furnaces, ca. 1950. The technique of heating silicate samples to known temperatures, then rapidly cooling the liquid to a glass and examining the resulting crystals was pioneered at the Lab by Shepherd and Rankin in 1909. The method remains a staple of experimental igneous petrology to this day, and is used to determine the melting points and sequence of crystallization of rock-forming minerals.

G. W. Morey with early pressure regulators and "Morey bomb," ca. 1950. Morey and Ingerson's design of a new type of high pressure vessel ("bomb") more the four decades earlier had opened the path to successful laboratory investigation of hydrothermal mineral formation.

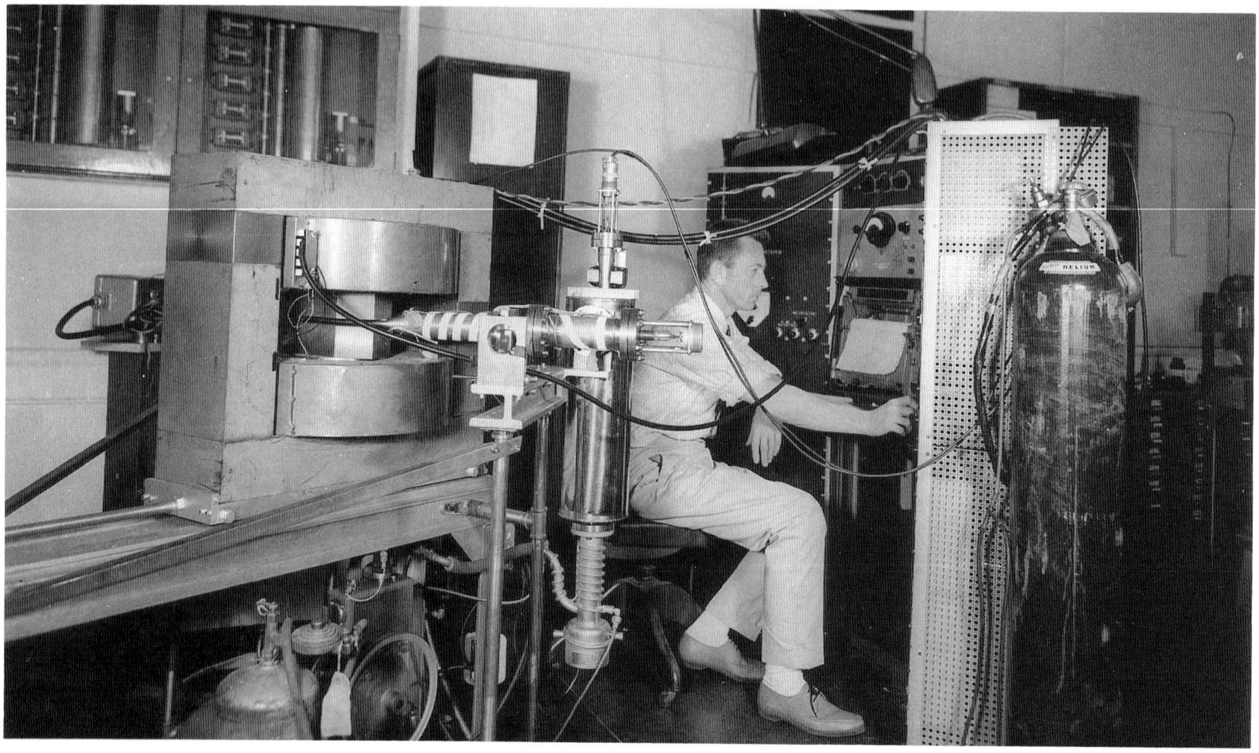

GL staff member G. L. Davis, with the mass spectrometer at DTM, ca. 1960. By 1952, mass spectrometry had progressed sufficiently to allow the accurate dating of rocks from the relative amounts of parent and daughter isotopes in radioactive decay. DTM and the Geophysical Laboratory established a successful collaboration in building instruments and developing techniques in the young discipline of geochronology. [See articles by Le Grand and Aldrich, this volume.]

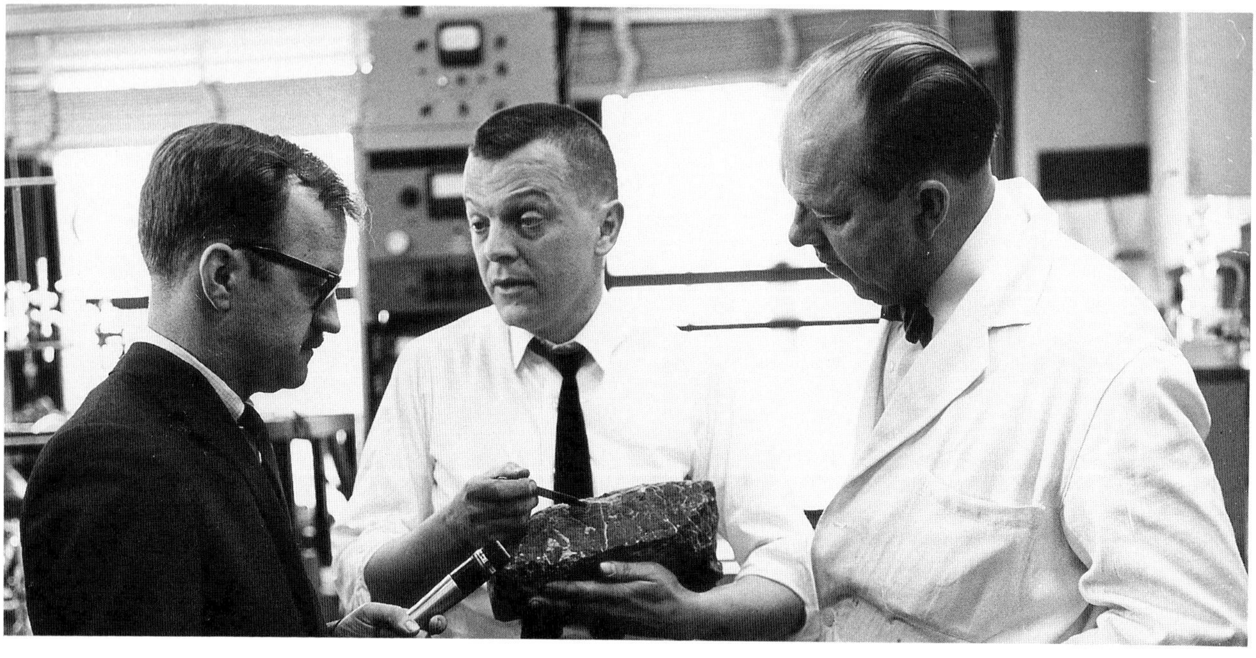

P. H. Abelson (right) and T. H. Hoering (center) describing evidence of life in a 3 billion year old rock to John Henahan of American Chemical Society News Service, 1963. Abelson's demonstration in the 1950's that fossils retained amino acids from some of their original proteins opened a new research agenda at the Geophysical Laboratory——the application of organic chemistry to geological problems. [Reprinted with permission from *Chemical and Engineering News 41 (3),* pp. 100-101, January 21, 1963. Copyright 1963 American Chemical Society.]